International Series in Operations Research & Management Science

Volume 279

More information about this series at http://www.springer.com/series/6161

Guillermo Gallego • Huseyin Topaloglu

Revenue Management and Pricing Analytics

Springer

Guillermo Gallego
Clearwater Bay
Kowloon, Hong Kong

Huseyin Topaloglu
ORIE, Cornell University
New York, NY, USA

ISSN 0884-8289 ISSN 2214-7934 (electronic)
International Series in Operations Research & Management Science
ISBN 978-1-4939-9604-9 ISBN 978-1-4939-9606-3 (eBook)
https://doi.org/10.1007/978-1-4939-9606-3

This Springer imprint is published by the registered company Springer Science+Business Media, LLC, part of Springer Nature.
The registered company address is: 233 Spring Street, New York, NY 10013, U.S.A.

To my wife, Mengqian.

Guillermo Gallego

To my parents, Gülşen and Ali Topaloğlu,
for their unconditional love and support.

Huseyin Topaloglu

Foreword

When Kalyan Talluri and I began to work on our book *The Theory and Practice of Revenue Management* (Talluri and van Ryzin 2004b) around the millennial year of 2000, the task was as clear-cut as it was daunting, attempting for the first time to synthesize a field that was young yet highly impactful, spanned multiple scientific disciplines, and whose core contributions lay scattered throughout both the academic and industrial communities.

The boundaries of what was and was not revenue management were unclear. Basic concepts, terminology, and notation were muddled. The audience for the work was ambiguous. Was it for scientists or business people? Researchers or practitioners? Students or experienced professionals? And even the format was not entirely clear: textbook, monograph, or reference book?

The end result published in 2004 was an imperfect compromise. It was overly expansive in places and not deep enough in others. Parts were accessible to general readers while others only to postdoctoral researchers. Thanks to our naivete in dealing with academic publishers, it was a pricing book that was embarrassingly overpriced.

Still, it managed to serve its purpose of defining the field for what it was: a productive blend of scientific theory with industry practice, enabled by a creative merging of business, engineering, and economic mindsets. It cataloged both the business context that leads to the practice of revenue management and the fundamental mathematical models and methods behind that practice. It helped expand the field, serving as an entry point for both students and experienced professionals alike. In these ways, it accomplished our goals in writing it.

However, 15 years have passed since *The Theory and Practice of Revenue Management* was published. Research and practice have exploded since then. Entirely new industries like ridesharing have emerged where dynamic pricing is central to the entire service model. And the need for an updated book on the field has been growing.

Revenue Management and Pricing Analytics by Guillermo Gallego and Huseyin Topaloglu is that book—and more. For starters, Gallego and Topaloglu have done more to advance the state of the art in revenue management over the past 15 years than any other researchers in the field, so the reader is in expert hands. Both are meticulous mathematicians too. And the book is clearly focused on three core topic areas central to modern revenue management: (1) traditional revenue management (capacity control and overbooking), (2) choice-based models and assortment optimization, and (3) pricing analytics (price-based controls).

Along with this increased focus comes increased depth. In particular, in the section on traditional capacity controls, the authors present in detail many new methods and results on these traditional problems, including performance bounds, new approximate dynamic programming methods, and probabilistic admission methods. The extensive section on choice-based revenue management and assortment optimization is almost entirely new, with many recent results on, for example, Markov chain choice models and the authors' signature results on the sales-based linear programming model. This section will also prove valuable to the many applied scientists working on retailing and recommendation system problems which involve assortment optimization. Lastly, the section on pricing analytics contains an updated treatment of classical price-based revenue management problems and its impact on consumers, new results on the learning-earning (explore-exploit) trade-off in combining pricing estimation and optimization, as well as results on competitive pricing and assortment optimization.

In addition to the updated and greatly expanded coverage of these core revenue management areas, the book provides an extensive collection of problems at the end of each chapter. It therefore serves as an invaluable resource for educating the next generation of revenue management scientists.

In short, it has all the qualities one could hope for as the new standard reference in the field of revenue management.

Lastly, on a personal note, I can think of no two colleagues I would rather see write this book. Guillermo has been a close friend since we carpooled together as assistant professors at Columbia and wrote our first joint paper on dynamic pricing. That paper was the start of both our careers in the field. Huseyin and I are currently colleagues together at Cornell Tech, and before that, I watched with admiration as his research career has blossomed over the past decade.

It makes me smile to see what they have accomplished together.

Cornell Tech, New York, NY, USA
Lyft Inc., San Francisco, CA, USA
March 2019

Garrett van Ryzin

Preface

Revenue management can be defined as a data-driven, computerized system to support the tactical pricing of perishable assets at the micro-market level to maximize expected revenues from sunk investments in capacity. More broadly, revenue management includes product design at the strategic level (e.g., coming up with travel restrictions and ancillary services that comprise an airline fare class as well as upgrade and upsell policies to be used) and inventory control of fare classes at the operational level (e.g., opening and closing fare classes in response to remaining capacities). While inventory control for the fare classes is the main lever in revenue management, dynamic pricing is about selecting optimal price paths over the sales horizon. Pricing analytics goes beyond dynamic pricing to include the study of consumer surplus and welfare and issues related to competition and on-line learning.

Revenue management originated in the United States after the deregulation of the airline industry in the late 1970s. The often-heard story of its origins is the dilemma faced by American Airlines (AA) when People Express (PE) offered deeply discounted fares in markets that were crucial to AA, resulting in unsustainable capacity spoilage. The alternative of matching the fares of PE was not feasible for AA due to its more expensive cost structure. The solution AA came up with was to offer non-refundable super-saver fares designed to undercut PE, while restricting availability to protect inventory for consumers willing to pay full fares. The saga ends a few months later with the bankruptcy of PE. While competition drove AA to a market segmentation strategy, competition is not essential for segmentation to be profitable. Indeed, market segmentation policies have existed for centuries in monopolies, ranging from canal and river tolls in China's Ming dynasty to Danish Sound tolls in the sixteenth century. What was new in the AA-PE saga were the restrictions (e.g., time of sale and booking limits) that went into the design of the super-saver fares, which made these fares available in limited quantities and to consumers who were willing to book tickets in advance. Interestingly, the use of different fare classes can improve expected revenues even if there is only one market segment.

Revenue management and dynamic pricing are essential tools for firms in the travel and leisure industry including airlines, hotels, car rental companies, art performances, and sports events. Revenue management systems were developed by some airlines, while others acquired customizable systems from vendors. A similar development has occurred for the hotel industry, car rental companies, and other segments of the leisure industry.

Applications of revenue management and dynamic pricing have expanded to other areas including e-commerce, health care, media, telecom, and financial services. In e-commerce, for example, retailers have to decide which products to offer and how to display them on their webpages. These are assortment optimization and product framing problems that require a deep understanding of consumer choice models. Firms offering new products with a short life cycle need to simultaneously learn and earn during the product sales horizon. Many of these new applications require modern revenue management and pricing analytics that go beyond the models developed in the travel and leisure industries.

There are several books in revenue management available to the reader. Here, we briefly mention three books and explain how the book in your hands (or on your screen) is different. *The Theory and Practice of Revenue Management* (Talluri and van Ryzin 2004b) has a comprehensive treatment of the subject that artfully mixes theory and practice including chapters on demand estimation and the economics of pricing. The book is aimed at a core audience with at least a master's degree or higher in engineering, operations research, statistics, or economics. The scope of our book is not as broad as Talluri and van Ryzin (2004b), as it does not cover demand estimation, a subject for which there are many excellent references. While also a technical book, our book aims at a broader audience that includes advance undergraduates in operations research, statistics, economics, and other related fields. Our book provides an up-to-date treatment of the subject containing many results that were not available in 2004.

The book *Pricing and Revenue Optimization* (Phillips 2005) provides a high-level treatment of the subject with many interesting managerial insights. It is the textbook of choice for courses in business schools, executives, and practitioners that are less analytically tolerant. It remains relevant today for both technical and nontechnical readers. Our book is almost orthogonal to Phillips (2005), as the focus is not on making the concepts accessible to executive or nontechnical readers.

Lastly, the book *The Oxford Handbook of Pricing Management* (Ozer and Phillips 2012) focuses on showcasing the recent research in revenue management, along with a detailed discussion of the industries in which revenue management tools find applications. The book is a comprehensive edited volume with contributed chapters. As such, it does not lend itself easily to classroom use at the master's or introductory PhD level. Our book is intended to be useful as a textbook.

The book is divided into three parts: traditional revenue management, revenue management under customer choice, and pricing analytics. Each part is approximately of the same length and written in a self-contained way, so readers can read them independently, although reading the first part may make the second part easier to understand. Each chapter ends with bibliographical notes where the reader can find the sources of the material covered as well as many useful references. Proofs of some important technical results can be found in the appendix of each chapter. Solving the end-of-chapter problems helps reinforce the material in the book, with some of the questions expanding on the subject.

The first part of the book, on traditional revenue management, has a rigorous treatment of the most important models developed in industry where time is treated implicitly and those developed in academia where time is considered explicitly. Coverage includes Littlewood's rule, bounds and heuristics for single resource and network models, overbooking, and some extensions such as the generalized newsvendor problem with convex costs. Academic models treat time explicitly which allows for arbitrary fare arrival sequences. Heuristics are developed, and conditions under which simple heuristics are asymptotically optimal are discussed.

The second part of the book, on revenue management under customer choice, is a modern treatment of revenue management that largely gives up on Littlewood's rule and its extensions in favor of a dynamic programming framework based on discrete choice models and the machinery of assortment optimization. Bounds, heuristics, and asymptotic properties are presented. This section starts with a chapter on discrete choice models, followed by a chapter on assortment optimization as these are the foundation of revenue management models under customer choice. A large class of discrete choice models can be approximated by the Markov chain choice model, and this model can in turn be used to develop bounds and heuristics for the network revenue management model that have the same complexity as the models for the independent demand model.

The third part of the book is on pricing analytics. It starts with an in-depth treatment of pricing models with linear costs (that arise as dual prices of capacity constraints) and a rigorous treatment of the necessary and sufficient conditions for existence and uniqueness of optimal prices. We show that firms prefer input costs that are random and that consumers also prefer random prices. We establish conditions on the pass-through rate under which randomness in inputs for the firm benefits both the firm and the consumer. We also provide treatments of a number of extensions including call options on capacity and models that allow for bargaining and concise treatments of the peak-load pricing and priority pricing. The chapter on dynamic pricing extends known formulations to allow for inventory replenishments, holding costs, discounted costs, compound Poisson demands, and dynamic non-linear pricing. The chapter on online learning presents some of the most relevant results in learning and earning, while the final chapter on competition presents models relevant to revenue management and dynamic pricing.

There is enough material in the book for a full-semester course for advanced undergraduate or master's students. Parts I and II can be covered in about 9 weeks and Part III in about 4 weeks excluding the last two chapters on online learning and competition, which can be assigned as independent readings.

The book has also been used for PhD seminars on revenue management and pricing analytics. These seminars would meet once a week for 3 hours at a time. Graduate students were asked to read the chapters in advance, with the instructor covering the material at a high level in 6 weeks and with students presenting papers during the next 4 weeks, leaving the last 3 weeks for project presentations.

Clearwater Bay, Hong Kong
New York, NY, USA
January 2019

Guillermo Gallego
Huseyin Topaloglu

Acknowledgments

We would like to thank our institutions, research collaborators, colleagues, and family who together made this book possible.

We would like to jointly thank the following colleagues for providing us feedback on earlier versions of the book: Gerardo Berbeglia at the University of Melbourne, Jing Dong at Columbia University, Ming Hu at the University of Toronto, Sang Jo Kim at the Shanghai University of Finance and Economics, Anran Li at London School of Economics, Lin Li at Kennesaw State University, Irene Lo at Stanford University, Preston Mcafee at Caltech, Robert Phillips at Amazon, Richard Ratliff at Sabre Corporation, Özge Şahin at Johns Hopkins University, Feng Youyi at Sichuan University, and Rachel Zhang at Hong Kong University of Science and Technology (HKUST).

We also want to recognize feedback from graduate students, including Wentau Lu, Zhuodong Tang, and many others who endured earlier drafts of the book and provided us with their feedback.

Guillermo Gallego is grateful to Columbia University and to Mr. M. C. Liu for providing the funds for the Liu Family Professorship which he held from 2012 to 2017. Professor Gallego is also grateful to HKUST and to James E. Thompson for providing the funds for the Crown Worldwide Professorship, which he holds at HKUST since 2016. The funding for these chairs gave him the extra time he needed to devote to this book.

Professor Gallego is also grateful to his academic collaborators including Jose Blanchet, Tamer Boyacı, Youyi Feng, Vineet Goyal, Ming Hu, Garud Iyengar, Anran Li, Lin Li, Michael Li, Yan Liu, Özalp Özer, Robert Phillips, Catalina Stefenescu, Kalyan Talluri, Özge Şahin, Garrett van Ryzin, and Ruxian Wang. It has been a pleasure to work with all of them. Special thanks are due to Garrett van Ryzin for early discussions about pricing that culminated in two joint papers in the mid-1990s. Special thanks are also due to Robert Phillips for many insightful discussions about pricing at our favorite tapas bar in NYC. Last, but not least, I am indebted to Huseyin Topaloglu, the coauthor of this book and several papers, for his unbounded energy and faith in this project.

Professor Gallego is also grateful to his industrial collaborators including Jose Luis Beltran, Kay-Yut Chen, Enis Kayış, Shelen Jain, and Julie Ward at Hewlett Packard, Aliza Heching and Kaan Katırcıoğlu at IBM, and Ross Darrow, Richard Ratliff, Barry Smith, Sergei Shebalov, and Ben Vinod at Sabre. Special thanks are due to Jose Luis Beltran for many fruitful discussions that culminated in several papers and the implementation of decision support systems at Hewlett Packard. Professor Gallego is particularly indebted to Richard Ratliff with whom he had numerous discussions about the theory and practice of revenue management that led to implementations at Sabre. His feedback and support have always been valued and appreciated.

Huseyin Topaloglu is grateful to Cornell University for providing a wonderful working environment. He greatly acknowledges the interactions with the Cornell colleagues Bob Bland, Jim Dai, Itai Gurvich, Shane Henderson, Kris Iyer, Peter Jackson, Nathan Kallus, Adrian Lewis, Mark Lewis, Jack Muckstadt, Uma Prabhu, Jim Renegar, Sid Resnick, Robin Roundy, Gena Samorodnitsky, David Shmoys, Les Trotter, Garret van Ryzin, and David Williamson.

Professor Topaloglu has had a delightful working relationship with the coauthor of this book, Guillermo Gallego. Professor Gallego's vision and enthusiasm for the book were unmatched, and his patience was the kindest. Working on this book and several other papers with Professor Gallego was a source of great joy. Very special thanks go to Professor Topaloglu's long-time collaborator and former Cornell colleague, Paat Rusmevichientong. He has had a big impact on Professor Topaloglu's approach to research. Professor Topaloglu fondly remembers the hours spent working on research with Professor Rusmevichientong, and he is looking forward to many more!

Professor Topaloglu's interactions with graduate students were a constant source of inspiration. Working with James Davis, Chao Ding, Alexander Erdelyi, Jake Feldman, Angela Gao, Sumit Kunnumkal, Venus Lo, Yuhang Ma, Zach Rayfield, and Mika Sumida shaped his view of revenue management. Other graduate students, Collin Chan, Matt Maxwell, and Mateo Restrepo, were also influential. Fortunately, Professor Topaloglu has had many other opportunities to collaborate with Jake Feldman and Sumit Kunnumkal after graduation. He hopes to continue this relationship and to form others.

Working with Nurşen Aydın, İlker Birbil, Hans Frenk, Shane Henderson, Ravi Kumar, Yuri Levin, Guang Li, Nan Liu, Jeff McGill, Mikhail Nediak, Nilay Noyan, Alice Paul, Warren Powell, Serdar Şimşek, Kalyan Talluri, Heng Zhang, and Serhan Ziya was a great opportunity to learn. Professor Topaloglu very deeply appreciates the time and effort these wonderful people spent with him.

Professor Topaloglu is grateful for Theresa Schmidt's love and support.

The work of Professor Topaloglu was funded by the National Science Foundation and Schmidt Sciences.

This book would not have been possible without the support of our friends and family.

Contents

Part I
Traditional Revenue Management

Chapter 1
Single Resource Revenue Management with Independent Demands

1.1 Introduction

In this chapter, we consider the single resource, independent demand revenue management problem with multiple fare classes. This problem arises in the airline industry where different fares for the same cabin are designed to cater to different market segments. As an example, a low fare may have advance purchase and length of stay restrictions and exclude ancillary services such as advance seat selection, luggage handling, and priority boarding. This low fare may target price-conscious consumers who travel for leisure on restricted budgets. On the other hand, a high fare designed for business consumers may be unrestricted, include ancillary services and be designed to be frequently available for late bookings. If requests for the low fare arrive first, the airline risks selling all of its capacity before seeing requests for the high fare. A key decision in revenue management is how much capacity to reserve for higher fare classes, or equivalently how much capacity to make available for lower fare classes. Throughout the chapter, we will refer to airline applications, but the reader should keep in mind that the models apply more generally.

We assume that the set of fare classes is given. This set of fare classes corresponds to a menu of prices, restrictions, and ancillary services. Demands for the different fare classes are assumed to be independent random variables. In particular, we assume that if a consumer finds that her preferred fare class is unavailable, then she will leave the system without purchasing anything. This assumption holds approximately when the difference in fares is large enough that demands for the different fare classes are decoupled or when consumers can find alternative sources of capacity for their preferred fare class, perhaps on a different flight or with a different carrier. In some cases, the demand for a fare class that is closed may be recaptured by other open fare classes. Demand recapture makes the independent assumption untenable. We will address this issue in a separate chapter on dependent demands based on discrete choice models.

G. Gallego, H. Topaloglu, *Revenue Management and Pricing Analytics*, International Series in Operations Research & Management Science 279, https://doi.org/10.1007/978-1-4939-9606-3_1

We will also assume that the capacity available for sale is fixed and cannot be modified or replenished during the selling horizon. Later in the chapter, we will discuss how to optimally select the initial capacity in situations where it can be purchased at an increasing convex cost. Unsold capacity is assumed to have a zero salvage value. This assumption is without loss of generality as any problem with positive salvage value can be transformed into an equivalent problem with zero salvage value.

The objective is to maximize the total expected revenue from all fare classes by dynamically choosing the fare classes to offer for sale during the selling horizon. In practice this is done by selectively denying requests for lower fare classes with the hope of using capacity for requests from higher fare classes. Consequently, the revenue curve that monitors the accumulation of the total revenue during the selling horizon may start low, but later catch up and hopefully exceed the revenue curve corresponding to a policy that accepts all requests.

In Sect. 1.2, we study the two fare class revenue management problem, where the capacity provider has to decide how much capacity to make available to the low-fare class before seeing the demand for the high fare class. In this formulation time is treated implicitly by assuming that requests for different fare classes arrive *sequentially*. The solution is given by Littlewood's rule. In Sect. 1.3, we present a dynamic programming formulation for multiple fare classes under the assumption that the requests for different fare classes arrive sequentially in a *low-before-high* order. Structural results such as concavity of the value function are also presented. While arrival order can be relaxed, the sequential arrival assumption cannot. In Sect. 1.4, we study the problem of setting initial capacity and its connections to the newsvendor problem. In Sect. 1.5, we present commonly used heuristics for the multiple fare class revenue management problem. In Sect. 1.6, we provide bounds on the optimal total expected revenue. In Sect. 1.7, we introduce a model that allows for non-sequential arrival patterns by modeling demands as independent Poisson processes with time-dependent arrival rates. In Sect. 1.8, we study models that do not allow fare classes to reopen once they are closed. This restriction may be helpful to cope with strategic consumers as it deters them from waiting for lower fares. In Sect. 1.9, we extend the analysis to the case of compound Poisson demands, where arriving consumers may request more than one unit of capacity. In Sect. 1.10, we conclude the chapter by comparing the performance of the multi-fare revenue problem under the sequential, low-before-high fare arrival pattern, to the performance of the more flexible model that allows for compound Poisson demands. Not surprisingly, the latter model outperforms the former when the sequential fare arrival pattern fails to hold.

1.2 Two Fare Classes

Consider an airline flight with c units of capacity. The capacity can be sold either at full fare for a price of p_1 or at a discounted fare for a price of $p_2 < p_1$. The discounted fare typically has advanced purchasing and usage restrictions. As an example, a round trip discounted fare may need to be purchased three weeks in advance and may require a Saturday night stay. We assume that all booked consumers will actually travel. This assumption avoids the need to overbook the capacity. We discuss how to deal with cancellations before the departure day and no-shows on the departure day in a separate chapter on overbooking models.

We assume a low-before-high fare class arrival order, which implies that demand for the discounted fare, say D_2, books before the demand for the full fare, say D_1. This arrival pattern holds approximately in practice and it is encouraged by the advance purchase restrictions imposed on the lower fare. Notice that low-before-high is a worst case arrival pattern in terms of revenues. Indeed, if the full fare consumers arrived first, then we would accept them up to the available capacity and use the residual capacity, if any, to satisfy demand from the discounted fare consumers. When the arrival pattern is low-before-high, it is critical to impose booking limits on the low-fare consumers, as otherwise the low-fare consumers may exhaust capacity and force the provider to deny capacity to consumers willing to pay higher fares.

Let c be the capacity on the flight. Suppose that we protect $y \in \{0, 1, \ldots, c\}$ units of capacity for the full fare before observing the actual demand for the discounted fare. This leaves $c - y$ units of capacity available to satisfy the demand for the discounted fare. We refer to $c - y$ as the booking limit for the discounted fare. Consequently, sales at the discounted fare are given by $\min\{c - y, D_2\}$. The remaining capacity is equal to $c - \min\{c - y, D_2\} = \max\{y, c - D_2\}$, so sales at the full fare are given by $\min\{\max\{y, c - D_2\}, D_1\}$. The total expected revenue from the two fare classes is therefore

$$W(y, c) := p_2\, \mathbb{E}\{\min\{c - y, D_2\}\} + p_1\, \mathbb{E}\{\min\{\max\{y, c - D_2\}, D_1\}\}. \tag{1.1}$$

The objective is to find the protection level y that maximizes the expected revenue $W(y, c)$. The extreme strategies $y = 0$ and $y = c$ correspond, respectively, to the case where no capacity is protected, and to the case where all of the capacity is protected for the full fare consumers. We will later discuss when these extreme strategies are optimal. In most cases, however, an intermediate strategy is optimal.

The fare ratio $r := p_2/p_1$ plays an important role in determining the optimal protection level. If the ratio is close to zero, then the full fare is substantially larger than the discounted fare and we would be inclined to protect more capacity for the full fare demand. If the ratio is close to one, then the discounted fare is close to the full fare and we would be inclined to accept more requests for the discounted fare since we can get almost the same revenue per unit of capacity. The distribution of the full fare demand is also important in determining the optimal protection level. If $\mathbb{P}\{D_1 \geq c\}$ is very large, then the full fare demand exceeds the available capacity with high probability, so it makes sense to protect the entire capacity for the full fare consumers as it is likely that the provider can sell all of the capacity at the full fare. However, if $\mathbb{P}\{D_1 \geq c\}$ is very small, then it is unlikely that all the capacity can be sold at the full fare, so fewer units should be protected for the full fare consumers. As we demonstrate shortly, the demand for the discounted fare has no influence on the optimal protection level.

We can use marginal analysis to study the tradeoff between accepting and rejecting a request for the discounted fare when we have y units of capacity available. If we accept this request, then we obtain a revenue of p_2 for the marginal unit. If we reject the request and close down the discount fare, then we will sell the y-th unit at fare p_1 only if the full fare demand D_1 is at least as large as y. Thus, it is intuitively optimal to reject the request for the discounted fare when $p_1 \mathbb{P}\{D_1 \geq y\} > p_2$. This argument suggests that an optimal protection level y_1^* is given by

$$y_1^* = \max\{y \in \mathbb{N}_+ : \mathbb{P}\{D_1 \geq y\} > r\}, \tag{1.2}$$

where $\mathbb{N}_+ = \{0, 1, \ldots\}$ is the set of non-negative integers. The formula for the optimal protection level in (1.2) is known as Littlewood's rule. Later we will show that (1.2) is a maximizer of (1.1).

Example 1.1 Suppose that D_1 is a Poisson random variable with mean 80, the full fare is $p_1 = \$100$ and the discounted fare is $p_2 = \$60$, so $r = 60/100 = 0.6$. To compute the optimal protection level y_1^*, we are interested in the cumulative tail distribution $\mathbb{P}\{D_1 \geq y\} = 1 - \mathbb{P}\{D_1 \leq y - 1\}$. Since most statistical software packages return the value of the cumulative distribution $\mathbb{P}\{D_1 \leq y - 1\}$, rather than the value of the tail distribution $\mathbb{P}\{D_1 \geq y\}$, we see that y_1^* should satisfy $\mathbb{P}\{D_1 \leq y_1^* - 1\} < 1 - r \leq \mathbb{P}\{D_1 \leq y_1^*\}$. Since $\mathbb{P}\{D_1 \leq 77\} < 0.4 \leq \mathbb{P}\{D_1 \leq 78\}$, we conclude that $y_1^* = 78$. Consequently, it is optimal to protect 78 seats for the full fare consumers. If $c = 200$, then the booking limit for the discounted fare consumers is $c - y_1^* = 122$, which indicates that it is optimal to allow at most 122 bookings for the discounted fare class. If $c \leq y_1^*$, then all of the capacity should be protected for the full fare consumers, resulting in a booking limit of zero for the discounted fare class.

Remark 1.2 The following remarks are immediately derived from Littlewood's rule.

- The optimal protection level y_1^* is independent of the distribution of the discounted fare demand D_2.
- If $\mathbb{P}\{D_1 \geq y_1^* + 1\} = r$, then $y_1^* + 1$ is also optimal protection level, so both y_1^* and $y_1^* + 1$ result in the same total expected revenue. Protecting the $y_1^* + 1$ units of capacity increases the variance of the revenue, but it reduces the probability of rejecting requests from the full fare consumers.

From Littlewood's rule in (1.2), we see that the extreme strategy $y_1^* = 0$ is optimal when $\mathbb{P}\{D_1 \geq 1\} \leq r$ and $y_1^* = c$ is optimal when $\mathbb{P}\{D_1 \geq c\} > r$.

1.2.1 Continuous Demand Distributions

Although the demand in revenue management models is inherently a discrete quantity, it can be easier to model the demand with a continuous random variable. If the demand D_1 from fare class 1 is a continuous random variable with cumulative distribution function $F_1(y) = \mathbb{P}\{D_1 \leq y\}$, then the optimal protection level is

$$y_1^* = F_1^{-1}(1 - r),$$

where $F_1^{-1}(\cdot)$ denotes the inverse of $F_1(\cdot)$. In particular, if D_1 is a normal random variable with mean μ_1 and standard deviation σ_1, then we have

$$y_1^* = \mu_1 + \sigma_1 \Phi^{-1}(1 - r),$$

where $\Phi(\cdot)$ is the cumulative distribution function of the standard normal random variable. This formula can be used to understand how the protection level changes as a function of the demand parameters. Notice that if $r < 1/2$, then $\Phi^{-1}(r) < 0$, so $y_1^* < \mu_1$ and y_1^* decreases with σ_1. Similarly, if $r > 1/2$, then $\Phi^{-1}(r) > 0$, so $y_1^* > \mu_1$ and y_1^* increases with σ_1. If $r = 1/2$, then $\Phi^{-1}(r) = 0$ so that $y_1^* = \mu_1$ and it does not depend on σ_1.

Example 1.3 Suppose that D_1 is a normal random variable with mean 80 and standard deviation 9, the full fare is $p_1 = \$100$ and the discount fare is $p_2 = \$60$. Thus, we have $y_1^* = 80 + 3 \times \Phi^{-1}(1 - 0.6) = 77.72$. Since $r < 1/2$, we have $y_1^* < 80$. Notice that the solution is quite close to that of Example 1.1. This is because a Poisson random variable with mean 80 can be well approximated by a normal random variable with mean 80 and standard deviation $9 \approx \sqrt{80}$.

1.2.2 Quality of Service, Salvage Values, and Callable Products

As we will now see, Littlewood's rule can result in poor service to consumers who prefer the high fare when the fare ratio is high. The probability of denying service to at least one high fare consumer is known as the full fare spill rate. Since at least one consumer is denied service when $D_1 > \max\{y_1^*, c - D_2\}$, we have

$$\mathbb{P}\{D_1 > \max\{y_1^*, c - D_2\}\} \leq \mathbb{P}\{D_1 > y_1^*\} \leq r < \mathbb{P}\{D_1 \geq y_1^*\},$$

where the last two inequalities follow from (1.2). We call $\mathbb{P}\{D_1 > y_1^*\}$ the maximal spill rate. Notice that if the inequality $y_1^* \geq c - D_2$ holds with high probability, as it typically does in practice when the discount fare demand D_2 is large relative to c, then the spill rate approaches the maximal spill rate which is, by design, close to the ratio r.

High spill rates may lead to the loss of full fare consumers to competition. To see this, imagine two airlines, each offering a discounted fare and a full fare in the same market, where the fare ratio r is large and there is high demand for the discounted fare. In this situation, the maximal spill rate is high, indicating that with high probability at least one fare class 1 consumer will be denied capacity. Suppose Airline A uses Littlewood's rule with spill rates close to r, which implies that Airline A turns down fare class 1 consumers with high probability. Airline B can protect more seats for the full fare consumers than recommended by Littlewood's rule. By doing so, Airline B sacrifices revenue in the short run but can attract some of the fare class 1 consumers spilled by Airline A. Over time, Airline A may see a decrease in class fare 1 demand as a secular change and protect even fewer seats for fare class 1 consumers. Meanwhile, Airline B will see an increase in class fare 1 consumers. At this point, Airline B can switch to the optimal protection level recommended by Littlewood's rule, deriving larger revenues in the long run. In essence, Airline B has strategically traded discounted fare consumers for full fare consumers with Airline A.

One way to cope with high spill rates and their adverse strategic consequences is to impose a penalty cost of ρ for each unit of full fare demand that is not served. This penalty is supposed to measure the ill will incurred when service is denied to a full fare consumer. Imposing a penalty ρ for each unit of full fare demand that is not served can be shown to be equivalent to increasing the fare of fare class 1 to $p_1 + \rho$ in the corresponding optimization problem. This leads to a modification of Littlewood's rule, resulting in optimal protection level given by

$$y_1^* = \max\left\{y \in \mathbb{N}_+ : \mathbb{P}\{D_1 \geq y\} > \frac{p_2}{p_1 + \rho}\right\}. \tag{1.3}$$

Since $p_2/(p_1+\rho) < p_2/p_1$, we get lower maximal spill rates by imposing a penalty for each unit of fare class 1 demand that is not served. Obviously, this adjustment in maximal spill rates comes at the expense of having larger protection levels and lower total expected revenues. This is, in essence, a sacrifice in expected revenues to keep a stream of consumers from defecting to competitors.

To improve the spill rate without sacrificing sales at the discount fare, the airline can modify the discount fare offering by adding a restriction that allows the airline to recall or buy back capacity when needed. This approach leads to revenue management with callable products. Callable products can be sold either by giving consumers an upfront discount or by giving them a compensation if and when capacity is recalled. When managed correctly, callable products lead to better capacity utilization, provide better service to full fare consumers, and induce demand from consumers who are attracted either to the upfront discount or to the compensation when the capacity is recalled. Callable products are common in the secondary market for event tickets as they provide a hedge against uncertainty in the supply in the sense that the liability for failing to deliver capacity is limited to the monetary compensation tied to the callable product.

We can also account for salvage values for the capacity that is not sold at the end of the selling horizon. Suppose there is a salvage value $s < p_2$ on excess capacity after the arrival of the full fare demand. This salvage value can be interpreted as the revenue from standby tickets or last minute travel deals. A salvage value of s for each unit of unsold capacity is equivalent to decreasing the fare of the fare classes 1 and 2, respectively, to $p_1 - s$ and $p_2 - s$. Therefore, using Littlewood's rule, the optimal protection level is given by

$$y_1^* = \max\left\{y \in \mathbb{N}_+ : \mathbb{P}\{D_1 \geq y\} > \frac{p_2 - s}{p_1 - s}\right\}. \tag{1.4}$$

1.3 Multiple Fare Classes

In this section, we present an exact solution to the multiple fare class problem using dynamic programming. The analysis is somewhat technical and readers may prefer to first focus on the dynamic programming formulation in (1.6) and the main results in Proposition 1.5, Theorem 1.6, and Corollary 1.7 before going over the details of the analysis.

We assume that the capacity provider has c units of perishable capacity to be allocated among n fare classes, where the fares are indexed so that $p_n < \ldots < p_1$. Lower fares typically have severe time-of-purchase and traveling restrictions. Given the time-of-purchase restriction, it is natural to assume, as we do, a low-before-high fare arrival order, with fare class n arriving first and fare class 1 arriving last. We use $N = \{1, \ldots, n\}$ to denote the set of fare classes. Let D_j denote the random demand for fare class j. We assume that the demand random variables $D_1, \ldots, D_n$ are independent of each other with finite means $\mu_j := \mathbb{E}\{D_j\} < \infty$ for all $j \in N$.

Let $V_j(x)$ denote the optimal total expected revenue from fare classes $j, j-1 \ldots, 1$ given x units of remaining capacity just before facing the demand for fare class j. To write a dynamic program, we first review the sequence of events in stage j (just before the arrival of demand for fare class j):

- Given x units of remaining capacity select protection level $y \in \{0, \ldots, x\}$ for fare classes $j-1, j-2, \ldots, 1$ and make $x - y$ units of capacity available for sale to fare class j.
- Observe demand for fare class j. The capacity sold to fare class j is given by $\min\{x-y, D_j\}$, and the revenue generated is $p_j \min\{x-y, D_j\}$.
- The remaining capacity before facing demand for fare class $j-1$ is given by $x - \min\{x-y, D_j\} = \max\{y, x-D_j\}$.

Let $W_j(y, x)$ be the optimal expected revenue from fare classes $j, j-1, \ldots, 1$ assuming that we protect $y \leq x$ units of capacity for the fare classes $j-1, j-2, \ldots, 1$. In this case, the functions $\{W_j(\cdot,\cdot) : j = n, \ldots, 1\}$ and $\{V_j(\cdot) : j = n, \ldots, 1\}$ satisfy the relationship

$$W_j(y,x) = p_j\, \mathbb{E}\{\min\{x-y, D_j\}\} + \mathbb{E}\{V_{j-1}(\max\{y, x-D_j\})\}. \tag{1.5}$$

In the expression above, $p_j\, \mathbb{E}\{\min\{x-y, D_j\}\}$ is the expected revenue obtained from fare class j given that we have x units of remaining capacity when facing the demand for fare class j and we protect y units of capacity for fare classes $j-1, j-2, \ldots 1$. On the other hand, $V_{j-1}(\max\{y, x-D_j\})$ is the optimal expected revenue obtained from fare classes $j-1, j-2, \ldots, 1$, given that we have $\max\{y, x-D_j\}$ units of remaining capacity before facing the demand from fare class $j-1$. Given x units of remaining capacity, we maximize $W_j(y,x)$ over $y \in \{0, \ldots, x\}$ resulting in the dynamic program:

$$\begin{aligned} V_j(x) &= \max_{y \in \{0,\ldots,x\}} W_j(y,x) \\ &= \max_{y \in \{0,\ldots,x\}} \Big\{ p_j \mathbb{E}\{\min\{x-y, D_j\}\} + \mathbb{E}\{V_{j-1}(\max\{y, x-D_j\})\} \Big\}. \end{aligned} \tag{1.6}$$

By convention, we have $V_0(x) = 0$, since we do not collect any revenue when there are no fare classes left to arrive. By definition, $V_n(c)$ is the optimal total expected revenue for the multiple fare class problem with n fare classes and an initial capacity of c units. Assuming that computing each one of the expectations in (1.6) takes constant time, we can solve the dynamic program in $O(c^2)$ operations.

1.3.1 Structure of the Optimal Policy

For any function f with integer domain, let $\Delta f(x) = f(x) - f(x-1)$.

The following lemma is important in establishing structural results.

Lemma 1.4 *Let $g(x) := \mathbb{E}\{G(\min\{X, x\})\}$, where X is an integer valued random variable with $\mathbb{E}X < \infty$, and $G(\cdot)$ is a function over the integers. Then,*

$$\Delta g(x) = \Delta G(x)\,\mathbb{P}\{X \geq x\}.$$

Similarly, let $h(x) := \mathbb{E}\{H(\max\{X, x\})\}$, where X is an integer valued random variable with $\mathbb{E}X < \infty$, and $H(\cdot)$ is a function over the integers. Then,

$$\Delta h(x) = \Delta H(x)\,\mathbb{P}\{X < x\}.$$

The next two results provide the key results of this section.

Proposition 1.5 *The value functions computed through* (1.6) *satisfy the following properties.*

- *$\Delta V_j(x)$ is decreasing in $x \in \{1, \ldots, c\}$.*
- *$\Delta V_j(x)$ is increasing in $j \in \{1, \ldots, n\}$.*

Since $\Delta V_j(x)$ is decreasing in x, the value function $V_j(\cdot)$ is concave. The following theorem speaks to the monotonicity of the protection levels.

Theorem 1.6 *For all $j = n, \ldots, 1$, the function $W_j(y, x)$ is unimodal in y and the maximizer of $W_j(y, x)$ over $y \in \{0, \ldots, x\}$ is given by* $\min\{y^*_{j-1}, x\}$, *where*

$$y^*_{j-1} = \max\{y \in \mathbb{N}_+ : \Delta V_{j-1}(y) > p_j\}. \tag{1.7}$$

Moreover, $y^*_{n-1} \geq y^*_{n-2} \geq \cdots \geq y^*_1 \geq y^*_0 = 0$. Thus, optimal protection levels are monotone in the number of stages left until the end of the selling horizon.

For the case of $n = 2$, we get a formal proof of Littlewood's rule.

Corollary 1.7

$$y^*_1 = \max\{y \in \mathbb{N}_+ : p_1 \mathbb{P}\{D_1 \geq y\} > p_2\}.$$

Likewise, it is possible to show (1.3) and (1.4) by appropriately modifying $V_0(x)$.

Table 1.1 Optimal total expected revenues $V_j(c)$ for the values of capacity $c \in \{50, 100, 150, 200, 250, 300, 350\}$ and $j = 1, 2, 3, 4, 5$

c	Load factor	$V_1(c)$	$V_2(c)$	$V_3(c)$	$V_4(c)$	$V_5(c)$
50	560%	1500	3427	3427	3427	3427
100	280%	1500	3900	5441	5441	5441
150	187%	1500	3900	5900	7189	7189
200	140%	1500	3900	5900	7825	8159
250	112%	1500	3900	5900	7825	8909
300	93%	1500	3900	5900	7825	9564
350	80%	1500	3900	5900	7825	9625

Remark 1.8 The following remarks apply for the structure of the optimal policy:

- Let x_j be the remaining capacity just before facing the demand for fare class j. Then capacity is allocated to fare class j only if $x_j > y^*_{j-1}$ with at most $[x_j - y^*_{j-1}]^+$ bookings allowed.
- The protection level y^*_{j-1} is independent of the distribution of the demand from fare classes $n, n-1, \ldots, j$.
- The policy is implemented as follows: At stage n, we start with $x_n = c$ units of inventory and we protect $y_{n-1}(x_n) = \min\{y^*_{n-1}, x_n\}$ units of capacity for fares $n-1, n-2, \ldots, 1$. Therefore, we allow up to $[x_n - y^*_{n-1}]^+$ units of capacity to be sold to fare class n. We sell $\min\{[x_n - y^*_{n-1}]^+, D_n\}$ units of capacity to fare class n and we have a remaining capacity of $x_{n-1} = x_n - \min\{[x_n - y^*_{n-1}]^+, D_n\}$ at stage $n-1$. We protect $y_{n-2}(x_{n-1}) = \min\{y^*_{n-2}, x_{n-1}\}$ units of capacity for fares $n-2, n-1, \ldots, 1$. Therefore, we allow up to $[x_{n-1} - y^*_{n-2}]^+$ units of capacity to be sold to fare class $n-1$. We continue until we reach stage 1 with x_1 units of capacity, allowing $(x_1 - y_0)^+ = (x_1 - 0)^+ = x_1$ to be sold to fare class 1.

Example 1.9 Suppose that there are five fare classes. The demand for all fare classes is a Poisson random variable. The fares and the expected demand for the five fare classes are given by $(p_5, p_4, p_3, p_2, p_1) = (15, 35, 40, 60, 100)$ and $(\mathbb{E}\{D_5\}, \mathbb{E}\{D_4\}, \mathbb{E}\{D_3\}, \mathbb{E}\{D_2\}, \mathbb{E}\{D_1\}) = (120, 55, 50, 40, 15)$. For this problem instance, the optimal protection levels are $y^*_4 = 169$, $y^*_3 = 101$, $y^*_2 = 54$, and $y^*_1 = 14$. In Table 1.1, we show the total expected revenue $V_j(c)$ obtained from fare classes $j, j-1, \ldots, 1$ when we have c units of remaining capacity at the beginning of stage j, as well as the corresponding load factors $\sum_{j=1}^{5} \mathbb{E}\{D_j\}/c = 280/c$.

The effect of restricting capacity for low fares is apparent in the pattern of total expected revenues across fare classes. For example, the total expected revenues $V_2(50)$, $V_3(50)$, $V_4(50)$, and $V_5(50)$ are all equal to \$3427 since fare classes 5, 4, and 3 are rationed when $c = 50$. On the other hand, $V_1(350)$ through $V_5(350)$ vary from \$1500 to \$9625 since there is enough capacity to accommodate all or nearly all demand from the five fare classes. In Fig. 1.1, we show the marginal value of capacity $\Delta V_j(x)$ when we have x units of remaining capacity at the beginning of stage j. The marginal value of capacity increases as we have fewer units of capacity and as we have more stages left until the end of the selling horizon.

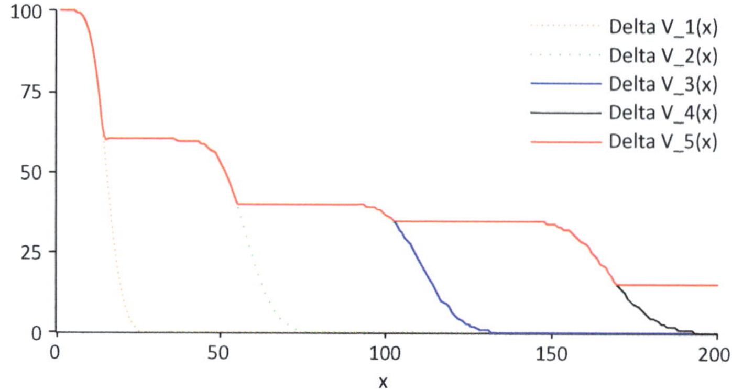

Fig. 1.1 Marginal value of capacity $\Delta V_j(x)$ as a function of x for $j = 1, 2, 3, 4, 5$

Although computing the optimal policy is not numerically onerous, some computations can be streamlined to obtain a more efficient implementation. It can be shown that

$$\Delta V_j(x) = p_j \, \mathbb{P}\{D_j \geq x - y^*_{j-1}\} + \sum_{k=0}^{x-y^*_{j-1}-1} \Delta V_{j-1}(x-k) \, \mathbb{P}\{D_j = k\} \qquad (1.8)$$

for all $x > y^*_{j-1}$, and that $\Delta V_j(x) = \Delta V_{j-1}(x)$ if $x \leq y^*_{j-1}$. The proof of this result is left as an exercise.

1.3.2 *Nonmonotone Fares*

It is possible to relax the low-before-high assumption while retaining the assumption that requests for the different fare classes arrive sequentially. Proposition 1.5 holds as stated. The optimal protection level y^*_{j-1} is computed as stated in Theorem 1.6, but optimal protection levels are not necessarily monotone. Clearly $y^*_{j-1} = 0$ whenever $p_j > \max\{p_{j-1}, p_{j-2}, \ldots p_1, \}$ since it is optimal to accept all requests for fare class j up to capacity since there is no point in protecting capacity to sell it later at a lower fare! As an example, suppose that $p_3 < p_2$ and $p_2 > p_1$. Since no fare classes are arriving after fare class 1, it is optimal to serve the demand from fare class 1 as much as possible, which implies that $V_1(x) = p_1 \, \mathbb{E}\{\min\{x, D_1\}\}$. Using Lemma 1.4, we obtain $\Delta V_1(x) = p_1 \, \mathbb{P}\{D_1 \geq x\} < p_2$, where the inequality uses the fact that $p_2 > p_1$. In this case, (1.7) implies that $y^*_1 = 0$. Since $y^*_1 = 0$, we see that $V_2(x) = p_2 \, \mathbb{E} \min\{x, D_2\} + \mathbb{E}\{V_1([x - D_1]^+)\} = p_2 \, \mathbb{E} \min\{x, D_2\} + p_1 \, \mathbb{E}\{\min\{D_1, [x - D_2]^+\}\}$. One can check that

$$\Delta V_2(x) = p_2 \, \mathbb{P}\{D_2 \geq x\} + p_1 \, \mathbb{P}\{D_2 < x \leq D[1,2]\} \geq p_2 \, \mathbb{P}\{D_2 \geq x\},$$

where we use $D[i, j] := \sum_{k=i}^{j} D_k$. The optimal amount of capacity to protect for fare classes 2 and 1 is given by $y_2^* = \max\{y \in \mathbb{N}_+ : \Delta V_2(y) > p_3\}$. On the other hand, if there were no fare classes arriving after fare class 2, then by Littlewood's rule, the optimal amount of capacity to protect for fare classes 2 would be $\max\{y \in \mathbb{N}_+ : \mathbb{P}\{D_2 \geq y\} > p_3/p_2\}$. Since $\Delta V_2(x) > p_2 \mathbb{P}\{D_2 \geq x\}$, the capacity protected for fare classes 2 and 1 is larger than it would be when there was no demand for fare class 1.

1.4 The Generalized Newsvendor Problem

Consider the problem of selecting c to maximize $\Pi_n(c) := V_n(c) - K(c)$, where $V_n(c)$ is the solution to the multi-fare revenue management problem and $K(c)$ is the cost of procuring c units of capacity. This problem is relevant in revenue management when capacity decisions are made. We will assume that $K(c)$ is increasing convex. Two plausible models are the linear model: $K(c) = k\,c$ for some unit cost k, or the fixed cost model with a capacity limit: $K(0) = 0$, $K(c) = k$ for all $0 < c \leq \bar{c}$ and $K(c) = \infty$ for $c > \bar{c}$.

The smallest maximizer of $\Pi_n(c)$, say c^*, is characterized in the following proposition.

Proposition 1.10 *The smallest optimal procurement quantity is given by*

$$c^* = \max\{c \in \mathbb{N}_+ \ : \ \Delta V_n(c) > \Delta K(c)\}.$$

For the linear model, we can write c^* as a function of k, yielding

$$c(k) = \max\{c \in \mathbb{N}_+ \ : \ \Delta V_n(c) > k\}.$$

The order quantities at the price points are the corresponding protection levels of the corresponding revenue management problem, so

$$c(p_{j+1}) = y_j \quad \forall \ j \in \{1, \ldots, n-1\},$$

with $c(k)$ increasing in k.

For the fixed cost model, $c^* = \bar{c}$ if $V_n(\bar{c}) > k$ and $c^* = 0$ otherwise.

Figure 1.2 depicts $c(k)$ for the data of Example 1.9, with $c(60) = 14$, $c(40) = 54$, $c(35) = 101$, and $c(15) = 169$.

In a retail setting, the low-before-high arrival pattern is unlikely to hold, and a more useful model is to seek c to maximize $\Pi_n(c) = V_n(c) - k\,c$, where

$$V_n(c) := \sum_{j=1}^{n} p_j \, \mathbb{E}\{\min\{D_j, (c - D[1, j-1])^+\}\}$$

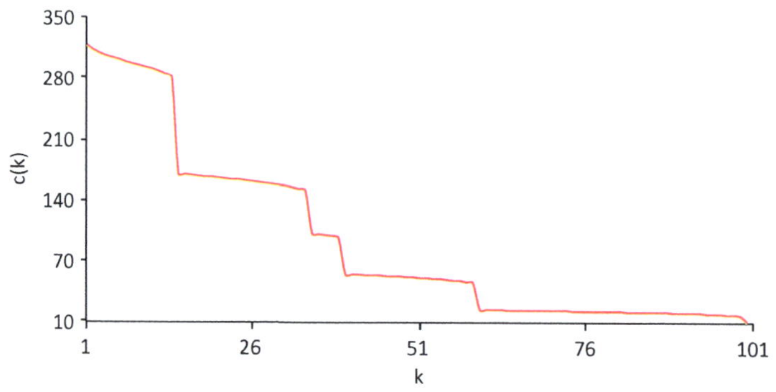

Fig. 1.2 Optimal capacity as a function of cost for Example 1.9

is the expected revenue corresponding to the high-before-low arrival pattern: $D_1, D_2, \ldots, D_n$ at prices $p_1 > p_2 > \ldots > p_n$. The marginal value of capacity can be written as

$$\Delta V_n(c) = \sum_{j=1}^{n} (p_j - p_{j+1}) \mathbb{P}\{D[1, j] > c\},$$

where for convenience we define $p_{n+1} = 0$. This leads directly to the following result.

Corollary 1.11 *In a retail setting, with a high-before-low arrival pattern, an optimal order quantity for the linear cost model $K(c) = k\,c$ is given by*

$$c(k) = \max\{c \in \mathbb{N}_+ : \sum_{j=1}^{n} (p_j - p_{j+1}) \mathbb{P}\{D[1, j] > c\} > k\}.$$

The classical newsvendor problem corresponds to the case $n = 2$ with $p = p_1 > p_2 = s$, where s is the salvage value and demand at the salvage value is infinite (an implicit assumption that is seldom discussed in the context of the newsvendor model). The model presented here is more general, even if $n = 2$, as it allows for random demand at the salvage value. In fact, it allows for random demand at all discounted prices $p_2, \ldots, p_n$.

Integrating dynamic pricing into the newsvendor problem allows for multiple price discounts, and this results in higher profits as more sales occur at price points above the salvage value.

Example 1.12 Consider a newsvendor problem with four prices and independent normally distributed demands as given in Table 1.2. At $k = 30$, the optimal order quantity is $c = 121$, resulting in an expected profit of \$4,101.05. Suppose instead, that the retailer used only price $p_1 = 100$ and salvage value $s = 20$ at which all residual capacity can be sold. In this case, the retailer will order $c = 30$ units and an expected profit of \$1,666.67.

Table 1.2 Problem parameters for Example 1.12

j	p_j	μ_j	σ_j
1	100	125	5
2	70	36	6
3	50	50	10
4	25	100	20

Further research is needed in terms of selecting a multi-price schedule $p_1 > p_2 > \ldots > p_n$ to maximize expected profits taking into account strategic behavior as consumers may want to wait for a lower price at the cost of deriving lower utility from the product and at the risk of being rationed.

1.5 Heuristics for Multiple Fare Classes

Several heuristics for the multiple fare class problem were developed in the 1980s. These heuristics are essentially extensions of Littlewood's rule. The most important heuristics are known as EMSR-a and EMSR-b, where EMSR stands for expected marginal seat revenue. For a while, some of these heuristics were even thought to be optimal by their proponents, until comparisons with optimal policies based on dynamic programming were carried out in the 1990s. By that time heuristics were already part of implemented systems and industry practitioners were reluctant to replace them with the solutions provided by dynamic programming. There are several reasons for this reluctance. People feel more comfortable with something they understand. Also, the performance gap between the heuristics and the optimal policies tends to be small. Finally, there is a feeling among some users that the heuristics may be more robust to demand estimation errors.

EMSR-a is based on the idea of adding protection levels produced by applying Littlewood's rule to each pair of fare classes. Suppose that we are at stage j and we need to decide how much capacity to protect for fare classes $j-1, j-2, \ldots, 1$. We can use Littlewood's rule to decide how much capacity to protect for fare class k demand against fare class j for all $k = j-1, \ldots, 1$. More precisely, we compute y_{kj}^* as

$$y_{kj}^* = \max\left\{ y \in \mathbb{N}_+ : \mathbb{P}\{D_k \geq y\} > \frac{p_j}{p_k} \right\}.$$

so that y_{kj}^* is the amount of capacity that we protect for fare class k when serving the demand for fare class j. In this case, EMSR-a heuristic protects

$$y_{j-1}^a := \sum_{k=1}^{j-1} y_{kj}^*$$

units of capacity for fare classes $j-1, \ldots, 1$ when serving the demand for fare class j. In particular, if the demand D_k for fare class k is a normal random variable with mean μ_k and standard deviation σ_k, then $y^*_{jk} = \mu_k + \sigma_k \Phi^{-1}(1 - p_j/p_k)$, so

$$y^a_{j-1} = \mu[1, j-1] + \sum_{k=1}^{j-1} \sigma_k \Phi^{-1}(1 - p_j/p_k),$$

we use $\mu[i, j] := \sum_{k=i}^{j} \mu_k$. Notice that EMSR-a heuristic involves $j - 1$ calls to Littlewood's rule to compute the protection level y^a_{j-1}.

EMSR-b heuristic is based on a single call to Littlewood's rule to compute each protection level. Suppose that we are at stage j and we need to decide how much capacity to protect for fare classes $j-1, j-2, \ldots, 1$. At this point in the problem, we assume that there are two fare classes, one fare class corresponding to fare class j and another fare class corresponding to the aggregation of fare classes $j-1, j-2, \ldots, 1$. The demand from fare class j is given by D_j, and the demand from the fare class that corresponds to the aggregation of fare classes $j-1, j-2, \ldots, 1$ is $D[1, j-1]$. The fare associated with fare class j is p_j. To compute the fare associated with the fare class that corresponds to the aggregation of fare classes $j-1, j-2, \ldots, 1$, we use a weighted sum of the fares of the aggregated fare classes and set the fare associated with the aggregated fare class as

$$\bar{p}_{j-1} = \sum_{k=1}^{j-1} p_k \frac{\mu_k}{\mu[1, j-1]}.$$

When serving the demand for fare class j, to decide how many units of capacity to protect for fare classes $j-1, j-2, \ldots, 1$, we compute the protection level in a two fare class problem, where the demand random variables for the two fare classes are D_j and $D[1, j-1]$, whereas the fares for the two fare classes are p_j and $\bar{p}_{j-1}$. Thus, when serving the demand for fare class j, EMSR-b heuristic protects

$$y^b_{j-1} = \max\left\{ y \in \mathbb{N}_+ : \mathbb{P}\{D[1, j-1] \geq y\} > \frac{p_j}{\bar{p}_{j-1}} \right\}$$

units of capacity for fare classes $j-1, j-2, \ldots, 1$. We note that using EMSR-b heuristic requires the distribution of $D[1, j-1] := \sum_{k=1}^{j-1} D_k$. Computing the distribution of $D[1, j-1]$ requires a convolution, but in some cases, such as the case where the demand has normal or Poisson distribution, the distribution of $D[1, j-1]$ can be easily obtained, since sums of independent normal or Poisson random variables are, respectively, also normal or Poisson random variables. In the special case where the demand for fare class j is a normal random variable with mean μ_j and standard deviation σ_j, we obtain

$$y^b_{j-1} = \mu[1, j-1] + \sigma[1, j-1] \Phi^{-1}(1 - p_j/\bar{p}_{j-1}),$$

where $\sigma[1, j-1]$ is the standard deviation of the demand $D[1, j-1]$ from aggregated fare class. More specifically, we have $\sigma[1, j-1] = \sqrt{\sum_{k=1}^{j-1} \sigma_k^2}$.

Once we compute protection levels either by using EMSR-a or EMSR-b heuristic, we use these protection levels to make capacity allocation decisions in the same way we use the optimal protection levels that are computed through the dynamic programming formulation of the problem. In particular, using $y_{n-1}^h, y_{n-2}^h, \ldots, y_1^h$ to denote the protection level computed by any heuristic, when serving the demand for fare class j, we protect y_{j-1}^h units of capacity for fare classes $j-1, j-2, \ldots, 1$. Thus, if we have x units of remaining capacity and $x \leq y_{j-1}^h$, then we do not make any capacity available for fare class j, so we do not serve any demand from fare class j and the remaining capacity at the next stage is still x. If $x > y_{j-1}^h$, then we make $x - y_{j-1}^h$ unit of capacity available for fare class j, so sales to fare class j equal $\min\{x - y_{j-1}^h, D_j\}$ and the remaining capacity that we have at the next stage is $x - \min\{x - y_{j-1}^h, D_j\} = \max\{y_{j-1}^h, x - D_j\}$. Let $V_j^h(x)$ be the total expected revenue obtained from the fare classes $j, j-1, \ldots, 1$ with x units of remaining capacity at stage j using heuristic protection levels $y_{n-1}^h, y_{n-2}^h, \ldots, y_1^h$, we can compute $\{V_j^h(\cdot) : j = n, \ldots, 1\}$ by using the recursion:

$$V_j^h(x) = \begin{cases} V_{j-1}^h(x) & \text{if } x \leq y_{j-1}^* \\ p_j\, \mathbb{E}\{\min\{x - y_{j-1}^*, D_j\}\} & \\ \qquad + \mathbb{E}\{V_{j-1}^h(\max\{y_{j-1}^*, x - D_j\})\} & \text{if } x > y_{j-1}^* \end{cases}$$

with the boundary condition that $V_0^h(x) = 0$.

Alternatively, the values $V_j^h(x)$ can be estimated using Monte Carlo simulation. Using $D_n^k, D_{n-1}^k, \ldots, D_1^k$ to denote the k-th sample, we can simulate the decisions made by using the protection levels $y_{n-1}^h, y_{n-2}^h, \ldots, y_1^h$. We start with a capacity of $x_n^k = c$ at stage n. Given that we have x_j^k units of remaining capacity at stage n, if $x_j^k \leq y_{j-1}^h$, then we do not make any capacity available for fare class j, so $s_j^k = 0$. If $x_j^k > y_{j-1}^h$, then we make $x_j^k - y_{j-1}^h$ units of capacity available for fare class j, so $s_j^k = \min\{x_j^k - y_{j-1}^h, D_j^k\}$. The remaining capacity at stage $j+1$ is $x_{j+1}^k = x_j^k - s_k^k$. For the k-th sample, the total revenue is $\sum_{j=1}^n p_j s_j^k$. Averaging the total revenue over many demand samples provides an estimate of the total expected revenue obtained by a set of protection levels.

Example 1.13 We have applied the EMSR-a and EMSR-b heuristics to the problem instance in Example 1.9. Recall that the optimal protection levels for this problem instance are $y_4^* = 169$, $y_3^* = 101$, $y_2^* = 54$, and $y_1^* = 14$. The protection levels provided by EMSR-a heuristic are $y_4^a = 171$, $y_3^a = 97$, $y_2^a = 53$, and $y_1^a = 14$. The protection levels provided by EMSR-b heuristic are $y_4^b = 166$, $y_3^b = 102$, $y_2^b = 54$, and $y_1^b = 14$. In Table 1.3, we show the total expected revenues obtained by the two heuristics and the optimal total expected revenue for different values of initial

Table 1.3 Performance of EMSR-a and EMSR-b heuristics for Example 1.13

c	Load factor	$V_5^a(c)$	$V_5^b(c)$	$V_5(c)$
50	560%	3427	3427	3427
100	280%	5432	5441	5441
150	187%	7181	7189	7189
200	140%	8157	8151	8159
250	112%	8907	8901	8909
300	93%	9564	9563	9564
350	80%	9625	9625	9625

Table 1.4 Problem parameters for Example 1.14

j	p_j	μ_j	σ_j	y_j^*	y_j^a	y_j^b
1	1050	17.3	5.8	16.7	16.7	16.7
2	567	45.1	15.0	42.5	38.7	50.9
3	534	39.6	13.2	72.3	55.7	83.2
4	520	34.0	11.3			

capacity. In particular, $V_5^a(c)$ and $V_5^b(c)$ correspond to the total expected revenues obtained by EMSR-a and EMSR-b heuristics with c units of initial capacity, whereas $V_5(c)$ corresponds to the optimal total expected revenue.

As seen in Table 1.3, the performance of both heuristics is close to optimal. We recall that this problem instance involves Poisson demand random variables and a low-before-high fare arrival order. The heuristics continue to perform well if the demand random variables are compound Poisson and the aggregate demand is approximated by a gamma random variable. However, the model constructed in this chapter makes strong assumptions about the arrival order of the fares and the heuristics may not perform well when the arrival order assumption does not hold.

Two more examples are presented below.

Example 1.14 There are four fare classes. The demand for each fare class j is normally distributed with mean μ_j and standard deviation σ_j. Table 1.4 shows the problem parameters and the protection levels computed by EMSR-a and EMSR-b heuristics, as well as the optimal protection levels. There are considerable discrepancies between the protection levels computed by the different approaches. The discrepancies are less severe in the later stages, since we essentially deal with a problem with a small number of fare classes in the later stages.

The total expected revenues obtained by the optimal policy, estimated through a simulation study with 500,000 replications, are shown in Table 1.5. Capacity is varied from 80 to 160 to create load factors in the range 1.7–0.85. The percent suboptimality of the two heuristics is also reported. For this problem instance, EMSR-a performs slightly better than EMSR-b, but both perform quite well, despite the discrepancies in the protection levels.

Example 1.15 This problem instance is similar to the one in Example 1.14. The only difference is in the fares of fare classes 2 and 3. The problem parameters and the protection levels are shown in Table 1.6. The total expected revenues

Table 1.5 Performance of EMRS-a and EMSR-b heuristics for Example 1.14

c	Load factor	$V_n(c)$	EMSR-a % Sub	EMSR-b % Sub
80	1.70	49,642	0.33%	0.43%
90	1.51	54,855	0.24%	0.52%
100	1.36	60,015	0.13%	0.44%
110	1.24	65,076	0.06%	0.34%
120	1.13	69,801	0.02%	0.21%
130	1.05	73,926	0.01%	0.10%
140	0.97	77,252	0.00%	0.04%
150	0.91	79,617	0.00%	0.01%
160	0.85	81,100	0.00%	0.00%

Table 1.6 Problem parameters and protection levels for Example 1.15

j	p_j	μ_j	σ_j	y_j^*	y_j^a	y_j^b
1	1050	17.3	5.8	9.7	9.8	9.8
2	950	45.1	15.0	54.0	50.4	53.3
3	699	39.6	13.2	98.2	91.6	96.8
4	520	34.0	11.3			

Table 1.7 Performance of EMRS-a and EMSR-b heuristics for Example 1.15

c	Load factor	$V_n(c)$	EMSR-a % Sub	EMSR-b % Sub
80	1.70	67,505	0.10%	0.00%
90	1.51	74,003	0.06%	0.00%
100	1.36	79,615	0.40%	0.02%
110	1.24	84,817	0.35%	0.02%
120	1.13	89,963	0.27%	0.01%
130	1.05	94,860	0.15%	0.01%
140	0.97	99,164	0.06%	0.01%
150	0.91	102,418	0.01%	0.00%
160	0.85	104,390	0.00%	0.00%

obtained by the optimal policy, estimated through a simulation study with 500,000 replications, are shown in Table 1.7, as well as the percent suboptimality gaps of the two heuristics. For this problem instance, both heuristics continue to perform well and EMSR-b has a slight edge.

1.6 Bounds on Optimal Expected Revenue

In this section, we show how to quickly compute lower and upper bounds on $V_n(c)$. It is natural to ask why we need bounds on $V_n(c)$ when we can compute this quantity exactly in $O(c^2)$ operations. Although we can compute $V_n(c)$ for the single resource, multiple fare problem, we will later encounter problems where exact computations are either not possible or very time consuming. In such cases, having bounds on the optimal total expected revenue becomes useful. The techniques that we develop in this section form a stepping stone for computing bounds on the optimal total expected revenue for more complicated revenue management problems.

To obtain an upper bound on $V_n(c)$, consider the perfect foresight problem where the demand vector $D = (D_n, \ldots, D_1)$ is known in advance. Having access to the demand for all fare classes in advance allows us to optimally allocate the capacity to the different fare classes by solving the knapsack type problem

$$\bar{V}(c \mid D) := \max \left\{ \sum_{j=1}^{n} p_j x_j : \sum_{j=1}^{n} x_j \leq c,\ 0 \leq x_j \leq D_j \ \forall\, j = 1, \ldots, n \right\}. \tag{1.9}$$

For each realization D of the demand random variables, advance knowledge results in a total revenue that is at least as large as the total revenue collected by the optimal policy with sequential arrivals and unknown demands. As a result, $\mathbb{E}\{\bar{V}(c \mid D)\} \geq V_n(c)$. For convenience, we will denote this upper bound as $V_n^U(c) := \mathbb{E}\{\bar{V}(c \mid D)\}$.

The solution to problem (1.9) can be written in explicit form as it is optimal to serve the demand from fare class 1 as much as possible before serving the demand from fare class 2. Therefore, the optimal value of the decision variable x_j is given by $\min\{D_j, (c - D[1, j-1])^+\}$. In this expression, we note that $(c - D[1, j-1])^+$ is the remaining capacity after we satisfy the demand from fare classes $1, 2, \ldots, j-1$ as much as possible. Therefore, the optimal objective value of problem (1.9) is given by

$$\bar{V}(c \mid D) = \sum_{j=1}^{n} p_j \min\{D_j, (c - D[1, j-1])^+\}.$$

Taking expectations, results in $V_n^U(c)$ as

$$\begin{aligned} V_n^U(c) &= \sum_{j=1}^{n} p_j\, \mathbb{E}\{\min\{D_j, (c - D[1, j-1])^+\}\} \\ &= \sum_{j=1}^{n} p_j\, (\mathbb{E}\{\min\{D[1, j], c\}\} - \mathbb{E}\{\min\{D[1, j-1], c\}\}) \\ &= \sum_{j=1}^{n} (p_j - p_{j+1})\, \mathbb{E}\{\min\{D[1, j], c\}\}, \\ &= \sum_{j=1}^{n} (p_j - p_{j+1}) \sum_{k=1}^{c} \mathbb{P}\{D[1, j] \geq k\} \end{aligned} \tag{1.10}$$

where we define $p_{n+1} \equiv 0$. The second equality follows from the fact that $\min\{D_j, (c - D[1, j-1])^+\} = \min\{D[1, j], c\} - \min\{D[1, j-1], c\}$. Computing this upper bound requires the evaluation of $\mathbb{E}\{\min\{D[1, j], c\}\}$. If $D[1, j]$ is a non-negative integer random variable, then $\mathbb{E}\{\min\{D[1, j], c\}\} = \sum_{k=1}^{c} \mathbb{P}\{D[1, j] \geq k\}$ and this justifies the last equality.

To make the upper bound more tractable, notice that $\bar{V}(c \mid D)$ is concave in D, so by Jensen's inequality $V_n^U(c) = \mathbb{E}\{\bar{V}(c \mid D)\} \leq \bar{V}(c \mid \mathbb{E}\{D\}) := \bar{V}_n(c)$. Letting $\mu_j := \mathbb{E}\{D_j\}$, we see that

$$\bar{V}_n(c) = \max\left\{\sum_{j=1}^{n} p_j x_j \,:\, \sum_{j=1}^{n} x_j \leq c, \quad 0 \leq x_j \leq \mu_j \ \forall j = 1, \ldots, n\right\} \tag{1.11}$$

is the solution to a linear program (1.11) known as the fluid model or the deterministic capacity allocation problem. It is a knapsack problem. Similar to problem (1.9), the optimal value of the decision variable x_j is given by $\min\{\mu_j, (c - \mu[1, j-1])^+\}$. In this case, the optimal objective value of problem (1.11) is

$$\bar{V}_n(c) = \sum_{j=1}^{n} (p_j - p_{j+1}) \min\{\mu[1, j], c\},$$

where $\mu[1, j] := \sum_{i=1}^{j} \mu_i$.

A lower bound on the optimal total expected revenue can be obtained by following a policy that uses zero protection levels. In this case, since the demand from fare class n arrives before the demand from fare class $n-1$, we serve the demand from fare class n as much as possible before servicing the demand from fare class $n-1$. Thus, the sales for fare class j are given by $\min\{D_j, (c - D[j+1, n])^+\}$, where $c - D[j+1, n])^+$ captures the remaining capacity after we satisfy the demand from fare classes $n, n-1, \ldots, j+1$. Therefore, the total expected revenue obtained by the policy that uses zero protection levels is $V^L(c) = \sum_{j=1}^{n} p_j \, \mathbb{E}\{\min\{D_j, (c - D[j+1, n])^+\}\}$. Similar to our approach in (1.10), we can simplify this expression as

$$\begin{aligned} V_n^L(c) &= \sum_{j=1}^{n} p_j \, \mathbb{E}\{\min\{D_k, (c - D[j+1, n])^+\}\} \\ &= \sum_{j=1}^{n} p_j (\mathbb{E} \min\{D[j, n], c\}\} - \mathbb{E} \min\{D[j+1, n], c\}\}) \\ &= \sum_{j=1}^{n} (p_j - p_{j-1}) \, \mathbb{E}\{\min\{D[j, n], c\}\}, \end{aligned} \tag{1.12}$$

where we define $p_0 = 0$. Notice that all of the terms in the sum are negative except for the term with $j = 1$. By the preceding discussion, it follows that $V_n^L(c) \leq V_n(c)$, since the total expected revenue $V_n^L(c)$ is computed under the possible suboptimal protection levels, which are all equal to zero. The above arguments justify the following proposition.

Proposition 1.16 *For the multiple fare class problem, we have*

$$V_n^L(c) \leq V_n(c) \leq V_n^U(c) \leq \bar{V}_n(c).$$

1.6.1 Revenue Opportunity Model

The bounds presented here can help with the so-called revenue opportunity model. The revenue opportunity is the spread between the ex-post optimal revenue using the estimated uncensored demand, and the revenue that results from not applying booking controls. Demand uncensoring refers to a statistical technique that attempts to estimate actual demand from the observed sales, which may be constrained by booking limits. The ex-post optimal revenue is a hindsight optimization and is equivalent to our perfect foresight model, resulting in revenue $\bar{V}(c|D)$, where D is the uncensored demand. On the other hand, the revenue that results from not applying booking controls is given by $V_n^L(c|D)$, which corresponds to the expression in (1.12) before taking the expectation. So for a given realization of demand, a measure of the revenue opportunity is defined as is $\bar{V}(c|D) - V_n^L(c|D)$. The achieved revenue opportunity is the difference between the actual revenue from applying optimal or heuristic controls and the lower bound. The ratio of the achieved revenue opportunity to the revenue opportunity is often called the percentage achieved revenue opportunity. The revenue opportunity $\bar{V}(c|D) - V_n^L(c|D)$ is sometimes approximated by its expectation $V_n^U(c) - V_n^L(c)$. Tables 1.8 and 1.9 show there is a significant revenue opportunity, particularly for $c \leq 140$. Thus, one use for the revenue opportunity model is to identify situations where revenue management has the most potential so that more effort can be put where it is most needed. The revenue opportunity model has also been used to show the benefits of using network-based controls versus using leg-based controls in networks.

Example 1.17 Tables 1.8 and 1.9 report $V_n^L(c)$, $V_n(c)$, $V_n^U(c)$, and $\bar{V}_n(c)$ for the data of Examples 1.14 and 1.15, respectively. Notice that $V_n^U(c)$ represents a significant improvement over the better known bound $\bar{V}_n(c)$, particularly for intermediate values of capacity. The spread $V_n^U(c) - V_n^L(c)$ between the lower and upper bound is a gauge of the potential improvements in revenues from using an optimal or heuristic admission control policy. If capacity is scarce relative to the potential demand, then the relative gap is large, and the potential for applying revenue management solutions is also relatively large. This is because significant improvements in revenues can be obtained from rationing capacity to lower fares. As capacity increases, the relative gap decreases indicating that less can be gained by rationing capacity. At very high levels of capacity, it is optimal to accept all requests, so there is nothing to be gained from the use of an optimal admission control policy.

Table 1.8 Optimal expected revenue and bounds for Example 1.14

c	$V_n^L(c)$	$V_n(c)$	$V_n^U(c)$	$\bar{V}_n(c)$
80	42,728	49,642	53,039	53,315
90	48,493	54,855	58,293	58,475
100	54,415	60,015	63,366	63,815
110	60,393	65,076	68,126	69,043
120	66,180	69,801	72,380	74,243
130	71,398	73,926	75,923	79,443
140	75,662	77,252	78,618	82,563
150	78,751	79,617	80,456	82,563
160	80,704	81,100	81,564	82,563

Table 1.9 Optimal expected revenue and bounds for Example 1.15

c	$V_n^L(c)$	$V_n(c)$	$V_n^U(c)$	$\bar{V}_n(c)$
80	52,462	67,505	72,717	73,312
90	61,215	74,003	79,458	80,302
100	70,136	79,615	85,621	87,292
110	78,803	84,817	91,122	92,850
120	86,728	89,963	95,819	98,050
130	93,446	94,869	99,588	103,250
140	98,630	99,164	102,379	106,370
150	102,209	102,418	104,251	106,370
160	104,385	104,390	105,368	106,370

1.7 General Fare Arrival Patterns with Poisson Demands

So far we have suppressed the time dimension. The order of the arrivals has provided us with stages that are a proxy for time. In this section, we consider models where time is considered explicitly. There are advantages of including time as part of the model as this allows for a more precise formulation of the consumer arrival process. For example, we can relax the low-before-high fare arrival assumption and allow for interleaved arrivals for different fare classes. On the other hand, the advantage of flexibility comes at the cost of estimating arrival rates for each of the fare classes over the sales horizon. If arrival rates are not estimated accurately, then adding the time dimension may hurt rather than help performance. In addition, formulations where time is handled explicitly usually assume that the demand for each fare class follows a Poisson process, whereas our earlier models based on sequential fare arrivals do not have this restriction. Here, we will provide formulations for both the Poisson and the compound Poisson cases. The compound Poisson model is flexible enough to fit any mean and variance of demand for each fare class. In Sect. 1.10, we compare the performance of the formulation with sequential fare arrivals to the formulation that allows for compound Poisson demands. Not surprisingly, optimal policies designed for arbitrary fare arrival patterns are superior.

1.7.1 Model

The length of the selling horizon is T, and time t will represent the time left until the end of the selling horizon. As before, there are n fare classes indexed by $\{1, \ldots, n\}$. We assume that consumers will leave the system when their preferred class is not available. Consumers requesting fare class j may be rejected because the fare is intentionally not made available in the hope of selling the capacity at a higher fare. There may also be time-of-purchase restrictions on some fares. We use M_t to denote the set of valid fares at time-to-go t. Typically $M_t = \{1, \ldots, n\}$ for large t, but low fares are dropped from M_t as time-of-purchase restrictions become binding. Consumers requesting fare class j arrive according to a Poisson process with arrival rate function $\{\lambda_{jt} : 0 \leq t \leq T\}$. The number of consumers that arrive during the last t units of time and request fare class j, say N_{jt}, is Poisson with parameter $\Lambda_{jt} := \int_0^t \lambda_{js} 1(j \in M_s)\, ds$, where $1(j \in M_s) = 1$ if $j \in M_s$ and 0 otherwise. We will use the shorthand notation $\Lambda_j := \Lambda_{jT}$ to denote the total expected demand for fare class $j \in \{1, \ldots, n\}$. We assume, without loss of generality, that $p_1 > p_2 > \ldots > p_n$.

Let $V(t, x)$ denote the maximum expected revenue that can be attained over the last t units of time from x units of capacity. We will develop both discrete and continuous time dynamic programs to compute $V(t, x)$. We now argue that the probability that there is exactly one request for fare class j over the interval $(t - \delta t, t]$ is $\lambda_{jt}\delta t + o(\delta t)$. Let $N_j(t - \delta t, t]$ denote the number of requests for fare class j over the interval $(t - \delta t, t]$. This is a Poisson random variable with mean $\int_{t-\delta t}^t \lambda_{js}\, ds = \lambda_{jt}\, \delta t + o(\delta t)$. Then,

$$\mathbb{P}\{N_j(t - \delta t, t] = 1\} = \lambda_{jt}\, \delta t\, \exp(-\lambda_{jt}\, \delta t) + o(\delta t) = \lambda_{jt}\, \delta t + o(\delta t),$$

while the probability that there are no requests for the other fare classes over the same interval is

$$\begin{aligned}\mathbb{P}\{N_k(t - \delta t, t] = 0, \forall k \neq j\} &= \exp\Big(-\sum_{k \neq j} \lambda_{kt}\delta t\Big) + o(\delta t) \\ &= 1 - \sum_{k \neq j} \lambda_{kt}\delta t + o(\delta t).\end{aligned}$$

Multiplying the two terms above and rearranging, we obtain $\lambda_{jt}\delta t + o(\delta t)$, as claimed.

Let $\Delta V(t, x) = V(t, x) - V(t, x - 1)$ for $x \geq 1$ and $t \geq 0$ and consider time steps of size $\delta t \ll 1$. Notice that

$$V(t,x) = \sum_{j \in M_t} \lambda_{jt}\, \delta t \, \max\{p_j + V(t-\delta t, x-1), V(t-\delta t, x)\}$$
$$+ \Big(1 - \sum_{j \in M_t} \lambda_{jt}\, \delta t\Big) V(t-\delta t, x) + o(\delta t)$$
$$= V(t-\delta t, x) + \delta t \sum_{j \in M_t} \lambda_{jt}[p_j - \Delta V(t-\delta t, x)]^+ + o(\delta t) \qquad (1.13)$$

with boundary conditions $V(t,0) = 0$ and $V(0,x) = 0$ for all $x \geq 0$. In the first equality, the first term on the right-hand side corresponds to the arrival of one request for fare class j, so a decision must be made between accepting the request earning $p_j + V(t-\delta t, x-1)$ or rejecting it and earning $V(t-\delta t, x)$, since we move to the next time period with the capacity x in the latter case. The second term on the right-hand side of the first equality corresponds to the case where no request arrives in the interval $(t-\delta t, t]$, resulting in expected revenue $V(t-\delta t, x)$. The second equality follows by arranging the terms. Subtracting $V(t-\delta t, x)$ from both sides of the equality in (1.13), dividing by δt and taking the limit as $\delta t \downarrow 0$, we obtain the Hamilton–Jacobi–Bellman (HJB) equation

$$\frac{\partial V(t,x)}{\partial t} = \mathcal{R}_t(\Delta V(t,x)). \qquad (1.14)$$

where

$$\mathcal{R}_t(z) := \sum_{j \in M_t} \lambda_{jt}\, [p_j - z]^+$$

is a decreasing convex function of z. The boundary conditions are as before $V(t,0) = V(0,x) = 0$. The equation tells us that the rate at which $V(t,x)$ grows with t is the sum of the arrival rates times the positive part of the fares net of the marginal value of capacity $\Delta V(t,x)$ at state (t,x). We can think of the right-hand side as the profit rate when the marginal cost is set equal to marginal value $\Delta V(t,x) = V(t,x) - V(t,x-1)$ of the xth unit of capacity.

1.7.2 Optimal Policy and Structural Results

Notice that fare j is accepted at state (t,x) if and only if $p_j \geq \Delta V(t,x)$. Thus, if fare j is accepted, then all fares $k \leq j$ are accepted since $p_k \geq p_j \geq \Delta V(t,x)$. This suggest that we find the index for the lowest acceptable fare by letting, for each time t and capacity x,

$$a(t,x) := \max\{j : p_j \geq \Delta V(t,x)\}.$$

In this case, if we are at time t with capacity x, then it is optimal to accept all fares in the active set

$$A(t,x) := \{j \in M_t : j \leq a(t,x)\},$$

and to reject all fares in the complement $R(t,x) := \{j \in \{1,\ldots,n\} : j \notin A(t,x)\}$. Note that the active set $A(t,x)$ essentially defines an admission control policy. The following theorem provides some structural results about the value function $V(t,x)$, its increments $\Delta V(t,x)$, and the admission control policy $A(t,x)$.

Theorem 1.18 *The value function $V(t,x)$ is increasing in t and in x. The increment of the value function $\Delta V(t,x)$ is decreasing in x and increasing in t. Moreover, $a(t,x)$ and $A(t,x)$ are increasing in x.*

If the arrival rates and the set of valid fares are stationary so that $\lambda_{jt} = \lambda_j > 0$ for all j and $M_t = M = \{1,\ldots,n\}$ for all t, then $a(t,x)$ and $A(t,x)$ are decreasing in t and $V(t,x)$ is strictly increasing and concave in t.

Notice that we can also express the optimal policy in terms of dynamic protection levels $y_j(t)$, $j = 1,\ldots,n-1$, $0 \le t \le T$, which are given by

$$y_j(t) := \max\{x : a(t,x) = j\},$$

Thus, if $x \le y_j(t)$, then fares $k > j$ should be closed. This observation follows because $a(t,x)$ is increasing in x.

1.7.3 Discrete-Time Formulation

The value function in (1.14) can be accurately computed by solving and pasting the HJB equation over a discrete mesh. Alternatively, $V(t,x)$ can be approximately computed by using a discrete time dynamic programming formulation. A discrete time dynamic programming formulation emerges from (1.13) by rescaling time, setting $\delta t = 1$, and dropping the $o(\delta t)$ term. This can be done by selecting $k > 1$, so that kT is an integer, and setting $\lambda_{jt} \leftarrow \frac{1}{k}\lambda_{j,t/k}$, for $t \in [0, kT]$. The scale factor k should be sufficiently large so that after scaling, we have $\sum_{j\in M_t}\lambda_{jt} \ll 1$, e.g., $\sum_{j\in M_t}\lambda_{jt} \le 0.01$ for all $t \in [0,T]$, with $T \leftarrow kT$. The resulting discrete time dynamic program is given by

$$\begin{aligned} V(t,x) &= \sum_{j\in M_t}\lambda_{jt}\max\{p_j + V(t-1,x-1), V(t-1,x)\} \\ &\qquad + \Big(1 - \sum_{j\in M_t}\lambda_{jt}\Big)V(t-1,x) \\ &= V(t-1,x) + \sum_{j\in M_t}\lambda_{jt}[p_j - \Delta V(t-1,x)]^+ \\ &= V(t-1,x) + \mathcal{R}_t(\Delta V(t-1,x)), \end{aligned} \tag{1.15}$$

with the same boundary conditions. Computing $V(t,x)$ via (1.15) is quite easy and fairly accurate if time is scaled appropriately. For each time period t and capacity x,

Table 1.10 Optimal total expected revenues in Example 1.19

c	50	100	150	200	250	300	350
$V(T, c)$	3553.6	5654.9	7410.1	8390.6	9139.3	9609.6	9625.0

the complexity is order $O(n)$, so the overall computational complexity to compute $V(t, x)$ for all t and x is $O(ncT)$.

From the dynamic program in (1.15), it is optimal to accept a request for fare class j when we have $p_j \geq \Delta V(t-1, x)$, or equivalently, when $p_j + V(t-1, x-1) \geq V(t-1, x)$. The latter condition compares the immediate revenue from fare class j plus the value of being in a state with one less unit of capacity at the next period with the value of not accepting the request and being in a state with the same capacity at the next period. Letting $a(t, x) := \max\{j : p_j \geq \Delta V(t-1, x)\}$, if we are at time period t with capacity x, then it is optimal to accept all fares in the active set

$$A(t, x) := \{j \in M_t : j \leq a(t, x)\},$$

and to reject all fares in the complement $R(t, x) := \{j \in \{1, \ldots, n\} : j \notin A(t, x)\}$. All of the structural results of Theorem 1.18 continue to hold for the discrete time model.

Example 1.19 Consider Example 1.9 with five fare classes with fares $p_1 = \$100$, $p_2 = \$60$, $p_3 = \$40$, $p_4 = \$35$, and $p_5 = \$15$. We also assume that the arrival rates are uniform over the horizon $[0, T]$, i.e., $\lambda_j = \Lambda_j/T$, and independent Poisson demands with means $\Lambda_1 = 15$, $\Lambda_2 = 40$, $\Lambda_3 = 50$, $\Lambda_4 = 55$, and $\Lambda_5 = 120$ and $T = 1$. The scaling factor was selected so that $\sum_{i=1}^{5} \Lambda_i/k < 0.01$ resulting in $T \leftarrow kT = 2800$. In Table 1.10, we present the expected revenues $V(T, c)$ for $c \in \{50, 100, 150, 200, 250, 300, 350\}$.

1.8 Monotonic Fare Offerings

The dynamic programs in (1.14) and (1.15) implicitly assume that fares can be opened and closed at any time. To see how a closed fare may reopen, suppose that $a(t, x) = j$ so set $A(t, x) = \{k \in M_t : k \leq j\}$ is offered at state (t, x), but an absence of sales may trigger fare $j+1$ to open as $a(s, x)$ increases and as the time-to-go s decreases. This can lead to the emergence of strategic consumers or third parties that specialize in exploiting inter-temporal fare arbitrage opportunities, where one waits for a lower fare class to be available. To avoid such strategic behavior, the capacity provider may commit to a policy of never opening fares once they are closed. Handling monotonic fares requires modifying the dynamic programming formulation into something akin to the dynamic program where time was handled implicitly through prefixed arrival order of the fare classes. In particular, let $V_j(t, x)$

be the maximum expected revenue from state (t, x) that can be obtained by offering any set $S_{it} = \{k \in M_t, k \leq i\}$ with $i \leq j$, so that we do not open any fares cheaper than fare class j. Let $W_k(t, x)$ be the expected revenue from accepting fares S_{kt} at state (t, x) and then following an optimal policy. More precisely,

$$\begin{aligned} W_k(t,x) &= \sum_{i \in S_{kt}} \lambda_{it}[p_i + V_k(t-1, x-1)] + (1 - \sum_{i \in S_{kt}} \lambda_{it}) V_k(t-1, x) \\ &= V_k(t-1, x) + \lambda_t \, [r_{kt} - \pi_{kt} \Delta V_k(t-1, x)] \\ &= V_k(t-1, x) + \lambda_t \, \pi_{kt} \, [q_{kt} - \Delta V_k(t-1, x)], \end{aligned}$$

where $\Delta V_k(t, x) = V_k(t, x) - V_k(t, x-1)$, and the quantities λ_t, π_{kt}, and q_{kt} are as defined as $\lambda_t := \sum_{j \in M_t} \lambda_{jt}, \pi_{jt} := \sum_{k \in S_{jt}} \lambda_{kt}/\lambda_t$, and $r_{jt} := \sum_{k \in S_{jt}} p_k \lambda_{kt}/\lambda_t$. Then, $V_j(t, x)$ satisfies

$$V_j(t, x) = \max_{k \leq j} W_k(t, x) \tag{1.16}$$

with the boundary conditions $V_j(t, 0) = V_j(0, x) = 0$ for all $t \geq 0$, $x \in \mathbb{N}_+$ and $j = 1, \ldots, n$. It follows immediately that $V_j(t, x)$ is monotone increasing in j. The complexity to compute each $V_j(t, x)$ is $O(1)$, so the complexity to compute $V_j(t, x)$ for all $j = 1, \ldots, n$, $x = 1, \ldots, c$ is $O(nc)$. Since there are T time periods, the overall complexity is $O(ncT)$. While computing $V_j(t, x)$ numerically is fairly simple, it is satisfying to know more about the structure of optimal policies as this gives both managerial insights and can simplify computations. The proof of the structural results are intricate and subtle, but they parallel the results for the dynamic programs in (1.14) and (1.15).

Lemma 1.20 *The value functions $\{V_j(t, x) : j = 1, \ldots, n,\ x = 1, \ldots, c,\ t = 1, \ldots, T\}$ computed through the dynamic program in (1.16) satisfy the following properties.*

- *$\Delta V_j(t, x)$ is decreasing in $x \in \{1, \ldots, c\}$, so the marginal value of capacity is diminishing.*
- *$\Delta V_j(t, x)$ is increasing in $j \in \{1, \ldots, n\}$, so the marginal value of capacity increases when we have more flexibility in terms of opening and closing fare classes.*
- *$\Delta V_j(t, x)$ is increasing in t, so the marginal value of capacity increases as the time-to-go increases.*

Let

$$a_j(t, x) := \max\{k \leq j : W_k(t, x) = V_j(t, x)\}.$$

In words, $a_j(t, x)$ is the index of the lowest open fare that is optimal to post if we are at time t with a capacity of x and we are allowed to use any fares in S_{jt}. Also, let

$$A_j(t, x) := \{k \in M_t : k \leq a_j(t, x)\}.$$

Then, it follows that $A_j(t,x)$ is the optimal set of fares to open at state (j,t,x). Clearly $V_i(t,x) = V_j(t,x)$ for all $i \in \{a_j(t,x),\ldots,j\}$. The following lemma asserts that $a_j(t,x)$ is increasing in x and in j and gives conditions for $a_j(t,x)$ to be decreasing in t.

Lemma 1.21 *The index $a_j(t,x)$ is increasing in x and in j. Furthermore, $a_j(t,x)$ is decreasing in t if the arrival rates λ_{jt} are time invariant and $M_t = M$ for all t. Moreover, $a_j(t,x) = k < j$ implies $a_i(t,x) = k$ for all $i \geq k$.*

It is possible to think of the policy in terms of protection levels, as well as in terms of stopping sets. Indeed, let $T_j := \{(t,x) : V_j(t,x) = V_{j-1}(t,x)\}$. We can think of T_j as the stopping set for fare j as it is optimal to close down fare j upon entering set T_j. For each t, let $y_j(t) := \max\{x \in \mathbb{N}_+ : (t,x) \in T_{j+1}\}$. We can think of $y_j(t)$ as the protection level for fares in S_j against higher fares. The following theorem is the counterpart to Theorem 1.6.

Theorem 1.22 *The optimal policy computed through the dynamic program in (1.16) satisfies the following properties.*

- *$A_j(t,x)$ is increasing in x and j. Furthermore, $A_j(t,x)$ is decreasing in t if the problem parameters is time invariant.*
- *$T_1 \subseteq T_2 \subseteq \ldots \subseteq T_n$.*
- *$y_j(t)$ is increasing in t and in j.*
- *If $x \leq y_j(t)$ then $V_i(t,x) = V_j(t,x)$ for all $i > j$.*

The policy is implemented as follows. The starting state is (n,T,c) as we can use any of the fares $\{1,\ldots,n\}$, we have T units of time-to-go and c is the initial inventory. At any state (j,t,x), we post fares $A_j(t,x) = \{1,\ldots,a_j(t,x)\}$. If a unit is sold during period t, then the state is updated to $(a_j(t,x-1),t-1,x-1)$ since all fares in the set $A_j(t,x)$ are allowed, the time-to-go is $t-1$ and the inventory is $x-1$. If no sales occur during period t, the state is updated to $(a_j(t,x),t-1,x)$. The process continues until either $t=0$ or $x=0$.

Example 1.23 Considering the same data in Example 1.19, in Table 1.11, we give the expected revenues $V_j(T,c)$, $j = 1,\ldots,5$ and $V(T,c)$ for $c \in \{50,100,150,200,250\}$. The first row is $V_5(T,c)$ from Example 1.19. Notice that $V(T,c) \geq V_j(T,c)$, since the optimal total expected revenue $V(T,c)$ is computed under the assumption that a closed fare class can be opened again. The difference in optimal total expected revenues $V(T,c) - V_5(T,c)$ due to the flexibility of opening and closing fares may be significant for some small values of c. For example, the difference is 1.7% for $c = 50$. However, the difference is small for larger values of c, and attempting to go for the extra revenue by opening an already closed fare may invite strategic consumers to wait for lower fares or for third parties to arbitrage the system by pre-selling capacity and then optimizing the time-of-purchase to exploit predictable price dynamics.

We close this section with a remark on mark-up and mark-down policies. Let us go back to the broader pricing interpretation coupled with the monotonic fare

Table 1.11 Optimal total expected revenues with monotonic fare offerings in Example 1.23

c	50	100	150	200	250	300	350
$V(T, c)$	3553.6	5654.9	7410.1	8390.6	9139.3	9609.6	9625.0
$V_5(T, c)$	3494.5	5572.9	7364.6	8262.8	9072.3	9607.2	9625.0
$V_4(T, c)$	3494.5	5572.9	7364.6	7824.9	7825.0	7825.0	7825.0
$V_3(T, c)$	3494.5	5572.9	5900.0	5900.0	5900.0	5900.0	5900.0
$V_2(T, c)$	3494.5	3900.0	3900.0	3900.0	3900.0	3900.0	3900.0
$V_1(T, c)$	1500.0	1500.0	1500.0	1500.0	1500.0	1500.0	1500.0

formulation in (1.16). In many applications the price menu $p_{jt}, j = 1, \ldots, n$ is time invariant, but the associated sales rates $\pi_{jt}, j = 1, \ldots, n$ are time varying. In addition, we will assume that there is a price p_{0t} such that $\pi_{0t} = 0$ for all t. This technicality helps with the formulation as a means of turning off demand when the system runs out of inventory. Recalling that we focus on monotonic fare offerings, the case $p_{1t} \geq p_{2t} \geq \ldots \geq p_{nt}$ and $\pi_{1t} \leq \pi_{2t} \leq \ldots \leq \pi_{nt}$ is known as the mark-up problem, while the case $p_{1t} \leq p_{2t} \leq \ldots \leq p_{nt}$ and $\pi_{1t} \geq \pi_{2t} \geq \ldots \geq \pi_{nt}$ is known as the mark-down problem. The former model is relevant in revenue management while the second is relevant in retailing.

For the revenue management formulation, the problem can be viewed as determining when to mark-up, i.e. switch from action j to $j-1$. The optimal mark-up times are random as they depend on the evolution of sales under the optimal policy. Suppose that the current state is (j, t, x), so the last action was j, the time-to-go is t and the inventory is x. We want to determine whether we should continue using action j or switch to action $j-1$. We know that if $x > y_{j-1}(t)$, then we should keep action j and if $x \leq y_{j-1}(t)$ then we should close fare class j. Let $T_j := \{(t, x) : x \leq y_{j-1}(t)\}$, so it is optimal to stop action j upon first entering set T_j. Notice that a mark-up occurs when the current inventory falls below a curve, so low inventories trigger mark-ups, and mark-ups are triggered by sales. The retailing formulation also has a threshold structure, but this time a mark-down is triggered by inventories that are high relative to a curve, so the optimal timing of a mark-down is triggered by the absence of sales.

1.9 Compound Poisson Demands

The formulations of the dynamic programs in (1.14) and (1.15) implicitly assume that each request is for a single unit of capacity. Instead, suppose that each arrival is for a random number of units. More specifically, suppose that size of a request for fare class j is a random variable Z_j, and that the probability mass function $P_j(z) := \mathbb{P}\{Z_j = z\}$ is known for fare class each j. As before, we assume independent demands for the different fare classes. We seek to generalize the dynamic programs in (1.14) and (1.15) so that at each state (t, x) we can decide whether or not to accept a fare request for a certain fare class and of a certain size. If we have a request for fare class j and for size z, then the expected revenue from accepting the request is $z p_j + V(t-1, x-z)$ and the expected revenue from rejecting the request is $V(t-1, x)$. Let $\Delta_z V(t,x) = V(t,x) - V(t,x-z) = \Delta V(t,x) + \Delta V(t,x-1) + \ldots + \Delta V(t,x-z+1)$ for all $z \leq x$. We set $\Delta_z V(t,x) = \infty$ if $z > x$. With this notation, the difference between accepting and rejecting a request for z units at state (t,x) is given by $z p_j - \Delta_z V(t,x)$, and it is optimal to accept the request whenever this quantity is non-negative. Notice that any request for $z > x$ is rejected as capacity is insufficient. (A different model is needed if a fraction of the consumers are willing to take partial orders.) The dynamic program in (1.14) with compound Poisson demands is given by

$$\frac{\partial V(t,x)}{\partial t} = \sum_{j \in M_t} \lambda_{jt} \sum_{z=1}^{x} P_j(z)[z p_j - \Delta_z V(t,x)]^+, \tag{1.17}$$

with boundary conditions $V(t,0) = V(0,x) = 0$. The optimal policy is to accept a request of size $z \leq x$ for fare class j, if $z p_j \geq \Delta_z V(t,x)$ and to reject all requests of size $z > x$. For $z \leq x$, define

$$j(z|t,x) := \arg\max \left\{ j : p_j \geq \frac{\Delta_z V(t,x)}{z} \right\},$$

so if we are at time-to-go t with remaining capacity of x, then it is optimal to accept requests of size z for all fares in the set

$$A(z|t,x) := \{ j \in M_t : j \leq j(z|t,x) \}.$$

The discrete time dynamic program in (1.15) with compound Poisson demands is given by

$$V(t,x) = V(t-1,x) + \sum_{j \in M_t} \lambda_{jt} \sum_{z=1}^{x} P_j(z)\,[z p_j - \Delta_z V(t-1,x)]^+, \tag{1.18}$$

with the same boundary conditions, and the optimal controls are of the same form except that $\Delta V(t-1,x)$ is used in defining $j(z|t,x)$.

For compound Poisson demands, we can no longer claim that the marginal value of capacity $\Delta V(t,x)$ is decreasing in x, although it is still true that $\Delta V(t,x)$ is increasing in t. To see why $\Delta V(t,x)$ is not monotone in x, consider a problem where the majority of the requests are for two units and requests are seldom for one unit. Then the marginal value of capacity for even values of x may be larger than the marginal value of capacity for odd values of x. Consequently, some of the structure may be lost. For example, it may be optimal to accept a request of a single unit of capacity when x is odd, but not if x is even. However, even if some of the structure is lost, the computations involved to solve the dynamic program in (1.18)

Table 1.12 Value function $V(T, c)$ in Example 1.24 with compound Poisson demand

c	50	100	150	200	250	300
$V(T, c)$	3837	6463	8451	10,241	11,724	12,559

Table 1.13 The first differences $\Delta V(208, x)$ in Example 1.24 with compound Poisson demand

x	1	2	3	4	5	6
$\Delta V(208, x)$	70.05	66.48	59.66	60.14	54.62	50.41

are straightforward as long as the distribution of Z_j is known. Airlines, for example, have a very good idea of the distribution of Z_j for different fare classes.

Example 1.24 Consider the same data in Example 1.19 with fares $p_1 = \$100$, $p_2 = \$60$, $p_3 = \$40$, $p_4 = \$35$, and $p_5 = \$15$ and independent Poisson requests with means $\Lambda_1 = 15, \Lambda_2 = 40, \Lambda_3 = 50, \Lambda_4 = 55, \Lambda_5 = 120$ over the horizon $[0, 1]$. Now, we will assume that the distribution of the demand sizes is given by $\mathbb{P}\{Z_j = 1\} = 0.65$, $\mathbb{P}\{Z_j = 2\} = 0.25$, $\mathbb{P}\{Z_j = 3\} = 0.05$, and $\mathbb{P}\{Z_j = 4\} = 0.05$ for all fare classes $j = 1, \ldots, 5$. Notice that $\mathbb{E}[Z_j] = 1.5$ and $\mathbb{E}[Z_j^2] = 2.90$. We will assume that $M_t = \{1, \ldots, n\}$ for all $t \in [0, T]$. Our computations are based on the dynamic program in (1.18) with a rescaled time horizon $T \leftarrow kT = 2800$, and rescaled arrival rates $\lambda_j \leftarrow \lambda_j/k$ for all $j = 1, \ldots, n$. Table 1.12 provides $V(T, c)$ for $c \in \{50, 100, 150, 200, 250, 300, 350\}$. Table 1.13 provides $\Delta V(t, x)$ for $t = 207$ in the rescaled horizon for $x \in \{1, \ldots, 6\}$ to illustrate the behavior of the policy. The reader can verify that at state $(t, x) = (208, 3)$, it is optimal to accept a request for one unit at fare p_2, and to reject the request if the request is for two units. If we have one more unit of inventory, so the state is $(t, x) = (208, 4)$ then it is optimal to reject a request for one unit at fare p_2, and to accept the request if it is for two units. The reason for this behavior is that the value of $\Delta V(t, x)$ is not monotone decreasing at $x = 4$.

1.10 Sequential vs. Mixed Arrival Formulations

In this section, we compare the performance of sequential policies obtained using the dynamic program in (1.6) with the performance of formulation (1.18) that allows for mixed arrivals and compound Poisson demands. Allowing for arbitrary arrivals provides more flexibility so it should not be surprising that policies based on a more flexible model would do better. Computational studies should concentrate on measuring the gap between the two. The gap is fairly small when the arrival rates are sequential and the low-before-high assumption holds, and more generally when the arrival rates follow a prescribed order that is consistent with the computations of the protection levels.

Table 1.14 Sub-optimality of EMSR-b with standard nesting vs optimal dynamic policy for Example 1.25

c	50	100	150	200	250	300
$V^s(T, c)$	3653	6177	8187	9942	11,511	12,266
$V(T, c)$	3837	6463	8451	10,241	11,724	12,559
% Sub	4.8%	4.4%	3.1%	2.9%	1.8%	2.3%

Comparing sequential and dynamic policies when the fare arrival rates do not follow a specific pattern is more difficult because revenues depend heavily on how the protection levels from the sequential policy are implemented. Two possible implementations are possible. Under theft nesting a request of size z for fare class j is accepted if $x - z \geq y_{j-1}$, where x is the current inventory and y_{j-1} is the protection level for fares $\{1, \ldots, j-1\}$. This method is called theft nesting because the remaining inventory x at time-to-go t deducts all previous bookings regardless of fare class. In contrast, standard nesting is implemented by accepting a size z request for fare j if $x - z \geq (y_{j-1} - b[1, j-1])^+$, where $b[1, j-1]$ are the observed bookings of fares $[1, j-1]$ at state (t, x). In practice, standard nesting works much better than theft nesting when the fare arrival pattern is not low-to-high. This makes sense because standard nesting does not insist on protecting y_{j-1} units for fares $\{1, \ldots, j-1\}$ even though we have already booked $b[1, j-1]$ units of these fares. Consequently, we use standard nesting in comparing sequential policies versus dynamic policies to give sequential policies a fighting chance.

Example 1.25 Consider the data of Example 1.24. Let $V(T, c)$ be the value function at the beginning of the selling horizon with initial capacities computed through the dynamic program in (1.18). Thus, $V(T, c)$ is the optimal total expected revenue under the compound Poisson arrivals. Let $V^s(T, c)$ be the total expected revenue collected by the sequential EMSR-b policy under standard nesting. In Table 1.14, we compare $V(T, c)$ with $V^s(T, c)$. Part of the gap between $V^s(T, c)$ and $V(T, c)$ can be reduced by frequently recomputing the booking limits applying the EMSR-b heuristic during the sales horizon. However, this is not enough to overcome the disadvantage of the EMSR-b heuristic when applied to mixed arrival patterns.

We end by noticing that it is possible to show that the upper bound $V_n^U(c)$ for $V_n(c)$, developed in Sect. 1.6 for the model with fixed arrival order for fares, is still a valid upper bound for $V(T, c)$ computed under arbitrary arrival pattern.

1.11 End of Chapter Problems

1. A coffee shop gets a daily allocation of 100 bagels. The bagels can be either sold individually at \$1.00 each or can be used later in the day for sandwiches. Each bagel sold as a sandwich provides a revenue of \$1.50 independent of the other ingredients.

Table 1.15 Fare classes, fares and demand distributions

Class	Fare	Demand distribution
1	\$600	Poisson(25)
2	\$475	Poisson(30)
3	\$265	Poisson(29)
4	\$130	Poisson(30)

(a) Suppose that demand for bagel sandwiches is estimated to be Poisson with parameter 80. How many bagels would you reserve for sandwiches?

(b) Compare the expected revenue of the solution of part (a) to the expected revenue of the heuristic that does not reserve capacity for sandwiches assuming that the demand for individual bagels is Poisson with parameter 150?

(c) Answer Part (a) if the demand for bagel sandwiches is normal with mean 100 and standard deviation 20.

2. Suppose capacity is 120 seats and there are four fares. The demand distributions for the different fares are given in Table 1.15.

   Determine the optimal protection levels. [Hints: The sum of independent Poisson random variables is Poisson with the obvious choice of parameter to make the means match. If D is Poisson with parameter λ, then $\mathbb{P}\{D = k+1\} = \mathbb{P}\{D = k\}\lambda/(k + 1)$ for any non-negative integer k.

3. Consider a parking lot in a community near Manhattan. The parking lot has 100 parking spaces. The parking lot attracts both commuters and daily parkers. The parking lot manager knows that he can fill the lot with commuters at a monthly fee of \$180 each. The parking lot manager has conducted a study and has found that the expected monthly revenue from x parking spaces dedicated to daily parkers is approximated well by the quadratic function $R(x) = 300x - 1.5x^2$ over the range $x \in \{0, 1, \ldots, 100\}$. Note: Assume for the purpose of the analysis that parking slots rented to commuters cannot be used for daily parkers even if some commuters do not always use their slots.

   (a) What would the expected monthly revenue of the parking lot be if all the capacity is allocated to commuters?

   (b) What would the expected monthly revenue of the parking lot be if all the capacity is allocated to daily parkers?

   (c) How many units should the parking manager allocate to daily parkers and how many to commuters?

   (d) What is the expected revenue under the optimal allocation policy?

4. A fashion retailer has decided to remove a certain item of clothing from the racks in 1 week to make room for a new item. There are currently 80 units of the item and the current sale price is \$150 per unit. Consider the following three strategies assuming that any units remaining at the end of the week can be sold to a jobber at \$30 per unit.

Table 1.16 Fare classes, fares and demand distributions

Class	Fare	Demand distribution
1	\$500	Poisson(45)
2	\$380	Poisson(55)
3	\$215	Poisson(50)
4	\$180	Poisson(100)

(a) Keep the current price. Find the expected revenue under this strategy under the assumption that demand at the current price is Poisson with parameter 50.

(b) Lower the price to \$90 per unit. Find the expected revenue under this strategy under the assumption that demand at \$90 is Poisson with parameter 120.

(c) Keep the price at \$150 but e-mail a 40% discount coupon for the item to a population of price sensitive consumers that would not buy the item at \$150. The coupon is valid only for the first day and does not affect the demand for the item at \$150. Compute the expected revenue under this strategy assuming that you can control the number of coupons e-mailed so that demand from the coupon population is Poisson with parameter x for values of x in the set $\{0, 5, 10, 15, 20, 25, 30, 35\}$. In your calculations assume that demand from coupon holders arrives before demand from consumers willing to pay the full price. Assume also that you cannot deny capacity to a coupon holder as long as capacity is available (so capacity cannot be protected for consumers willing to pay the full price). What value of x would you select? You can assume, as in parts (a) and (b) that any leftover units are sold to the jobber at \$30 per unit.

5. Prove that Eq. (1.8) holds.
6. Suppose we have a capacity of 220 seats and four fare classes. The fares and demand distribution for each fare class are given in Table 1.16. In all cases, except where noted, we will assume a low-to-high fare class arrival pattern.

(a) Determine the optimal protection levels using dynamic programming.
(b) Determine the protection levels under the EMSR-a heuristic
(c) Determine the protection levels under the EMSR-b heuristic
(d) Use simulation or the exact method to estimate the expected sales for each fare class and the total expected revenues for the policies determined in parts (a)–(c).
(e) Find the expected revenue under a policy that does not protect inventory for higher fare classes assuming the arrival pattern is low-to-high.
(f) Find the expected revenue of the policy in part (e) if the fare class arrival pattern is high-to-low.
(g) Solve the linear programming described in class to obtain an upper bound on the expected revenue of the optimal policy.

Table 1.17 Dynamic booking control with booking limits and protection levels

	Booking limits				Protection levels					
	1	2	3	4	1	2	3	4	Request	Action
1	50	45	37	22	5	13	28	50	4 seats in Class 4	
2									5 seats in Class 3	
3									7 seats in Class 2	
4									5 seats in Class 4	
5									5 seats in Class 1	
6									5 seats in Class 4	
7									6 seats in Class 3	
8									3 seats in Class 2	
9									1 seats in Class 1	
10									2 seats in Class 3	
11										

7. Consider a flight with a capacity of 50 seats and four fare classes. Suppose that we implement *nested protection levels* starting with $(y_1, y_2, y_3, y_4) = (5, 13, 28, 50)$. Table 1.17 shows a series of booking requests. For this problem, each request must be accepted on all-or-none basis, i.e. given a request of m units, we can only sell m units or none at all. Determine whether each request would be accepted, and update the booking limits and protection levels accordingly.
8. Suppose that you are the capacity provider for a popular event. The face value of the tickets is \$100 per seat, and the venue can hold 350 individuals. The \$100 tickets go on sale a month before the event. Assume demand for \$100 tickets is at least 350. You estimate that demand from people willing to pay \$300 for a ticket the day of the event can be modeled as a negative binomial with parameters $r = 36$ and $p = 1/4$ (mean 144 and variance 432). More precisely, the probability mass function of demand for \$300 tickets is $\mathbb{P}\{D_1 = k\} = \binom{k-1}{35}(1/4)^{36}(3/4)^{k-36}$ for integer values of $k \geq 36$.
   (a) How many tickets should you reserve for sale at \$300?
   (b) Evaluate the expected revenue of the strategy of part (a) and determine the average number of unsold seats under the strategy of part (a).
   (c) Suppose now that you sell the \$100 tickets with a callable option that allows you to buy them back for \$130 if needed (you can assume consumers are willing to accept this deal). Suppose that you exercise the option of buying back \$100 tickets at \$130 when demand for \$300 tickets exceeds the number you reserved for them in part (a). Use simulation to evaluate the expected revenue of this strategy and determine the average number of unsold seats. You can continue to assume that demand for \$100 tickets exceeds the capacity of the venue for the purpose of your calculations.

Table 1.18 Fare classes, fares and demand distributions

j	p_j	$\mathbb{E}[D_j]$
1	\$75	8
2	\$100	21
3	\$75	31
4	\$60	20

(d) Consider now a refinement of the strategy in part (c) where you can fine tune the number of tickets that you reserve for sale at \$300. How many tickets would you reserve? Compute the expected profit under the new strategy and also the expected number of unsold seats.

9. Find the optimal protection levels for the data in Table 1.18 and compute the optimal expected revenues $V_1(c)$, $V_2(c)$, $V_3(c)$, and $V_4(c)$ for $c \in \{50, 55, 60, 65, 70, 75, 80\}$ assuming Poisson demands.
10. Modify Problem 9 so that $p_1 = \$125$ and compute optimal protection levels and the value function $V_4(c)$ for $c \in \{50, 55, 60, 65, 70, 75, 80\}$.
11. Compute the upper bound $V^H(c)$ and $\bar{V}(c)$ and the lower bound $V^L(c)$ and the spread $V^H(c) - V^L(c)$ for Problem 9 for $c \in \{50, 55, 60, 65, 70, 75, 80\}$.
12. Use the discrete time dynamic programs to compute $V(T, c)$ and $V_j(T, c)$, $j = 1, 2, 3, 4$ for the data of Problem 9, for the values of the capacity $c \in \{50, 55, 60, 65, 70, 75, 80\}$ for the following arrival rate models:

    (a) Uniform arrival rates, e.g. $\lambda_{tj} = \Lambda_j = E[D_j]$ for $0 \leq t \leq T = 1$. Be sure to rescale time so that $T = a$ is an integer large enough so that $\sum_{j=1}^{3} \mathbb{E}[D_j]/a \leq 0.01$. What accounts for the difference between $V(T, c)$ and $V_4(T, c)$? What accounts for the difference between $V_4(T, c)$ and $V_4(c)$?

    (b) Low-to-high arrival rates: Dividing the selling horizon $[0, T] = [0, 1]$ into 4 sub-intervals $[t_{j-1}, t_j]$, $j = 1, \ldots, 4$ with $t_j = j/4$, and set $\lambda_{jt} = 4\Lambda_j$ over $t \in [t_{j-1}, t_j]$ and $\lambda_{jt} = 0$ otherwise. Again, be sure to rescale the system so that $T = a$ is an integer large enough so that $\max_j \max_t \lambda_{jt}/a \leq 0.01$. What accounts for the difference between $V(T, c)$ and $V_4(T, c)$? What accounts for the difference between $V_4(T, c)$ and $V_4(c)$?

13. Show that the upper bound (1.11) holds for the model presented in Sect. 1.7 with $\mu_i = \Lambda_i$ for all $i \in N$. Find the dual for the formulation and show that you can reduce this to a single dimensional convex problem in the dual of the capacity constraint.

1.12 Bibliographic Remarks

Talluri and van Ryzin (2004b), Phillips (2005) and Ozer and Phillips (2012) are comprehensive reference books on revenue management and pricing. Weatherford and Bolidy (1992), McGill and van Ryzin (1999), van Ryzin and Talluri (2003) and van Ryzin and Talluri (2005) give reviews of the literature on the subject. For the two-fare class model, a formula for the optimal protection level as a function of $\mathbb{P}\{D_1 \geq y\}$ and r was proposed by Littlewood (1972). His arguments were not formal; however, they were later justified by Bathia and Prakesh (1973) and Richter (1982). Our discussion of quality of service and salvage values borrows from Brumelle et al. (1990). Gallego et al. (2008a) discuss how callable products can improve the quality of service. The two-fare class model has connections to the newsvendor problem. A coverage of the newsvendor problem can be found in Zipkin (2000). Cachon and Kok (2007a) study the newsvendor problem when the salvage value of the product depends on how much inventory is left over. Gallego and Moon (1993) and Perakis and Roels (2008) focus on the newsvendor problem when the demand distribution is not known fully. Boyaci and Ozer (2010) consider a problem of information acquisition for capacity planning. Levi et al. (2015) give bounds for sample average approximation solution to the newsvendor problem, which extend to the two-fare class model. Hu et al. (2016a) consider a newsvendor problem where the customers choose between products.

Wollmer (1992) uses dynamic programming to obtain the optimal policy for the multi-fare problem with discrete demands and fixed arrival rate for the fare classes. Curry (1990) derives optimality conditions when demands are assumed to follow a continuous distribution. Brumelle and McGill (1993) allow for either discrete or continuous demand distributions and make a connection with the theory of optimal stopping. The reader is referred to Robinson (1995) for the case where the arrival pattern of the fare classes is not necessary low-to-high. The papers by van Ryzin and McGill (2000) and Kunnumkal and Topaloglu (2009) give an algorithm for computing the optimal protection levels only by using samples of the random demand, instead of using the demand distributions. Ball and Queyranne (2009) and Ma et al. (2018) provide a competitive analysis for single-leg revenue management problems.

Credit for the EMSR heuristics is sometimes given to the American Airlines team working on revenue management problems shortly after deregulation. The first published account of these heuristics appears in Simpson (1985), Belobaba (1987) and Belobaba (1989). Ratliff (2005) reports that EMSR-b usually provides improved performance on real world problems, especially ones involving nested inventory controls. Diwan (2010) numerically compares various approaches for single-leg revenue management problems.

Examples 1.14 and 1.15 are from Wollmer (1992). The reader is referred to Chandler and Ja (2007) and Temath et al. (2010) for further information on the uses of the revenue opportunity model. Lee and Hersh (1993) first proposed a model that is equivalent to our discrete-time formulation with arbitrary arrival patterns. Brumelle and Walczak (2003) present more general results in this vein.

Weatherford et al. (1993) consider problems with diversion possibilities between the different fare classes. Revenue management problems have clear connections to dynamic packing problems studied in Kleywegt and Papastavrou (1998). Belobaba and Farkas (1999) focus on the interactions between revenue management decisions and estimating the spill rate between different classes. Zhao and Zheng (2001) consider the case with monotonic fare offerings but with only two fare classes. The sample path based proof technique that can be found in their paper becomes useful in numerous revenue management settings. In particular, this approach can be used to show the results related to monotonic fare offerings in this chapter. Mark-up and the mark-down problems can be studied from the point of view of stopping times. We refer the reader to Feng and Gallego (1995, 2000), and Feng and Xiao (2000) for mark-up and mark-down problems.

Gupta and Cooper (2005) and Cooper and Gupta (2006) provide comparisons between revenues in different systems with different demand distributions.

Appendix

Proof of Lemma 1.4 Taking expectations yields $g(x) = G(x)\mathbb{P}\{X \geq x\} + \sum_{j\leq x-1} G(j)\,\mathbb{P}\{X = j\}$ and $g(x-1) = G(x-1)\,\mathbb{P}\{X \geq x\} + \sum_{j\leq x-1} G(j)\,\mathbb{P}\{X = j\}$. Taking the difference yields $\Delta g(x) = \Delta G(x)\mathbb{P}\{X \geq x\}$. Similarly, taking expectations, we have $h(x) = H(x)\,\mathbb{P}\{X < x\}+\sum_{j\geq x} H(j)\,\mathbb{P}\{X = j\}$. Similarly, we also have $h(x-1) = H(x-1)\,\mathbb{P}\{X < x\}+\sum_{j\geq x} H(j)\,\mathbb{P}\{X = j\}$. Taking the difference, we see that $\Delta h(x)=\Delta H(x)\,\mathbb{P}\{X<x\}$. □

Proof of Proposition 1.5 We will prove the result by induction on j. The result holds for $j = 1$ since $\Delta V_1(y) = p_1\,\mathbb{P}\{D_1 \geq y\}$ is decreasing in y, and clearly $\Delta V_1(y) = p_1\,\mathbb{P}\{D_1 \geq y\} \geq \Delta V_0(y) = 0$. Assume that the result is true for V_{j-1}. It follows from the dynamic programming equation that

$$V_j(x) = \max_{y\leq x} \left\{W_j(y, x)\right\},$$

where, for any $y \leq x$,

$$W_j(y, x) = \mathbb{E}\{p_j \min\{D_j, x-y\}\} + \mathbb{E}\{V_{j-1}(\max\{x - D_j, y\})\}.$$

Directly using the definition of $W_j(y, x)$, for $y \in \{1, \ldots, x\}$, we can show that

$$\Delta W_j(y, x) = W_j(y, x) - W_j(y-1, x) = [\Delta V_{j-1}(y) - p_j]\,\mathbb{P}\{D_j > x - y\}.$$

Since $\Delta V_{j-1}(y)$ is decreasing in y by the inductive hypothesis, we see that $W_j(y, x) \geq W_j(y-1, x)$ if $\Delta V_{j-1}(y) > p_j$ and $W_j(y, x) \leq W_j(y-1, x)$ if $\Delta V_{j-1}(y) \leq p_j$. Consider the expression

$$y_{j-1} = \max\{y \in \mathbb{N}_+ : \Delta V_{j-1}(y) > p_j\},$$

where the definition of $\Delta V_j(y)$ is extended to $y = 0$ for all j by setting $\Delta V_j(0) = p_1$. If $x \geq y_{j-1}$, then

$$V_j(x) = \max_{y \leq x} W_j(y, x) = W_j(y_{j-1}, x).$$

On the other hand, if $x < y_{j-1}$, then

$$V_j(x) = \max_{y \leq x} W_j(y, x) = W_j(x, x).$$

In summary, we have

$$\begin{aligned} V_j(x) &= W_j(\min(x, y_{j-1}), x) \\ &= \begin{cases} V_{j-1}(x), & \text{if } x \leq y_{j-1} \\ \mathbb{E}\{p_j \min\{D_j, x - y_{j-1}\}\} & \\ \qquad + \mathbb{E}\{V_{j-1}(\max\{x - D_j, y_{j-1}\})\} & \text{if } x > y_{j-1}. \end{cases} \end{aligned}$$

Using the expression above, computing $\Delta V_j(x) = V_j(x) - V_j(x-1)$ for $x \in \mathbb{N}_+$ results in

$$\Delta V_j(x) = \begin{cases} \Delta V_{j-1}(x), & \text{if } x \leq y_{j-1} \\ \mathbb{E}\{\min\{p_j, \Delta V_{j-1}(x - D_j)\}\} & \text{if } x > y_{j-1}. \end{cases}$$

We will now use this result to show that $\Delta V_j(x)$ is itself decreasing in x. Since $\Delta V_j(x) = \Delta V_{j-1}(x)$ for $x \leq y_{j-1}$ and $\Delta V_{j-1}(x)$ is decreasing in x, we only need to worry about the case $x > y_{j-1}$. However, in this case, we have

$$\Delta V_j(x) = E \min(p_j, \Delta V_{j-1}(x - D_j))$$

is decreasing in x, since $\Delta V_{j-1}(x)$ is itself decreasing in x. Lastly, at y_{j-1}, using the expression for the first difference $\Delta V_j(x)$ above,

$$\begin{aligned} \Delta V_j(y_{j-1}) = \Delta V_{j-1}(y_{j-1}) > p_j & \\ \geq \mathbb{E}\{\min\{p_j, \Delta V_{j-1}(x - D_j)\}\} &= \Delta V_j(y_{j-1} + 1), \end{aligned}$$

showing that $\Delta V_{j-1}(x)$ is decreasing at $x = y_{j-1}$ as well.

Now, we show that $\Delta V_j(x) \geq \Delta V_{j-1}(x)$. For $x > y_{j-1}$, we have

$$\min\{p_j, \Delta V_{j-1}(x - D_j)\} \geq \min\{p_j, \Delta V_{j-1}(x)\} = \Delta V_{j-1}(x),$$

where the inequality follows since $\Delta V_{j-1}(x)$ is decreasing in x, and the equality holds since $x > y_{j-1}$. Taking expectations we see that $\Delta V_j(x) \geq \Delta V_{j-1}(x)$ on $x > y_{j-1}$. Lastly, note that $\Delta V_j(x) = \Delta V_{j-1}(x)$ on $x \leq y_{j-1}$. □

Proof of Theorem 1.6 By Lemma 1.4, we have

$$\mathbb{E}\{\min\{x-y, D_j\}\} - \mathbb{E}\{\min\{x-(y-1), D_j\}\}$$
$$= -\mathbb{E}\{\min\{x-y+1, D_j\}\} + \mathbb{E}\{\min\{x-y, D_j\}\}$$
$$= -\mathbb{P}\{D_j \geq x-y+1) = -\mathbb{P}\{D_j > x-y\}.$$

Similarly, $\mathbb{E}\{V_{j-1}(\max\{y, x-D_j\})\} - \mathbb{E}\{V_{j-1}(\max\{y-1, x-D_j\})\} = \Delta V_{j-1}(y) \times \mathbb{P}\{x-D_j < y\} = \Delta V_{j-1}(y)\,\mathbb{P}\{D_j > x-y\}$. This implies that

$$\Delta W_j(y,x) = W_j(y,x) - W_j(y-1,x) = (\Delta V_{j-1}(y) - p_j)\,\mathbb{P}\{D_j > x-y\},$$

thus, the sign of $\Delta W_j(y,x)$ is dictated by the sign of $\Delta V_{j-1}(y) - p_j$.

We now show that $W_j(y,x)$ is a unimodal in y. Letting y^*_{j-1} be as defined in (1.7), for all $y > y^*_{j-1}$, we have $\Delta V_{j-1}(y) \leq p_j$. Therefore, $\Delta W_j(y,x) \leq 0$ for all $y > y^*_{j-1}$. Similarly, we have $\Delta V_{j-1}(y^*_{j-1}) > p_j$, but since $\Delta V_j(x)$ is decreasing in x by the first part of Proposition 1.5, it follows that $\Delta V_{j-1}(y) > p_j$ for all $y \leq y^*_{j-1}$. Therefore, $\Delta W_j(y,x) \geq 0$ for all $y \leq y^*_{j-1}$. Having $\Delta W_j(y,x) \geq 0$ for all $y \leq y^*_{j-1}$ and $\Delta W_j(y,x) \leq 0$ for all $y > y^*_{j-1}$ implies that $W_j(y,x)$ is unimodal in y and its maximizer occurs at y^*_{j-1}. So, the maximizer of $W_j(y,x)$ over $y \in \{0,\ldots,x\}$ occurs at $\min\{y^*_{j-1}, x\}$. Third, we show that the optimal protection levels are monotone in the fare classes. By the definition of y^*_{j-1}, we have $\Delta V_{j-1}(y^*_{j-1}) > p_j$, and since $\Delta V_j(x) \geq \Delta V_{j-1}(x)$ by the second part of Proposition 1.5, we obtain $\Delta V_j(y^*_{j-1}) \geq \Delta V_{j-1}(y^*_{j-1}) > p_j > p_{j+1}$, which implies that $\Delta V_j(y^*_{j-1}) > p_{j+1}$. In this case, since y^*_j is given by $\max\{y \in \mathbb{N}_+ : \Delta V_j(y) > p_{j+1}\}$, it must be the case that $y^*_j \geq y^*_{j-1}$. □

Proof of Corollary 1.7 Let $G(x) = p_1 x$, then $V_1(x) = g(x) = \mathbb{E}\{G(\min\{D_1, x\})\}$, so $\Delta V_1(x) = \Delta g(x) = p_1\,\mathbb{P}\{D_1 \geq x\}$. Then, by Theorem 1.6,

$$y_1 = \max\{y \in \mathbb{N}_+ : p_1\,\mathbb{P}\{D_1 \geq x\} > p_2\}$$

which coincides with Littlewood's rule. □

Proof of Proposition 1.10 Since $\Pi_n(c,k)$ is the difference of a concave and a linear function, $\Pi_n(c,k)$ is itself concave. The marginal value of adding the c-th unit of capacity is $\Delta V_n(c) - k$ so the c-th unit increases profits as long as $\Delta V_n(c) > k$. Therefore, the smallest optimal capacity is given by $c(k)$. (Notice that $c(k)+1$ may be also optimal if $\Delta V_n(c(k)+1) = k$.) Note that $c(k)$ is decreasing in k since $\Delta V_n(c)$ is decreasing in c. Suppose that $k = p_{j+1}$. To establish $c(p_{j+1}) = y_j$, it is enough to show that $\Delta V_n(y_j) > p_{j+1} \geq \Delta V_n(y_j+1)$. By definition, $y_j = \max\{y \in \mathbb{N}_+ : \Delta V_j(y) > p_{j+1}\}$, so that we have $\Delta V_j(y_j) > p_{j+1} \geq \Delta V_j(y_j+1)$. Since it is optimal to protect up to y_j units of capacity for sale at fares $j, j-1, \ldots, 1$, it follows that $V_n(c) = V_j(c)$ for all $c \leq y_j$, and consequently $\Delta V_n(y_j) =$

$\Delta V_j(y_j) > p_{j+1}$. Now $\Delta V_n(y_j + 1)$ can be written as a convex combination of p_{j+1} and $\Delta V_j(y_j+1) \leq p_{j+1}$ which implies that $\Delta V_n(y_j+1) \leq p_{j+1}$, as desired. □

Proof of Theorem 1.18 Since $\mathcal{R}_t(z) \geq 0$, it follows from (1.14) that $V(t, x)$ is increasing in t, with strict inequality as long as there is a fare $k \in M_t$ such that $p_k > \Delta V(t, x)$ and $\lambda_{kt} > 0$. To show that $V(t, x+1) \geq V(t, x)$ consider a sample path argument where the system with $x+1$ units of inventory uses the optimal policy for the system with x units of inventory until either the system with x units runs out of stock or time runs out. If the system with x units of inventory runs out at time s, then the system with $x+1$ units of inventory can still collect $V(s, 1) \geq 0$. On the other hand, if time runs out the two systems collect the same revenue. Consequently, the system with $x+1$ units of inventory makes at least as much revenue resulting in $V(t, x+1) \geq V(t, x)$.

Clearly $\Delta V(t, 1) \leq \Delta V(t, 0) = \infty$. Assume as the inductive hypothesis that $\Delta V(t, y)$ is decreasing in $y \leq x$ for all $t \geq 0$. We want to show that $\Delta V(t, x+1) \leq \Delta V(t, x)$, or equivalently that

$$V(t, x+1) + V(t, x-1) \leq V(x) + V(x). \tag{1.19}$$

We will use a sample path argument to establish inequality (1.19). Consider four systems, one with $x+1$ units of inventory, one with $x-1$ units of inventory, and two with x units of inventory. Assume that we follow the optimal policy for the system with $x+1$ and for the system with $x-1$ that are on the left-hand side of inequality (1.19). For the two systems on the right, we use the sub-optimal policies designed for $x+1$ and $x-1$ units of inventory, respectively. We follow these policies until one of the following events occurs: time runs out, the difference in inventories for the systems on the left drops to 1, or the inventory of the system with $x-1$ units drops to zero. After that time we follow optimal policies for all four systems. To establish inequality (1.19), we will show that the revenues obtained for the systems in the right are at least as large as for the systems on the left, even though sub-optimal policies are used for the systems in the right. This is obviously true if we run out of time since the realized revenues of the two systems on the right are exactly equal to the realized revenues from the two systems on the left. Assume now that at time $s \in (0, t)$, the difference in inventories on the two systems on the left-hand side drops to 1, so that the states are $(s, y+1)$ and (s, y) for some $y < x$. This means that system on the left with $x+1$ units of inventory had $x-y$ units of sale and the system with $x-1$ units of inventory had $x-1-y$ units of sale. This implies that the system on the right that was following the policy designed for $x+1$ reaches state (s, y), while the system that was using the policy designed for $x-1$ reaches state $(s, y+1)$. Clearly, the additional optimal expected revenues over $[0, s]$ for each pair of systems is $V(s, y+1) + V(s, y) = V(s, y) + V(s, y+1)$, showing that the system on the right gets as much revenue as the system on the left even if sub-optimal polices are used for part of the horizon. Finally, if the inventory of the system with $x-1$ units of inventory drops to 0 at some time $s \in [0, t)$, so that state of the systems on the left are, respectively, (s, y) and $(s, 0)$ for some y, such that $1 < y \leq x$, while the systems on the left are $(s, y-1)$ and $(s, 1)$. From the inductive hypothesis, we know that $\Delta V(s, y) \leq \Delta V(s, 1)$ for all $y \leq x$ and all $s \leq t$. Consequently,

$$V(s, y) = V(s, y) + V(s, 0) \leq V(s, y-1) + V(s, 1),$$

and once again the pair of systems on the right result in at least as much revenue even though sub-optimal policies are used for part of the sales horizon.

We now show that $\Delta V(t, x)$ is increasing in t. This is equivalent to

$$\frac{\partial V(t, x)}{\partial t} = \mathcal{R}_t(\Delta V(t, x)) \geq \mathcal{R}_t(\Delta V(t, x-1)) = \frac{\partial V(t, x-1)}{\partial t},$$

but this is true on account of $\mathcal{R}_t(z)$ being decreasing in z and $\Delta V(t, x)$ being decreasing in x.

Notice that the set $a(t, x)$ is increasing in x since $\Delta V(t, x)$ is decreasing in x. Consequently, the set $A(t, x)$ is also increasing in x.

We now show that $V(t, x)$ is strictly increasing in t when $d_t(p) = d(p)$ is time invariant. This is because

$$\frac{\partial V(t, x)}{\partial t} = r(\Delta V(t, x)) \geq r(\Delta V(t, 1)) = r(V(t, 1)) > 0,$$

where the first inequality follows because r is decreasing and $V(t, 1) = \Delta V(t, 1) \geq \Delta V(t, x)$ for all $x \geq 1$. The strict inequality follows because $V(t, 1)$ must be below p_1 as otherwise if $V(t, 1) = p_1$, then the single unit of inventory must be priced at p_1 over the horizon $[0, t]$ and must sell with probability one over that interval. However, there is a positive probability equal to $e^{-\lambda_1 t}$ that the unit does not sell, so $V(t, 1) < p_1$, implying that $r(V(t, 1)) > 0$, so $V(t, x)$ is strictly increasing in t.

To show that $V(t, x)$ is concave, notice that since $r(z)$ is almost everywhere differentiable, then

$$\frac{\partial^2 V(t, x)}{\partial t^2} = r'(\Delta V(t, x)) \frac{\partial \Delta V(t, x)}{\partial t} \leq 0,$$

follows since $r'(z) \leq 0$, on account of $r(z)$ being decreasing in z, and from the fact that $\Delta V(t, x)$ is increasing in t. The fact that $r(z)$ is not differentiable at points $p_j \in M$ does not change the argument because we can take the right derivative of r and things work well given that $\Delta V(t, x)$ is increasing in t. □

Proof of Lemma 1.20 We will first show part that $\Delta V_j(t, x)$ is decreasing in x which is equivalent to showing that $2V_j(t, x) \geq V_j(t, x+1) + V_j(t, x-1)]$ for all $x \geq 1$. Let A be an optimal admission control rule starting from state $(t, x+1)$ and let B be an optimal admission control rule starting from $(t, x-1)$. These admission control rules are mappings from the state space to subsets $S_k = \{1, \ldots, k\}, k =$

$0, 1, \ldots, j$ where $S_0 = \emptyset$ is the optimal control whenever a system runs out of inventory. Consider four systems: two starting from state (t, x), using control rules A' and B', respectively, and one each starting from $(t, x+1)$ and $(t, x-1)$, using control rule A and B, respectively. Our goal is to specify heuristic control rules A' and B' that together make the expected revenues of the two systems starting with (t, x) at least as large as the expected revenues from the systems starting at $(t, x+1)$ and $(t, x-1)$. This will imply that $2V_j(t, x) \geq V_j(t, x+1) + V_j(t, x-1)$.

We will use the control rules $A' = A \cap B$ and $B' = A \cup B$ until the first time, if ever, the remaining inventory of the system (t, x) controlled by A' is equal to the remaining inventory of the system $(t, x+1)$ controlled by A. This will happen the first time, if ever, there is a sale under A and not under A', i.e. a sale under A but not under B. Let t' be the first time this happens, if it happens before the end of the horizon, and set $t' = 0$ otherwise. If $t' > 0$ then we apply policy $A' = A$ and $B' = B$ over $s \in [0, t')$. We claim that the expected revenue from the two systems starting with (t, x) is the same as the expected revenue from the other two systems. This is because the sales and revenues up to, but before t', are the same in the two systems. At t' sales occur only for the system (t, x) controlled by B' and the system $(t, x+1)$ controlled by A, and the revenues from the two sales are identical. After the sales at t', the inventory of the system (t, x) controlled by A' becomes identical to the inventory of the system $(t, x+1)$ controlled by A while the inventory of the system (t, x) controlled by B' becomes identical to the inventory of the system $(t, x-1)$ controlled by B. Since the policy switches to $A' = A$ and $B' = B$, then sales and revenues are the same over $[0, t')$. If $t' = 0$, then the sales of the two systems are the same during the entire horizon.

It remains to verify that inventories don't become negative. Prior to time t', the systems remain balance in the sense that system (t, x) governed by A' always has one unit of inventory less than system $(t, x+1)$ governed by A and system (t, x) governed by B' has one more unit of inventory than system $(t, x-1)$ governed by B. Thus the only two systems that could potential run out of inventory before t' are A' and B.

Since sales under $A' = A \cap B$ are more restricted than sales under B, the inventory of system (t, x) governed by A' will always be at least one unit since at most $x - 1$ units of sale are allowed under B. Therefore the only way the system can run out of inventory is if system $(t, x-1)$ runs out of inventory under B before t'. However, in this case, sales would stop under systems A' and B, while sales will continue under $B' = A$ and A so revenues will continue to be the same until the first sale under A at which point we reached t'. This shows that even if the system $(t, x-1)$ runs out of inventory under B the two systems continue to have the same revenues over the entire horizon. Consequently $2\Delta V_j(t, x) \geq V_j(t, x+1) + V_j(t, x-1)$ for all $x \geq 1$.

To show that $\Delta V_j(t, x)$ is increasing in j, it is enough to show that

$$V_j(t, x) + V_{j-1}(t, x-1) \geq V_j(t, x-1) + V_{j-1}(t, x).$$

To do this, we again use a sample path argument. Let A be an optimal admission control rule for the system $(j, t, x-1)$ and B be an admission control rule for the system $(j-1, t, x)$. Let A' and B' be heuristic admission rules applied, respectively, to the systems (j, t, x) and $(j-1, t, x-1)$. Our goal is to exhibit heuristics A' and B' such that when applied to the systems (j, t, x) and $(j-1, t, x-1)$ they generate as much revenue as the applying A to $(j, t, x-1)$ and B to $(j-1, t, x)$. This will imply that $V_j(t, x) + V_{j-1}(t, x-1) \geq V_j(t, x-1) + V_{j-1}(t, x)$.

Let $A' = A \cup B$ and $B' = A \cap B$ and let t' be the first time there is a sale under $A \cup B$ without a corresponding sale in A, so there is a sale under B but not under A. If $t' = 0$, then the revenues of the sets of two systems are equal. If $t' > 0$ switch at that point to the policy $A' = A$ and $B' = B$. Then sales and revenues under both sets of two systems are equal up to t'. At t' there are sales for the system (j, t, x) and $(j-1, t, x-1)$ that generate the same revenues. Moreover, the inventories of the two sets of two systems have the same inventories immediately after the sale at t'. Since the policy then switches to $A' = A$ and $B' = B$ then sales and revenues are the same for the two set of systems over $s \in [0, t')$. The only system in danger to run out of inventory is system (j, t, x) under $A' = A \cup B$, but that system has the same number of sales as the system $(j, t, x-1)$ under A up to t'. Therefore the system (j, t, x) has at least one unit of inventory up to t'.

To show that $\Delta V_j(t, x)$ is increasing in t it is enough to show that

$$V_j(t, x) + V_j(t-1, x-1) \geq V_j(t, x-1) + V_j(t-1, x).$$

To do this we again use a sample path argument. Let A be an optimal admission control rule for the system $(t, x-1)$ and B be an optimal admission control rule for the system $(t-1, x)$ Let A' and B' be heuristic admission rules applied, respectively, to the systems (t, x) and $(t-1, x-1)$. Our goal is to exhibit heuristics A' and B' such that when applied to the systems (t, x) and $(t-1, x-1)$ they generate as much revenue as the applying A to $(t, x-1)$ and B to $(t-1, x)$. This will imply that $V_j(t, x) + V_j(t-1, x-1) \geq V_j(t, x-1) + V_j(t-1, x)$. Let $A' = A \cup B$ and $B' = A \cap B$ and let t' be the first time there is a sale under A' without a corresponding sale in A, so there is a sale under B but not under A. If $t' = 0$ then the revenues of the sets of two systems are equal. If $t' > 0$ switch at that point to the policy $A' = A$ and $B' = B$. Then sales and revenues under both sets of two systems are equal up to t'. At t', there are sales for the system (t, x) and $(t-1, x)$ that generate the same revenues. Moreover, the inventories of the two sets of two systems have the same inventories immediately after the sale at t'. Since the policy then switches to $A' = A$ and $B' = B$, then sales and revenues are the same for the two set of systems over $s \in [0, t')$. The only system in danger to run out of inventory is system $(t-1, x-1)$ under $B' = A \cup B$, but that system has the same number of sales as the system $(t-1, x)$ under B up to t'. Therefore, the system $(t-1, x-1)$ has at least one unit of inventory up to t'. □

Proof of Lemma 1.21 We will first show that $a_j(t, x)$ can also be characterized as $a_j(t, x) = \max\{k \leq j : p_k \geq \Delta V_k(t-1, x)\}$. The result will then follow from

Lemma 1.20. First notice that if $a_j(t, x) = k < j$ then $V_i(t, x) = V_k(t, x)$ for all $i \in \{k, \ldots, j\}$. Moreover, $a_j(t, x) = k < j$ implies that $W_k(t, x) > W_{k+1}(t, x)$. Consequently, $0 > W_{k+1}(t, x) - W_k(t, x) = (p_{k+1} - \Delta V_{k+1}(t-1, x))\lambda_{k+1}$, so $p_{k+1} < \Delta V_{k+1}(t-1, x)$. Conversely, if $p_k \geq \Delta V_k(t-1, x)$ then $W_k(t, x) - W_{k-1}(t, x) \geq (p_k - \Delta V_k(t-1, x))\lambda_k \geq 0$ so $W_k(t, x) \geq W_{k-1}(t, x)$. With the new characterization, we now turn to the monotonicity of $a_j(t, x) = \max\{k \leq j : p_k \geq \Delta V_k(t-1, x)\}$. The monotonicity with respect to j is obvious because it expands the set over which we are maximizing. To see the monotonicity with respect to t, notice that $\Delta V_k(t, x) \geq \Delta V_k(t-1, x)$ so k is excluded from the set whenever $\Delta V_k(t-1, x) \leq p_k < \Delta V_k(t, x)$. To see the monotonicity with respect to x, notice that $\Delta V_k(t-1, x+1) \leq \Delta V_k(t, x) \leq p_k$ implies that k contributes positively at state $(t-1, x+1)$ whenever it contributes at $(t-1, x)$. □

Proof of Theorem 1.22 The properties of $A_j(t, x)$ follow from the properties of $a_j(t, x)$ established in Lemma 1.21. Note that $T_j = \{(t, x) : a_j(t, x) < j\}$. From Lemma 1.21, $a_j(t, x) < j$ implies that $a_i(t, x) < i$ for all $i > j$, so $T_j \subseteq T_i$ for all $i > j$. This implies that $y_j(t)$ is increasing in j for any $t \geq 0$. If $t' > t$, then $a_{j+1}(t', y_j(t)) \leq a_{j+1}(t, y_j(t)) < j+1$, so $y_j(t') \geq y_j(t)$. Note that $y_j(t) \leq y_i(t)$ for all $i > j$, then $x \leq y_j(t)$ implies $V_{i+1}(t, x) = V_i(t, x)$ for all $i \geq j$ and therefore $V_i(t, x) = V_j(t, x)$ for all $i > j$. □

Chapter 2
Network Revenue Management with Independent Demands

2.1 Introduction

In this chapter, we consider a firm that has finite capacities of several resources that can be instantly combined into different products with fixed prices. We assume that there is an independent demand stream for each of the products that arrives as a Poisson process. A requested product is purchased if available. The firm generates the revenue associated with the sale and updates the inventories of the resources consumed by the product. If the requested product is not available, then the customer leaves the system without purchasing. The objective of the firm is to decide which products to make available over a finite sales horizon to maximize the total expected revenue from fixed initial inventories that cannot be replenished during the sales horizon.

In the context of an airline, the resources are the flight legs that are combined to form itineraries from different origins to different destinations. Itineraries from different origins to different destinations can be sold at multiple price levels. Thus, the products are the origin-destination-fare (ODF) class combinations, which correspond to tickets from different origins to different destinations sold to customers at different fare levels. Lower fares are often more restricted in terms of the time-of-purchase, length-of-stay, and have limited ancillary services such as advanced seat selection, luggage handling, mileage accrual, and meals. In hotel applications, the resources are the room nights. The products are different sequences of night stays along with the prices associated with different sequences of night stays. In general, this problem setting captures situations where a set of resources are combined in different ways to generate products sold to customers at multiple price points. Although the problem setting is more general, we will adopt the airline vocabulary and we will use the terms flight leg and ODF to, respectively, refer to a resource and a product.

G. Gallego, H. Topaloglu, *Revenue Management and Pricing Analytics*, International Series in Operations Research & Management Science 279, https://doi.org/10.1007/978-1-4939-9606-3_2

Throughout this chapter, we assume that the set of possible ODF's is given and the demand for each ODF is independent of the demand for other ODF's. More precisely, we assume that an arriving customer has a particular ODF in mind and purchases that ODF if available and does not make a purchase otherwise. Under this model, a customer interested in a nonrestricted fare for a given itinerary does not change his mind and purchase a lower priced but more restricted fare. Alternatively, a customer interested in a low fare itinerary does not buy up when the desired fare is unavailable. Such an independent demand assumption, while restrictive, holds approximately when fares are well differentiated and customers can turn to a competitor when they fail to find the fare they want. The independent demand model performs poorly when fares are not well differentiated or alternative providers are not available. Indeed, when fares differ only in price, it makes sense to purchase the one with the lowest price. Moreover, when alternative low fares are not available, some customers may buy up to more expensive but less restricted fares. In such situations, one can abandon the independent demand model in favor of a dependent demand model where customer choices depend on the available fares. We treat the dependent demand case by using discrete choice models in a subsequent chapter.

In Sect. 2.2, we give a dynamic programming formulation of the network revenue management problem with independent demands. This formulation involves a high-dimensional state variable, so it is difficult to compute the optimal policy. In Sect. 2.3, we focus on obtaining an upper bound on the optimal expected revenue by using a linear program. In Sect. 2.4, we study bid-price and probabilistic admission heuristic control policies derived from the linear program. In Sect. 2.5, we dwell on refinements of the linear program, which helps us capture the randomness in the demand arrivals and extract bid-prices that are dependent on time. In Sect. 2.6, we study dynamic programming decomposition methods, which aim at approximately solving the dynamic programming formulation of the problem by decomposing it by resources. In Sect. 2.7, we give EMSR-like heuristics. In Sect. 2.8, we dwell on approximate dynamic programming methods. In particular, we show how to use the linear programming representation for the dynamic programming formulation of the network revenue management problem to calibrate the parameters of linear value function approximations. Out of this approach, we also obtain a bid-price policy.

2.2 Formulations

Suppose that there are m resources in the network and let $c := (c_1, \ldots, c_m) \in \mathbb{Z}_+^m$ denote the vector of initial capacities. We will measure time backwards, so t will represent the time-to-go. At the start of the sales horizon, we have $t = T$. At departure, we have $t = 0$. Customers arrive according to a Poisson or compound Poisson process with time-varying arrival rates. Expositionally, it helps to introduce the main ideas for the Poisson case and later take care of the changes needed to deal with the compound Poisson case. We index the itineraries by $1, \ldots, K$ and the possible fares for itinerary k by p_{kj} for $j \in \{1, \ldots, n_k\}$. Each ODF corresponds to an itinerary and fare combination. We use the double index kj to denote the ODF corresponding to itinerary k and fare j. We let λ_{tkj} denote the arrival rate at time t of customers interested in ODF kj. Let A_k be the column vector of resources utilized by itinerary k. That is, $A_k = (a_{1k}, \ldots, a_{mk})$ is an m-dimensional vector, where $a_{ik} \in \{0, 1\}$ with $a_{ik} = 1$ if resource i is consumed by itinerary k.

For any time-to-go t, we will let (t, x) represent the state of the system where $x \in \mathbb{Z}_+^m$ represents the vector of remaining inventory of resources. The initial state is (T, c). We will now develop the necessary notation to write a continuous time dynamic program for the maximum total expected revenue, say $V(t, x)$, that can be extracted from state (t, x). To capture the decisions at any time period, we use $u = \{u_{kj} : j = 1, \ldots, n_k,\ k = 1, \ldots, K\}$, where $u_{kj} = 1$ if we accept a request for ODF kj, and $u_{kj} = 0$ if we do not. The feasible set of decisions when the remaining inventory is x is given by $U(x) := \{u_{kj} \in \{0, 1\} : A_k u_{kj} \leq x,\ j = 1, \ldots, n_k,\ k = 1, \ldots, K\}$, which indicates that we cannot offer ODF jk if the current inventory x is insufficient to cover the requirements A_k of itinerary k.

Assume now that the state is (t, x) and consider a time increment δt that is small enough so that we can approximate the probability of an arrival of a request for fare j of itinerary k by $\lambda_{tkj}\,\delta t$. If $A_k \leq x$, then it is feasible to accept a request for itinerary k, and the problem is to maximize $p_{kj}u_{kj} + V(t - \delta t, x - A_k u_{kj})$ over $u_{kj} \in \{0, 1\}$. The choice $u_{kj} = 0$ results in expected revenue $V(t - \delta t, x)$, and the choice $u_{kj} = 1$ results in expected revenue $p_{kj} + V(t - \delta t, x - A_k)$. We can expand the optimization to all $x \geq 0$, by setting $V(t - \delta t, x - A_k) = -\infty$ whenever $A_k \leq x$ fails to hold. In this case, it is optimal not to offer itinerary k whenever there is insufficient capacity to do so. By following this convention, we can write

$$
\begin{aligned}
V(t, x) = & \sum_{k=1}^{K} \sum_{j=1}^{n_k} \lambda_{tkj}\delta t \max_{u_{kj} \in \{0,1\}} [p_{kj}u_{kj} + V(t - \delta t, x - A_k u_{kj})] \\
& + \left\{ 1 - \sum_{k=1}^{K} \sum_{j=1}^{n_k} \lambda_{tkj}\delta t \right\} V(t - \delta t, x) + o(\delta t),
\end{aligned}
$$

where $o(\delta t)$ is a quantity that goes to zero faster than δt. Subtracting $V(t - \delta t, x)$ from both side of the equation, dividing by δt, and using the notation $\Delta_k V(t, x) = V(t, x) - V(t, x - A_k)$, we obtain the Hamilton–Jacobi–Bellman (HJB) equation

$$
\frac{\partial V(t, x)}{\partial t} = \sum_{k=1}^{K} \sum_{j=1}^{n_k} \lambda_{tkj} [p_{kj} - \Delta_k V(t, x)]^+
$$

with boundary conditions $V(t, 0) = V(0, x) = 0$ for all $t \geq 0$ and all $x \geq 0$. Notice that term $[p_{kj} - \Delta V_k(t, x)]^+$ is equivalent to the maximum of $p_{kj}u_{kj} + V(t, x - A_k u_{kj}) - V(t - \delta t, x)$ over $u_{kj} \in \{0, 1\}$.

For any vector $z \geq 0$, let

$$R_t(u,z) := \sum_{k=1}^{K}\sum_{j=1}^{n_k} \lambda_{tkj}[p_{kj} - z_k]u_{kj},$$

and consider the optimization problem

$$\mathcal{R}_t(z) := \max_u R_t(u,z) = \sum_{k=1}^{K}\sum_{j=1}^{n_k} \lambda_{tkj} \max_{u_{jk}\in\{0,1\}} [p_{kj} - z_k]\, u_{kj}$$
$$= \sum_{k=1}^{K}\sum_{j=1}^{n_k} \lambda_{tkj}[p_{kj} - z_k]^+.$$

Notice that an optimal solution to this optimization problem is nested-by-revenue as it is optimal to accept all fares for itinerary k that exceed z_k. If a fare is equal to z_k then it does not matter whether the fare is included or excluded from the solution. With this notation, the HJB equation can be written as

$$\frac{\partial V(t,x)}{\partial t} = \mathcal{R}_t(\Delta V(t,x)), \tag{2.1}$$

where we interpret $\Delta V(t,x)$ as the vector $(\Delta_1 V(t,x), \ldots, \Delta_K V(t,x))$.

This continuous time formulation indicates that the marginal value of time is equal to the maximum expected revenue that can be obtained when the marginal cost for itinerary k is $\Delta_k V(t,x)$ and the available fares for itinerary k are $\{p_{kj}, j = 1, \ldots, n_k\}$. It is then optimal to accept ODF kj if $p_{kj} \geq \Delta_k V(t,x)$.

It is conceptually useful to think of $\lambda_{tk} = \sum_{j=1}^{n_k} \lambda_{tkj}$ as the arrival rate of requests for itinerary k and the requests for specific fares $\lambda_{tkj} = \lambda_{tk}\, q_{tkj}$ as thinned Poisson processes, where q_{tkj} is the probability of a customer selecting fare j given that she selected itinerary k. This is equivalent to having random fare P_{tk} at time t for itinerary k where P_{tk} takes value p_{kj} with probability q_{tkj}. Writing (2.1) in terms of random fares yields the formulation

$$\frac{\partial V(t,x)}{\partial t} = \sum_{k=1}^{K} \lambda_{tk}\, \mathbb{E}[P_{tk} - \Delta_k V(t,x)]^+ = \mathcal{R}_t(\Delta V(t,x)). \tag{2.2}$$

Formulation (2.2) makes explicit the arrival rate for each itinerary, and the fact that distribution of fares for an itinerary may change over time to reflect time preferences or fare restrictions. Formulation (2.2) is actually more general than the motivation that led to it as it is valid for any random fares P_{tk} with finite means. Viewing P_{tk} as a general random variable, not necessarily restricted to the set of offered fares $\{p_{k1}, \ldots, p_{kn_k}\}$, may be helpful to deal with net fares that may be different depending on the distribution channel costs. However, most of these benefits can be captured by enlarging, if necessary, the set of fares $\{p_{k1}, p_{k2}, \ldots, p_{kn_k}\}$ to allow to deal with commission netting and channel effects.

While formulation (2.2) is rich in detail, it is technically and notationally more convenient to work with a more streamlined formulation that aggregates ODF's into a single index and letting $n = \sum_{k=1}^{K} n_k$ be the number of ODF's. With $N = \{1, \ldots, n\}$, the single index formulation results in

$$\frac{\partial V(t,x)}{\partial t} = \mathcal{R}_t(\Delta V(t,x)) = \sum_{j \in M_t} \lambda_{tj}[p_j - \Delta_j V(t,x)]^+, \tag{2.3}$$

subject to the boundary conditions $V(t,0) = V(0,x) = 0$ for all $t \geq 0$ and all $x \geq 0$. Here $M_t \subseteq N$ is the set of valid fares at time-to-go t. In this formulation, λ_{tj} and p_j denote, respectively, the arrival rate and the fare of ODF j, while we let $\Delta_j V(t,x) = V(t,x) - V(t, x - A_j)$, where now A_j is vector of resources consumed by ODF j. Notice that there may be multiple identical columns corresponding to different fare classes for the same itinerary. In terms of the optimal controls, $u_j^*(t,x)$ we have

$$u_j^*(t,x) = \begin{cases} 1 & \text{if } j \in M_t,\ A_j \leq x, \text{ and } p_j \geq \Delta_j V(t,x) \\ 0 & \text{otherwise,} \end{cases} \tag{2.4}$$

so we accept a request for ODF j as long as j is a valid fare, we have enough capacity $x \geq A_j$, and the price p_j is greater than or equal to the displacement cost $\Delta_j V(t,x) = V(t,x) - V(t, x - A_j)$.

Since this dynamic program cannot be solved numerically for large-scale systems, we will often resort to heuristics that approximate the value function $V(t,x)$ by an easier way to compute function, say $\tilde{V}(t,x)$. We would then solve the assortment problem $\mathcal{R}(\Delta\tilde{V}(t,x))$ and use the heuristic control rule

$$\tilde{u}_j(t,x) = \begin{cases} 1 & \text{if } j \in M_t,\ A_j \leq x \text{ and } p_j \geq \Delta_j \tilde{V}(t,x) \\ 0 & \text{otherwise.} \end{cases}$$

We finalize this section by presenting a discrete-time dynamic programming formulation that emerges from the logic that leads to the construction of (2.3) by rescaling time, setting $\delta t = 1$, and dropping the $o(\delta t)$ term. This can be done by selecting $a > 1$, so that aT is an integer, and setting $\lambda_{jt} \leftarrow \frac{1}{a}\lambda_{j,t/a}$, for $t \in [0, aT]$. The scale factor a should be sufficiently large so that after scaling, $\sum_{j=1}^{n} \lambda_{jt} \ll 1$, e.g., $\sum_{j=1}^{n} \lambda_{jt} \leq 0.01$ for all $t \in [0, T]$, with $T \leftarrow aT$. The resulting discrete-time dynamic program is given by

$$V(t,x) = V(t-1,x) + \mathcal{R}_t(\Delta V(t-1,x)) \tag{2.5}$$

with the same boundary conditions $V(t,0) = V(0,x) = 0$.

The single index formulations (2.3) and (2.5) have the virtue of simplicity and generality at the cost of abstractness as it distances the model a bit from one of its potential origins. Its simplicity allows us to more easily treat the compound Poisson case, to discuss the nested by fare structure of optimal policies and randomness in the arrival rates. Next, we discuss several extensions of the basic model formulated so far.

2.2.1 Upgrades and Upsells

Let U_j be the set of products that can be used to fulfill a request for product j. The set U_j contains j and other more desirable substitutes. We assume that customers are willing to take any product $k \in U_j$ at the price of product p_j. The value function with upgrades is given by

$$\frac{\partial V(t,x)}{\partial t} = \sum_{j \in M_t} \lambda_j \max_{k \in U_j}(p_j - \Delta_k V(t,x))^+ = \sum_{j \in M_t} \lambda_j (p_j - \tilde{\Delta}_j V(t,x))^+$$

with boundary conditions $V(t,0) = V(0,x) = 0, x \geq 0$, where

$$\tilde{\Delta}_j V(t,x) = \min_{k \in U_j} \Delta_k V(t,x).$$

Now, consider an upsell model. Let U_j be as before, but assume now that r_{jk} is the revenue obtained from selling product j and fulfilling it with product $k \in U_j$. When $r_{jk} > p_j$, there is a probability, say π_{jk}, that a customer will accept the upgrade to product $k \in U_j$ as it now requires an additional payment of $r_{jk} - p_j$. Naturally, the probability π_{jk} that a customer agrees to pay for an upgrade from product j to product k is expected to decrease as r_{jk} increases. The upgrade model above assumes that $r_{jk} = p_j$ and $\pi_{jk} = 1$. Designing an upsell policy requires a careful tradeoff between a high r_{jk} and a low π_{jk}. A conservative approach would select r_{jk} to maximize the expected revenue $\pi_{jk} r_{jk}$. A bolder approach may allow for a higher r_{jk} even at the cost of a low probability π_{jk}, and this strategy may work if we have many candidates for upsells and we only need to convert a few of them.

In our upsell model, a customer may be asked to pay r_{jk} for product $k \in U_j$ instead of p_j for product j. If the customer accepts, the provider must fulfill with a product $l \in U_k$, and if he rejects, the request must be fulfilled with a product $l \in U_j$. This leads to the HJB equation:

$$\frac{\partial V(t,x)}{\partial t} = \sum_{j \in M_t} \lambda_j \max_{k \in U_j} \left[\pi_{jk}(r_{jk} - \tilde{\Delta}_k V(t,x))^+ + (1 - \pi_{jk})(p_j - \tilde{\Delta}_j V(t,x))^+ \right]$$

with the same boundary conditions as before. Notice that in this formulation that if a consumer is willing to upgrade from j to k, the provider can fulfill with any product in U_k, and if the customer rejects to upgrade to k, the provider can fulfill with any product in U_j. Throughout the rest of the chapter, for simplicity, we will assume that all fares are valid at all times so that $M_t = N$, helping us reduce the notational clutter.

Example 2.1 Suppose that $n = 3$, the prices are $(p_1, p_2, p_3) = (500, 300, 200)$. The aggregate demands $(\Lambda_1, \Lambda_2, \Lambda_3) = (1, 1, 7)$ and the capacities $(c_1, c_2, c_3) = (2, 2, 4)$. Noting that there are three resources and three ODF's, ODF j uses only resource j. Here, the arrival rates are stationary with $\lambda_j := \lambda_{tj}$ for all $t \in [0, T]$, so $\Lambda_j = \lambda_j T$. Suppose that $U_3 = \{1, 2, 3\}$, $U_2 = \{1, 2\}$, and $U_1 = \{1\}$. Then the expected revenue from the standard model is given by \$1493, while the model with upgrades results in \$1779. This represents a 19.12% improvement over the model without upgrades. Assume now that $r_{32} = 240$, $\pi_{32} = 0.80$, $r_{31} = 320$, $\pi_{31} = 0.40$, and $r_{21} = 380$, $\pi_{21} = 0.60$. The expected revenues with upsells result in \$1896 resulting in a 6.62% improvement over the model with free upgrades.

2.2.2 *Compound Poisson Process*

Our dynamic programming formulation is written under the implicit assumption that arriving requests for ODF j are for a single unit. If demand for an ODF j is a compound Poisson process with arrival rate λ_{tj} and demand size $Z_j = k$ with probability $q_j^k = \mathbb{P}\{Z_j = k\}$, then we can model this by using columns of the form $A_j^k = kA_j$ with corresponding fare p_j^k and arrival rate $\lambda_{tj}^k = \lambda_{tj} q_j^k$ for all k such that $q_j^k > 0$. Notice that the total fare p_j^k for a size k request for ODF j need not be equal to kp_j, thus allowing for economies or diseconomies of scale in group pricing. Under this formulation, a size k request for ODF j is accepted if $kA_j \le x$ and $p_j^k \ge \Delta_j^k V(t, x) = V(t, x) - V(t, x - A_j^k) = V(t, x) - V(t - kA_j)$. This allows us to treat ODF of each size as a different request in the single index model.

2.2.3 *Doubly Stochastic Poisson Process*

If the arrival rates themselves are random for each t, say λ_{tj}^i with probability θ_j^i, then the HJB Eq. (2.3) is of the same form with $\lambda_{tj} = \sum_i \lambda_{tj}^i \theta^i$. Consequently, the same value function is obtained regardless of the variance of the arrival rates.

2.3 Linear Programming-Based Upper Bound on $V(T, c)$

We will use a perfect foresight idea to obtain an upper bound on $V(T, c)$ given by (2.3). Let D_j be the aggregate demand for ODF j over the horizon $[0, T]$. Then D_j is Poisson with parameter $\Lambda_j = \int_0^T \lambda_{sj}\, ds$. Letting $M = \{1, \ldots, m\}$ be the set of resources, and recalling $N = \{1, \ldots, n\}$ is the set of ODF's, consider the linear program

$$\begin{aligned} \bar{V}(T, c) \;:=\; \max \quad & \sum_{j \in N} p_j\, y_j \\ \text{s.t.} \quad & \sum_{j \in N} a_{ij}\, y_j \le c_i \qquad \forall i \in M \\ & 0 \le y_j \le \Lambda_j \qquad \forall j \in N. \end{aligned} \tag{2.6}$$

Here, the decision variable y_j corresponds to the number of requests for ODF j that we plan to sell. The first constraint ensures that our sales do not violate the capacities of the resources. The second constraint ensures that our sales do not exceed the expected demand. Thus, the problem above can be viewed as an approximation formulated under the assumption that the demand for the ODF's takes on its expected value. We will show that the optimal objective value of the problem above, $\bar{V}(T, c)$, is an upper bound on the optimal total expected revenue. Also, letting $D = \{D_j : j \in N\}$, we will also show that an even sharper bound is given by $\mathbb{E}[\bar{V}(T, c \mid D)]$, which is the expected value of the perfect foresight model

$$\begin{aligned} \bar{V}(T, c \mid D) \;:=\; \max \quad & \sum_{j=1}^{n} p_j\, y_j \\ \text{s.t.} \quad & \sum_{j \in N} a_{ij}\, y_j \le c_i \qquad \forall i \in M \\ & 0 \le y_j \le D_j \qquad \forall j \in N. \end{aligned} \tag{2.7}$$

Note that perfect foresight model in (2.7) maximizes revenues by deciding sales after observing demand.

Theorem 2.2

$$V(T, c) \le \mathbb{E}[\bar{V}(T, c \mid D)] \le \bar{V}(T, c).$$

Proof For every instance of D, sales under the optimal dynamic policy (2.4) constitute a feasible solution to program (2.7), so the revenue obtained by the optimal policy under demand realizations D is at most $V(T, c|D)$. Taking expectations and noting that $V(T, c)$ is the total expected revenue of the optimal policy, we obtain $V(T, c) \le \mathbb{E}[\bar{V}(T, c \mid D)]$. Note that $\bar{V}(T, c \mid D)$ is concave in D. (This can be seen by observing that a convex combination of feasible solutions for two values of D is feasible for the convex combination of the two values of D, so the optimal solution is at least as large as the convex combination.) Thus, by Jensen's inequality, we get $\mathbb{E}[\bar{V}(T, c \mid D)] \leq \bar{V}(T, c \mid \mathbb{E}[D]) = \bar{V}(T, c \mid \Lambda) = \bar{V}(T, c)$, where the last equality follows from the definitions of $\bar{V}(T, c)$ and $\bar{V}(T, c \mid D)$. □

Both the primal and the dual of problem (2.6) are feasible and bounded so they have an optimal solution. To formulate the dual of problem (2.6), we associate the dual variables $z = (z_1, \ldots, z_m)$ with the first set of constraints and the dual variables $\beta = (\beta_1, \ldots, \beta_n)$ with the second set of constraints. By strong duality, they both result in the same value for the objective function and all complementary slackness conditions apply. The dual formulation is given by

$$\begin{aligned} \bar{V}(T, c) \;=\; \min \quad & \sum_{i \in M} c_i\, z_i + \sum_{j \in N} \Lambda_j\, \beta_j && \qquad (2.8) \\ \text{s.t.} \quad & \sum_{i \in M} a_{ij}\, z_i + \beta_j \geq p_j && \forall\, j \in N \\ & z_i \geq 0, \;\; \beta_j \geq 0 && \forall\, i \in M, \;\; j \in N. \end{aligned}$$

At optimality, the dual variable z_i^* is the marginal value of capacity of resource i, while the dual variable β_j^* is the marginal value of demand for ODF j. By complementary slackness, having $z_i^* > 0$ implies that the capacity constraint for resource i is tight, and having $\beta_j^* > 0$ implies that all the demand of ODF j is used.

We can simplify the dual problem (2.8) by solving for β in terms of z and then optimizing over z only. Notice that $\beta_j \geq p_j - \sum_{i \in M} a_{ij} z_i$ and $\beta_j \geq 0$ for all $j \in N$. Since the dual is a minimization problem, it follows that $\beta_j = (p_j - \sum_{i \in M} a_{ij} z_i)^+$ for all $j \in N$. Consequently,

$$\sum_{j \in N} \Lambda_j \beta_j = \sum_{j \in N} \Lambda_j (p_j - \sum_{i \in M} a_{ij} z_i)^+ = \int_0^T \mathcal{R}_t(A'z)dt,$$

so

$$\bar{V}(T, c) = \min_{z \geq 0} \left\{ \int_0^T \mathcal{R}_t(A'z)dt + c'z \right\}. \qquad (2.9)$$

Program (2.9) involves only the decision variable z. An optimal solution z^* provides us with an estimate z_i^* of the marginal value of the ith resource. This is equivalent to approximating $\Delta_j V(T, c)$, by $e_j' \, A' \, z^* = \sum_{i \in M} a_{ij} \, z_i^*$ for $j \in N$, where e_j is the unit vector with one in the j-th component.

Example 2.3 Consider a network with two connecting flight legs and assume that two fares are available for each origin-destination pair. Thus, there are six ODF's with capacity consumption vectors

$$A_1 = A_2 = \begin{pmatrix} 1 \\ 0 \end{pmatrix} \quad A_3 = A_4 = \begin{pmatrix} 0 \\ 1 \end{pmatrix} \quad A_5 = A_6 = \begin{pmatrix} 1 \\ 1 \end{pmatrix}.$$

The fares and probabilities of getting requests for the different ODF's are given by

$$(p_1, \ldots, p_6) = (150, 100, 120, 80, 250, 170)$$

$$(\lambda_{t1}, \ldots, \lambda_{t6}) = \begin{cases} (0.00, 0.12, 0.00, 0.16, 0.00, 0.08) & \text{if } t = 1000, 999, \ldots, 501 \\ (0.06, 0.00, 0.04, 0.00, 0.06, 0.00) & \text{if } t = 500, 499, \ldots, 1. \end{cases}$$

Notice that low fare requests arrive before high fare ones. The capacities on the flight legs are $c = (90, 90)$. For this problem instance, the optimal solution to problem (2.6) is $y^* = (30, 30, 20, 40, 30, 0)$ with objective value $\bar{V}(T, c) =$ \$20,600. Thus, the total expected revenue obtained by the optimal policy is no larger than \$20,600. The solution to the dual is given by $z^* = (100, 80)$.

Example 2.4 Consider the problem instance in Example 2.3. Solving formulation (2.7) for 100,000 demand realizations, we estimate $\mathbb{E}[\bar{V}(T, c \mid D)]$ to be \$20,600, which is the same as the upper bound from problem (2.6). Although formulation (2.7) does not improve the upper bound (2.6) for this problem instance, if the capacities on the flight legs were $c = (60, 60)$, the upper bounds from formulation (2.6) and (2.7) would be, respectively, \$15,200 and \$15,054.

The linear program in (2.6) is known as the deterministic linear program or the fluid approximation to the network revenue management problem. To make decisions in the network revenue management problem, one can design heuristics by using the optimal primal or the optimal dual solution to the deterministic linear program. Heuristics based on the primal solution require keeping n parameters, whereas heuristics based on the dual solution require keeping m parameters. In practice, the number of ODF's is usually much larger than the number of flight legs, which makes heuristics based on primal solutions significantly more cumbersome to implement. Nevertheless, heuristics based on primal solutions have the potential of performing well when implemented correctly. In particular, these heuristics have optimality guarantees in a certain asymptotic regime. In the next section, we explain how we can construct heuristics by using the optimal primal or the optimal dual solution to the deterministic linear program, compare these heuristics with each other, and discuss the asymptotic optimality properties of these heuristics.

2.4 Bid-Prices and Probabilistic Admission Control

Here, we discuss two heuristics for the network revenue management problem, one based on the solution to the dual of problem ((2.6) and the other based on the solution to the primal problem. Suppose $\{z_i^* : i \in M\}$ is the solution to the dual (2.8).

We construct a bid-price heuristic as follows. We make ODF j available if and only if $A_j \leq x$ and $p_j \geq \sum_{i \in M} a_{ij} z_i^*$. In other words, the bid-price heuristic makes a product available if there is enough capacity and the fare is higher than the opportunity cost of the capacities consumed (or the displacement costs). In network revenue management vocabulary, the opportunity cost of seats on a flight leg is referred to as the bid-price of this flight leg. Although the bid-price heuristic described above is based on the vector of bid-prices obtained from the dual solution to the deterministic linear program, but there are other approaches in the literature for obtaining bid-prices and we go over some of these approaches later in this chapter.

Let $y = (y_1^*, \ldots, y_n^*)$ denote an optimal solution to primal problem (2.6). A heuristic will make available ODF j with probability y_j^*/Λ_j whenever $A_j \leq x$. We refer to this heuristic policy as the probabilistic admission control (PAC) heuristic. Comparing the PAC heuristic with a bid-price heuristic, the PAC uses n parameters corresponding to the admission probabilities $\{y_j^*/\Lambda_j : j \in N\}$. In contrast, the bid-price heuristic uses m parameters corresponding to the bid-prices $(z_1^*, \ldots, z_m^*)$. In most applications, the number of resources m is much smaller than the number of ODF's n. Thus, bid-price heuristics can be characterized in a more parsimonious fashion.

There are interesting connections between the two heuristics when the bid-prices are obtained through the dual solution to the deterministic linear program. Let (z^*, β^*) be the optimal solution to problem (2.8) and consider the bid-price heuristic characterized by the bid-prices $z^* \in \Re^m$. Let

$$F := \{j \in N : p_j > \sum_{i \in M} a_{ij} z_i^*\} \qquad P := \{j \in N : p_j = \sum_{i \in M} a_{ij} z_i^*\}$$

$$R := \{j \in N : p_j < \sum_{i \in M} a_{ij} z_i^*\}.$$

As long as there is enough capacity, the bid-price heuristic makes the ODF's in F and P available for purchase. Also, let

$$F' := \{j \in N : y_j^* = \Lambda_j\} \qquad P' := \{j \in N : 0 < y_j^* < \Lambda_j\}$$

$$R' := \{j \in N : y_j^* = 0\}.$$

The PAC heuristic makes the ODF's in F' available with probability one and the ODF's in P' available with a probability that is strictly between zero and one. The next proposition characterizes a connection between the bid-price and PAC heuristics.

Proposition 2.5 $R \subseteq R'$, $P' \subseteq P$, $F \subseteq F'$.

If $P = \emptyset$, then noting that $F \cup P \cup R = F' \cup P' \cup R'$, it follows that $F = F'$ and $R = R'$, in which case, the PAC and bid-price heuristics coincide. If $P' \neq \emptyset$, then the PAC heuristic has a more refined method of dealing with the ODF's in P. This feature of the PAC heuristic is good news when the solution to the deterministic linear program has many ODF's in P, particularly when the request for the ODF's in P arrives before the requests for the ODF's in F. In practice, since the ODF's in F are expected to have larger revenues than the ODF's in P, it is expected that the requests for the ODF's in P arrive before the requests for the ODF's in F. The next example illustrates this situation.

Example 2.6 Consider the problem instance in Example 2.3. Using the optimal dual solution to problem (2.6), we can check that the six ODF's are partitioned as

$$F = F' = \{1, 3, 5\},\ P = P' = \{2, 4\} \text{ and } R = R' = \{6\}.$$

Notice also that fares in P arrive before fares in F. The total expected revenue obtained by the bid-price heuristic is about \$17,732, which was estimated by averaging the total revenues in 100,000 simulated sample paths of the bid-price heuristic. The PAC heuristic has $y_j^* = 0.5\Lambda_j$ for $j \in P$, so the PAC admits ODF's in P with probability 50%. This improves the expected revenues dramatically, from \$17,732 to \$19,386.

Under the assumption that the fares associated with the ODF's are uniformly bounded continuous random variables, it is possible to show that the bid-price heuristic based on deterministic linear program (2.6) is asymptotically optimal. More precisely, if $V^b(T, c)$ is the optimal total expected revenue for a system with arrival rates $b\lambda_{tj},\ j \in N,\ t \in [0, T]$, and capacities bc, and $\hat{\Pi}^b(T, c)$ is the total expected revenue of the bid-price heuristic for the same system based on the dual solution z^* to problem (2.8), then it is possible to show that

$$\lim_{b\to\infty} \frac{\hat{\Pi}^b(T, c)}{V^b(T, c)} \to 1.$$

Intuitively speaking, the assumption that the fares are continuous random variables guarantees that the set of ODF's in P' that are partially made available is empty. In this case, the bid-price heuristic works similarly to the PAC heuristic. Unfortunately, the set P' is rarely empty, so the asymptotic optimality of the bid-price heuristic is destroyed when the continuity assumption of the random fares is relaxed.

We can see what goes wrong by considering a single leg problem with capacity c_1 and two fares $p_1 > p_2$, with aggregate demands $\Lambda_1 = (1-\epsilon)c_1$ and Λ_2 such that $\Lambda_1 + \Lambda_2 > c_1$ for a small ϵ. The bid-price from the dual solution of problem (2.6) is $z_1^* = p_2$, so $F = F' = \{1\}$ and $P = P' = \{2\}$. The bid-price heuristic makes all ODF's in $F \cup P = \{1, 2\}$ available until capacity is exhausted. If arrivals are low-to-high and $\Lambda_2 \gg c_1$, then the entire capacity is consumed by low-fare customers resulting in revenue p_2c_1, while the optimal revenue is at least $p_1c_1(1-\epsilon)$. One may try to correct things by redefining the bid-price heuristic to allow bookings only if $p_j > z_1^*$, so only ODF's in set $F = \{1\}$ are admitted while capacity is available. But things go wrong again when $\Lambda_1 = \epsilon c_1$, as most of the capacity would get spoiled. Neither version of the heuristic improves as demand and capacity are scaled by $b \geq 1$. As a result, the bid-price heuristic is *not*, in general, asymptotically optimal. In practice, we have $P \neq \emptyset$ and demand for the marginal fares in P tends to arrive before demand for fares in F. Thus, the bid-price heuristic risks being too generous in accepting requests for fares in P at the cost of consuming capacity that would be better used by fares in F.

We now show that unlike the bid-price heuristic, the PAC heuristic is asymptotic optimality even if $P \neq \emptyset$.

Theorem 2.7 *Let $\Pi^b(T,c)$ be the total expected revenue from the PAC heuristic and let $V^b(T,c)$ be the optimal total expected revenue corresponding to instance $b \geq 1$ with capacity bc and arrival rates $b\lambda_{jt}$ for $j \in N$ and $t \in [0,T]$. Then $\lim_{b\to\infty} \Pi^b(T,c)/V^b(T,c) = 1$.*

Theorem 2.7 essentially tells us that for large enough systems the PAC heuristic generates total expected revenues that represent a very high percentage of the optimal total expected revenues.

2.5 Refinements of Heuristics

In this section, we describe refinements of the bid-price and PAC heuristics that are often used in practice.

2.5.1 Resolving the Deterministic Linear Program

The practical performance of the bid-price heuristics tends to improve if the deterministic linear program is resolved frequently during the sales horizon, for example, at reading dates when the demand forecasts may also be updated. The key insight about resolving is that fares in P tend to close over time, so resolving frequently prevents accepting too many marginal fares. To see how resolving works, suppose that the state is (t,x). Then, we resolve problem (2.6) after replacing the right side of the first set of constraints with $(x_1,\ldots,x_m)$ and the right side of the second set of constraints with $\Lambda_{tj} = \int_0^t \lambda_{sj} ds$, $j \in N$. Resolving can help update both the bid-price heuristic and the PAC heuristic. Resolving has provable asymptotic optimality properties. However, resolving may also hurt performance in some cases, particularly towards the end of the sales horizon when the deterministic linear program significantly suffers from ignoring randomness.

2.5.2 Randomized Linear Program

Heuristics can benefit from refinements that make them more sensitive to randomness in demand. One idea is to solve the dual of the perfect foresight model (2.7) with many sample demand realizations and using the average of the dual $z^*(D)$

values as bid-prices. The idea is to approximate $\mathbb{E}[z^*(D)]$ via simulation. There is an important advantage of working with problem (2.7) instead of problem (2.6), because as indicated in Theorem 2.2, $\mathbb{E}[\bar{V}(T, c \mid D)]$ is a better approximation to $V(T, c)$ than $\bar{V}(T, c)$, and heuristics based on sharper upper bounds tend to produce better results because the approximation to the value function is better. Likewise, by averaging the primal solutions from different demand realizations, we can construct admission probabilities for the PAC heuristic. One can show that the randomized linear program also has asymptotic optimality properties. In practice, using the randomized linear programming approach helps when the problem is solved only once during the horizon or when the number of resolves is low. However, as the deterministic linear program is resolved more frequently, the benefit of randomizing tends to disappear and in some instances may result in worse performance.

2.5.3 *Time-Dependent Bid-Prices*

To describe the time-dependent bid-price heuristic, we will work with the discrete-time approximation of the HJB Eq. (2.3) given by (2.5). Recall that the discrete-time approximation requires scaling time so that we can think of the arrival rate λ_{tj} for product j in period t as the probability of a request for product j in period t.

The deterministic linear program in (2.6) produces a single bid-price for each flight leg. By resolving the deterministic linear program, we can obtain bid-prices that depend on the remaining time in the sales horizon, but an interesting question is whether we can come up with a version of the deterministic linear program that naturally produces bid-prices that depend on the remaining time in the sales horizon. To that end, we use the decision variable y_{tj} to capture the probability that we sell a ticket for ODF j at time period t and the decision variable x_{ti} to denote the expected remaining capacity for resource i at time period t. A more sophisticated version of the deterministic linear program in (2.6) is given by

$$
\begin{aligned}
\tilde{V}(T, c) := \max \quad & \sum_{t=1}^{T} \sum_{j \in N} p_j \, y_{tj} && && (2.10) \\
\text{s.t.} \quad & x_{Ti} = c_i && \forall i \in M \\
& x_{t-1,i} = x_{ti} - \sum_{j \in N} a_{ij} \, y_{tj} && \forall i \in M, \; t = 2, \ldots, T \\
& a_{ij} \frac{y_{tj}}{\lambda_{tj}} \le x_{ti} && \forall i \in M, \; j \in N, \; t = 1, \ldots, T \\
& 0 \le y_{tj} \le \lambda_{tj}, \; x_{ti} \ge 0 && \forall i \in M, \; j \in N, \; t = 1, \ldots, T.
\end{aligned}
$$

The objective function of the problem above accumulates the total expected revenue obtained over the sales horizon. The first set of constraints initializes the remaining capacities of the resources. Noting that $\sum_{i \in N} a_{ij}\, y_{tj}$ is the expected capacity consumption of resource i at time period t, the second set of constraints keep track of the dynamics of the expected remaining capacities. In the third set of constraints, y_{tj}/λ_{tj} is the probability that we sell a ticket for ODF j at time period t given that there is a request for this ODF at time period t. Thus, $a_{ij}\, y_{tj}/\lambda_{tj}$ is the expected capacity consumption on flight leg i at time period t given that there is a request for ODF j at time period t. The third set of constraints ensure that the expected capacity consumption of resource i given that there is a demand for ODF j at time period t cannot exceed the expected remaining capacity of resource i. The last set of constraints ensure that the probability of selling a ticket for ODF j at time period t cannot exceed the probability of having a request for this ODF at time period t. The next theorem shows that $\tilde{V}(T, c)$ lies between $V(T, c)$ and $\bar{V}(T, c)$, where the latter quantity is the optimal objective value of our earlier deterministic linear program.

Theorem 2.8 $V(T, c) \leq \tilde{V}(T, c) \leq \bar{V}(T, c)$.

The problem of the theorem is based on using an optimal solution to problem (2.10) to construct a feasible solution to problem (2.6) such that the objective values obtained for the two problems match.

Example 2.9 For the problem instance in Example 2.3, the optimal objective value of problem (2.10) is \$20,510, improving the upper bound provided by the deterministic linear program by \$90.

One can use the dual solution to problem (2.10) to obtain time-dependent bid-prices. In particular, the first two sets of constraints in problem (2.10) capture the dynamics of the remaining capacities. We let $\{z^*_{1i} : i \in M\}$ be the optimal values of the dual variable associated with the first constraint and $\{z^*_{ti} : t = 2, \ldots, T,\ i \in M\}$ be the optimal values of the dual variables associated with the second constraint. Thus, we can use z^*_{ti} to capture the opportunity cost of a unit of capacity of resource i at time period t. To make the decisions at time period t, the bid-price heuristic can use the bid-prices $(z^*_{t1}, \ldots, z^*_{tm})$. Also, the primal solution to problem (2.10) can be used to obtain admission probabilities in the PAC heuristic. The decision variable y_{tj} corresponds to the probability of selling a ticket for ODF j at time period t. The PAC heuristic can make ODF j available for purchase at time period t with probability y_{tj}/λ_{tj}.

Problem (2.10) has $O(Tmn)$ constraints, which can get too many in practical applications. To deal with this difficulty, one idea is to aggregate the constraints after some time period s. In this case, the problem would have $O(smn)$ constraints and returns a value function that is still an upper bound, but that may not be as tight as the full model. One can then check for optimality iteratively, increasing s if necessary until optimality to the full model is achieved.

Lastly, the bid-prices and the admission probabilities obtained from problem (2.10) depend on the remaining time in the sales horizon. Nevertheless, there is still value in resolving problem (2.10) periodically at reading dates over the sales horizon.

Table 2.1 Performance of various bid-price and PAC heuristics

Control	1	4	10
Bid-price, problem (2.6)	\$17,732	\$18,519	\$19,582
PAC, problem (2.6)	\$19,386	\$19,438	\$19,554
Bid-price, problem (2.7)	\$17,736	\$19,372	\$19,448
PAC, problem (2.7)	\$18,867	\$19,257	\$19,429
Bid-price, problem (2.10)	\$17,733	\$18,325	\$19,583
PAC, problem (2.10)	\$19,337	\$19,457	\$19,569

Example 2.10 For the problem instance in Example 2.3, Table 2.1 provides the total expected revenues obtained by the heuristics discussed so far. We resolve problems (2.6), (2.7), and (2.10) 1, 4, and 10 times over the sales horizon, at evenly spaced time periods. Table 2.1 shows the total expected revenues obtained by the bid-price and PAC heuristics that we extract from the deterministic linear program in (2.6), the randomized linear program in (2.7), and the refined version of the deterministic linear program in (2.10). The total expected revenues are estimated through 100,000 simulated sample paths. The bid-price heuristic from problem (2.6) improves as we move to more frequent resolves, actually overtaking the performance of the PAC heuristic from the same problem. This observation is not unusual as the bid-price heuristic tends to gradually tighten capacity for fares in P. The heuristics from problem (2.10) do not provide a significant increase in total expected revenues in this particular instance, except for the PAC heuristic with 4 resolves. The bid-price heuristic from problem (2.7) performs better than (2.6) for 1 and 4 resolves, but lags behind for 10 resolves.

In Example 2.10, the heuristics from problem (2.10) perform only marginally better than the heuristics from the deterministic linear program in (2.6). Computational experiments on larger problem instances show that problem (2.10) can be beneficial, especially if the problems in consideration are not resolved frequently. Notice that the bid-price heuristic benefits from resolving, eventually overtaking the performance of the PAC heuristic, in Table 2.1, with 10 equally spaced reading dates during the sales horizon.

2.6 Dynamic Programming Decomposition

In this section, we describe two possible approaches for approximating the value functions $\{V(t,\cdot) : t = 1,\ldots,T\}$ for the discrete-time formulation (2.5). Once we have such approximations to the value functions, we can construct a heuristic policy by replacing $V(t-1,\cdot)$ on the right side of (2.4) with the value function approximation. To facilitate the exposition, we will write the formulation (2.5) in a slightly different form, where we create an artificial ODF, say ϕ, with revenue $p_\phi = 0$, consumption vector $A_\phi = 0$, and arrival probability $\lambda_{t\phi} = 1 - \sum_{j \in N} \lambda_{tj}$. By adding this fictitious product, and making the number of products $n \leftarrow n+1$, we can assume without loss of generality that $\sum_{j \in N} \lambda_{tj} = 1$, and write the dynamic program (2.5) as

$$V(t, x) = \max_{u \in U(x)} \left\{ \sum_{j=1}^{n} \lambda_{tj} [p_j \, u_j + V(t-1, x - u_j \, A_j)] \right\} \tag{2.11}$$

with the same boundary conditions $V(t, 0) = V(0, x) = 0$.

2.6.1 Exploiting the Deterministic Linear Program

One approach for approximating the value functions is based on the deterministic linear program given in (2.6). Assume that we solve problem (2.6), and $(z_1^*, \ldots, z_m^*)$ are the optimal values of the dual variables associated with the first set of constraints. We choose an arbitrary resource i and relax the first set of constraints for all of the resources except for resource i by associating the dual multipliers $(z_1^*, \ldots, z_m^*)$ with them. The objective function of problem (2.6) reads $\sum_{j \in N} p_j \, y_j + \sum_{k \neq i} z_k^* \, [c_k - \sum_{j \in N} a_{kj} \, y_j] = \sum_{j \in N} [p_j - \sum_{k \neq i} a_{kj} \, z_k^*] \, y_j + \sum_{k \neq i} z_k^* \, c_k$. Since we use the optimal values of the dual variables as dual multipliers, by linear programming duality, relaxing the first set of constraints in this fashion does not change the optimal objective value of the problem. Thus, problem (2.6) and the problem

$$\begin{aligned} \bar{V}(T, c) \;=\; \max \quad & \sum_{j \in N} \Big[p_j - \sum_{k \neq i} a_{kj} \, z_k^* \Big] \, y_j + \sum_{k \neq i} z_k^* \, c_k \\ \text{s.t.} \quad & \sum_{j \in N} a_{ij} \, y_j \leq c_i \\ & 0 \leq y_j \leq \Lambda_j \qquad \forall \, j \in N \end{aligned} \tag{2.12}$$

share the same optimal objective value and we denote this common optimal objective value by $\bar{V}(T, c)$. Ignoring the constant term $\sum_{k \neq i} z_k^* \, c_k$ in the objective function and comparing the problem above with problem (2.6), we observe that problem (2.12) is the deterministic linear program for a single-resource revenue management problem that takes place over resource i only. In this single-resource revenue management problem, if we accept a request for ODF j, then we generate a revenue of $p_j - \sum_{k \neq i} a_{kj} \, z_k^*$.

By Theorem 2.2, the optimal objective value of the deterministic linear program provides an upper bound on the optimal total expected revenue. Noting that the optimal objective value of problem (2.12) is $\bar{V}(T, c)$, it follows that $\bar{V}(T, c) - \sum_{k \neq i} z_k^* \, c_k$ provides an upper bound on the optimal total expected revenue for the single-resource revenue management problem that takes place over resource i. On the other hand, we can compute the optimal total expected revenue in the single-resource revenue management problem that takes place over resources i by solving the dynamic program

$$v_i(t, x_i) = \max_{u \in \mathcal{U}_i(x_i)} \left\{ \sum_{j \in N} \lambda_{tj} \left\{ \left[p_j - \sum_{k \neq i} a_{kj}\, z_k^* \right] u_j + v_i(t-1, x_i - u_j\, a_{ij}) \right\} \right\}. \tag{2.13}$$

In the dynamic program above, we only keep track of the remaining capacity of resource i. If we sell a ticket for ODF j, then we generate revenue $p_j - \sum_{k \neq i} a_{kj}\, z_k^*$. If ODF j uses resource i, then we consume capacity a_{ij} on flight leg i. The set of feasible decisions is given by $\mathcal{U}_i(x_i) = \{u \in \{0, 1\}^n : a_{ij}\, u_j \leq x_i\ \forall\, j = 1, \ldots, n\}$. Since the state variable is a scalar, solving the dynamic program above is tractable.

The optimal total expected revenue for the single-resource revenue management problem that takes place over resource i is given by $v_i(T, c_i)$. By the discussion at the beginning of the previous paragraph, $\bar{V}(T, c) - \sum_{k \neq i} z_k^*\, c_k$ provides an upper bound on the optimal total expected revenue for the single-resource revenue management problem that takes place over resource i, implying that $v_i(T, c_i) + \sum_{k \neq i} z_k^*\, c_k \leq \bar{V}(T, c)$. The next theorem shows that $v_i(T, c_i) + \sum_{k \neq i} z_k^*\, c_k \geq V(T, c)$, implying that $v_i(T, c_i) + \sum_{k \neq i} z_k^*\, c_k$ is an upper bound on the optimal total expected revenue and this upper bound is tighter than $\bar{V}(T, c)$. Since this is true for every i, the minimum of $v_i(T, c_i) + \sum_{k \neq i} z_k^*$ over i is also an upper bound.

Theorem 2.11 *For any $t = 1, \ldots, T$ and $x \in \mathbb{Z}_+^m$,*

$$V(t, x) \leq \min_{i \in M} \{ v_i(t, x_i) + \sum_{k \neq i} z_k^*\, x_k \}.$$

Proof We show the result by using induction over the time periods. It is not difficult to check that the result holds at time period 1. Assuming that the result holds at time period $t - 1$, we show that the result holds at time period t. Letting u^* be an optimal solution to the problem on the right side of (2.11), we have

$$\begin{aligned}
V(t, x) &= \sum_{j \in N} \lambda_{tj} \left\{ p_j\, u_j^* + V(t-1, x - u_j^*\, A_j) \right\} \\
&\leq \sum_{j \in N} \lambda_{tj} \left\{ p_j\, u_j^* + v_i(t-1, x_i - u_j^*\, a_{ij}) + \sum_{k \neq i} z_k^*\, [x_k - u_j^*\, a_{kj}] \right\} \\
&= \sum_{j \in N} \lambda_{tj} \left\{ \left[p_j - \sum_{k \neq i} a_{kj}\, z_k^* \right] u_j^* + v_i(t-1, x_i - u_j^*\, a_{ij}) \right\} + \sum_{k \neq i} z_k^*\, x_k \\
&\leq v_i(t, x_i) + \sum_{k \neq i} z_k^*\, x_k,
\end{aligned}$$

where the first inequality is by the induction assumption, the second equality uses the fact that $\sum_{i \in N} \lambda_{tj} = 1$ and the second inequality follows by noting that having $u^* \in \mathcal{U}(x)$ implies $u^* \in \mathcal{U}_i(x_i)$ so that u^* is a feasible, but not necessarily the optimal, solution to the problem on the right side of (2.13). Since the choice of resource i is arbitrary, taking the minimum overall $i \in M$ on the right side of the chain of inequalities above yields the desired result. □

Example 2.12 Consider the problem instance in Example 2.3. For all $i \in M$, we compute the value functions $\{v_i(t, \cdot) : t = 1, \ldots, T\}$ through the dynamic program in (2.13). The upper bound that we obtain on the optimal total expected revenue is given by $\min_{i \in M}\{v_i(T, c_i) + \sum_{k \neq i} z_k^* c_k\} = \$20{,}181$, which tightens the upper bound of \$20,600 provided by the deterministic linear program.

The approach does not give guidance on how to make the availability decisions. As a heuristic, we can solve the dynamic program in (2.13) for all $i \in M$ to obtain the value functions $\{v_i(t, \cdot) : t = 1, \ldots, T\}$ for all $i \in M$ and then approximate $V(t, x)$ by $\sum_{i \in M} v_i(t, x_i)$. Although we cannot claim this approximation is an upper bound, it can serve as refinement to the approximation suggested by the linear program (2.6). We report the results of applying this heuristic after we present another heuristic based on Lagrangian relaxation.

2.6.2 *Lagrangian Relaxation*

Another approach for approximating the value functions is based on Lagrangian relaxation. To demonstrate the Lagrangian relaxation strategy, we use decision variables $\{w_{ij} : i \in M,\ j \in N\}$ in the dynamic programming formulation of the network revenue management problem, where $w_{ij} = 1$ if we make ODF j available for purchase on flight leg i, otherwise $w_{ij} = 0$.

Let $e_i \in \mathbb{Z}_+^m$ denote the unit vector with a one in the component corresponding to resource i, and let ψ denote a fictitious resource with infinite capacity. With this notation, the dynamic program in (2.11) can be written as

$$
\begin{aligned}
V(t, x) \;=\; \max \quad & \sum_{j \in N} \lambda_{tj} \Big\{ p_j\, w_{\psi j} + V(t-1, x - \sum_{i \in M} w_{ij}\, a_{ij}\, e_i) \Big\} && (2.14)\\
\text{s.t.} \quad & a_{ij}\, w_{ij} \le x_i && \forall i \in M,\ j \in N\\
& w_{ij} = w_{\psi j} && \forall i \in M,\ j \in N\\
& w_{ij} \in \{0, 1\},\ w_{\psi j} \in \{0, 1\} && \forall i \in M,\ j \in N.
\end{aligned}
$$

To see that the dynamic program above is equivalent to the one in (2.11), we can use the second set of constraints to replace the decisions variables $\{w_{ij} : i \in M\}$ with the single decision variable $w_{\psi j}$. The expression $\sum_{i \in M} w_{ij}\, a_{ij}\, e_i$ in the objective function is equivalent to $A_j\, w_{\psi j}$, and the first set of constraints can be written as $A_j\, w_{\psi j} \le x$ for all $j \in N$. Identifying the decision variable $w_{\psi j}$ with the decision variable u_j in the dynamic program in (2.11) shows the equivalence. The Lagrangian strategy is based on relaxing the second set of constraints by associating the Lagrange multipliers $\alpha = \{\alpha_{tij} : t = 1, \ldots, T,\ i \in M,\ j \in N\} \in \Re^{Tmn}$ with them. This results in the relaxed dynamic program

$$\begin{aligned} V^\alpha(t, x) = \max \sum_{j \in N} \lambda_{tj} \Big\{ \sum_{i \in M} \alpha_{tij}\, w_{ij} + \Big[p_j - \sum_{i \in M} \alpha_{tij} \Big] w_{\psi j} \\ + V^\alpha(t-1, x - \sum_{i \in M} w_{ij}\, a_{ij}\, e_i) \Big\} \qquad (2.15) \\ \text{s.t.}\ \ a_{ij}\, w_{ij} \le x_i \qquad \forall\, i \in M,\ j \in N \\ w_{ij} \in \{0, 1\},\, w_{\psi j} \in \{0, 1\} \qquad \forall\, i \in M,\ j \in N. \end{aligned}$$

The superscript α in the value functions emphasizes the fact that the value functions $\{V^\alpha(t, \cdot) : t = 1, \ldots, T\}$ obtained from the dynamic program above depend on our choice of the Lagrange multipliers. Notice that the Lagrange multiplier α_{tij} in (2.15) is scaled by λ_{tj}, but this scaling is not a critical concern since if $\lambda_{tj} = 0$, then we do not have a request for ODF j in time period t and we can drop the Lagrange multipliers $\{\alpha_{tij} : i \in M\}$.

We let $\mathcal{U}_i(x_i) = \{w_i \in \{0, 1\}^n : a_{ij}\, w_{ij} \le x_i\ \forall\, j \in N\}$, where we use w_i to denote the vector $w_i = (w_{i1}, \ldots, w_{in}) \in \{0, 1\}^n$. The first set of constraints in the dynamic program in (2.15) can succinctly be written as $w_i \in \mathcal{U}_i(x_i)$ for all $i \in M$. The next theorem shows that the dynamic program in (2.15) decomposes by the resources.

Theorem 2.13 *Assume that the value functions* $\{v_i^\alpha(t, \cdot) : t = 1, \ldots, T\}$ *are computed through the dynamic program*

$$v_i^\alpha(t, x_i) = \max_{w_i \in \mathcal{U}_i(x_i)} \Big\{ \sum_{j \in N} \lambda_{tj} \Big\{ \alpha_{tij}\, w_{ij} + v_i^\alpha(t-1, x_i - w_{ij}\, a_{ij}) \Big\} \Big\}. \qquad (2.16)$$

Then

$$V^\alpha(t, x) = \sum_{i \in M} v_i^\alpha(t, x_i) + \sum_{\tau=1}^{t} \sum_{j \in N} \lambda_{\tau j} \Big[p_j - \sum_{i \in M} \alpha_{\tau ij} \Big]^+. \qquad (2.17)$$

Proof The proof is by induction. It is easy to verify that the result holds at time period 1. Assume that the result holds at time period $t-1$. By the induction assumption, the dynamic program in (2.15) is given by

$$V^\alpha(t, x) = \max \quad \sum_{j \in N} \lambda_{tj} \Big\{ \sum_{i \in M} \alpha_{tij}\, w_{ij} + \Big[p_j - \sum_{i \in M} \alpha_{tij} \Big] w_{\psi j}$$

$$+ \sum_{i \in M} v_i^\alpha(t-1, x_i - w_{ij}\, a_{ij}) \Big\} + \sum_{\tau=1}^{t-1} \sum_{j \in N} \lambda_{\tau j} \Big[p_j - \sum_{i \in M} \alpha_{\tau ij} \Big]^+$$

$$\text{s.t.} \ \ w_i \in \mathcal{U}_i(x_i), \ \ w_\psi \in \{0, 1\}^n \qquad \forall i \in M.$$

We note that the optimal value of the decision variable $w_{\psi j}$ in the problem above is one if $p_j - \sum_{i \in M} \alpha_{tij} \geq 0$, otherwise the optimal value of this decision variable is zero. Therefore, for each $j \in N$, we can replace the expression $[p_j - \sum_{i \in M} \alpha_{tij}]\, w_{\psi j}$ in the objective function above with $[p_j - \sum_{i \in M} \alpha_{tij}]^+$ and drop the decision variable $w_{\psi j}$. The desired result follows by noting that the objective function and the feasible solution set of the problem decompose by the resources. □

By Theorem 2.13, we can compute the value functions $\{V^\alpha(t, \cdot) : t = 1, \ldots, T\}$ by solving the dynamic program in (2.16). The dynamic program in (2.16) is simple to solve since it has a scalar state variable. In the next theorem, we show that the value functions $\{V^\alpha(t, \cdot) : t = 1, \ldots, T\}$ upper bound the value functions $\{V(t, \cdot) : t = 1, \ldots, T\}$.

Theorem 2.14 *For any set of Lagrange multipliers* α, *we have* $V(t, x) \leq V^\alpha(t, x)$ *for all* $x \in \mathbb{Z}_+^m$ *and* $t = 1, \ldots, T$.

Proof The proof follows from an argument similar to the one in the proof of Theorem 2.11 and it uses induction over the time periods. It is simple to check that the result holds at time period 1. Assuming that the result holds at time period $t-1$, we show that the result holds at time period t as well. We let $\{w_{ij}^* : i \in M, \ j \in N\}$ and $\{w_{\psi j}^* : j \in N\}$ be an optimal solution to the problem on the right side of (2.14). Then,

$$V(t, x) = \sum_{j \in N} \lambda_{tj} \Big\{ p_j\, w_{\psi j}^* + V(t-1, x - \sum_{i \in M} w_{ij}^*\, a_{ij}\, e_i) \Big\}$$

$$= \sum_{j \in N} \lambda_{tj} \Big\{ \sum_{i \in M} \alpha_{tij}\, w_{ij}^* + \Big[p_j - \sum_{i \in M} \alpha_{tij} \Big] w_{\psi j}^* + V(t-1, x - \sum_{i \in M} w_{ij}^*\, a_{ij}\, e_i) \Big\}$$

$$\leq \sum_{j \in N} \lambda_{tj} \Big\{ \sum_{i \in M} \alpha_{tij}\, w_{ij}^* + \Big[p_j - \sum_{i \in M} \alpha_{tij} \Big] w_{\psi j}^* + V^\alpha(t-1, x - \sum_{i \in M} w_{ij}^*\, a_{ij}\, e_i) \Big\}$$

$$\leq V^\alpha(t, x),$$

where the second equality uses the fact that the solution $\{w_{ij}^* : i \in M, \ j \in N\}$ and $\{w_{\psi j}^* : j \in N\}$ satisfies the second set of constraints in problem (2.14), the first inequality follows from the induction assumption, and the second inequality uses the fact that $\{w^*_{ij} : i \in M,\ j \in N\}$ and $\{w^*_{\psi j} : j \in N\}$ is a feasible, but not necessarily the optimal, solution to problem (2.15). □

Theorem 2.14 implies that for any set of Lagrange multipliers α, $V^\alpha(T, c)$ provides an upper bound on the optimal total expected revenue. To find the tightest possible upper bound, we can solve the problem

$$\min_{\alpha \in \Re^{Tmn}} V^\alpha(T, c). \tag{2.18}$$

Thus, we can obtain a reasonable set of Lagrange multipliers by solving the problem above. In the next lemma, we show that the objective function of problem (2.18) is convex, which implies that we can use standard convex optimization methods to solve this problem. We can show this lemma from first principles, but the material covered later in this chapter allows us to give a simpler proof. Thus, we defer the proof of this lemma to the appendix.

Lemma 2.15 *For any $t = 1, \ldots, T$ and $x \in \mathbb{Z}^m_+$, $V^\alpha(t, x)$ is a convex function of α.*

Let α^* denote an optimal solution to problem (2.18) and with the corresponding value functions $\{V^{\alpha^*}(t, \cdot) : t = 1, \ldots, T\}$ obtained by solving the dynamic program in (2.15). A heuristic can now be developed based on using $V^{\alpha^*}(t, \cdot)$ as an approximation to $V(t, \cdot)$ on the right side of (2.4). Furthermore, $V^{\alpha^*}(T, c)$ provides an upper bound on the optimal total expected revenue.

Example 2.16 Consider the problem instance in Example 2.3. We solve problem (2.18) to obtain the optimal solution α^*, in which case, $V^{\alpha^*}(T, c)$ provides an upper bound on the optimal total expected revenue. This upper bound comes out to be \$19,988. Among all the upper bounds that we compute in this chapter, this upper bound is the tightest.

The two approaches discussed in this section can benefit from periodic resolving at reading dates. The next example illustrates the performance of these methods.

Example 2.17 Table 2.2 provides the total expected revenues obtained by the value function approximation methods discussed in this section. We use 1, 4, and 10 resolves over the sales horizon. The first line corresponds to the value function approximation approach that exploits the deterministic linear program, whereas the second line corresponds to the value function approximation approach that uses Lagrangian relaxation. The value function approximation methods can provide good revenues without resolving too many times. Furthermore, Example 7.7 shows that the optimal total expected revenue is no larger than \$19,988. The total expected revenues in Table 2.2 are quite close to \$19,988.

Table 2.2 Performance of the value function approximation approaches

Control	1	4	10
Deterministic linear program	\$19,842	\$19,817	\$19,837
Lagrangian relaxation	\$19,802	\$19,876	\$19,895

2.7 Heuristics That Take Randomness into Account

In addition to bid-price and PAC heuristics, researchers and practitioners have also devised heuristics that incorporate randomness. Here, we describe two such heuristics. If the number of itineraries is not overwhelmingly large, one can try to use the EMSR-b heuristic or a similar heuristic for each itinerary to find protection levels for higher fares relative to lower fares. If the number of itineraries is large, an alternative is to aggregate ODF's into buckets for each resource and to use EMSR-b or a similar heuristic for each resource. We will now explain these two heuristics.

2.7.1 Managing Itineraries

We use the two index notation to denote our ODF's. Thus, we represent an ODF by the pair (k, j), where k indexes the itinerary corresponding to the ODF and j is the corresponding fare p_{kj}, $j = 1, \ldots, n_k$. Similarly, we let λ_{tkj} be the probability of observing a request for ODF (j, k) at time period t and use $\Lambda_{Tkj} := \sum_{t=1}^{T} \lambda_{tkj}$ to denote the total expected demand for this ODF. Finally, we let $A_k := (a_{1k}, \ldots, a_{mk})$ be the vector that captures the set of resources used by itinerary k. In particular, we have $a_{ik} = 1$ if itinerary k uses resource i; otherwise $a_{ik} = 0$.

If the number of itineraries is not too large, or at least the number of itineraries with significant demand is not too large, then we can try to obtain strong heuristics by using a combination of capacity allocation and single-resource heuristics. More specifically, let $\bar{y}_k := \sum_{j=1}^{n_k} \bar{y}_{kj}$ be the total capacity that the deterministic linear program allocates to itinerary k and consider the problem of allocating that capacity among fares, $p_{k1} \geq p_{k2} \geq \ldots \geq p_{k,n_k}$ with Poisson demands $\Lambda_{Tk1}, \Lambda_{Tk2}, \ldots, \Lambda_{Tk,n_k}$ assuming low-to-high arrivals. The output will be protection levels and nested booking limits for the different fares. The heuristic allows bookings of itinerary k at fare p_{kj} at state (t, x) if $A_k \leq x$ and the current bookings for the itinerary are within the nested booking limits. If the fare is accepted, then the sale is recorded and this sale is counted against the nested booking limits.

Example 2.18 For Example 2.3, we have three itineraries. Itinerary 1 consumes one unit of resource 1 and is allocated 60 units of capacity for fares $p_{11} = 150$ and $p_{12} = 100$. The expected demands for these two fares over the entire horizon are 30 and 60, respectively. An EMSR-b analysis shows that we should protect 28 units of capacity for fare p_{11}, resulting in nested booking limits 60 and $60 - 28 = 32$, respectively, for fares p_{11} and p_{12}. Itinerary 2 consumes one unit of resource 2 and is allocated 60 units of capacity for fares $p_{21} = 120$ and $p_{22} = 80$. The expected demands for these two fares over the entire horizon are 20 and 80, respectively. An EMSR-b analysis shows that we should protect 18 units of capacity for fare p_{21}, resulting in nested booking limits 60 and $60-18 = 42$, respectively, for fare p_{21} and p_{22}. Finally, itinerary 3 consumes one unit of each resource and is allocated 30 units of capacity for fares $p_{31} = 250$ and $p_{32} = 170$. The expected demands for these fares are 30 and 40. An EMSR-b analysis shows that we should protect 27 units for fare p_{31} resulting in nested booking limits 30 and $30 - 27 = 3$, respectively, for fares p_{31} and p_{32}. The total expected revenue of the heuristic was estimated, through simulation, to be \$19,658 or 1.4% better than the PAC heuristic applied once at the beginning of the sales horizon.

2.7.2 *Managing Resources*

We return here to the single index model, where ODF's are indexed by j. Let $I_j := \{i \in M : a_{ij} > 0\}$ be the set of resources used by ODF j. The net contribution of ODF j to resource i is defined by subtracting the value of the capacity used by ODF j on all other resources consumed by ODF j. To capture this net contribution, for any vector of bid-prices $z^* \in \Re^m$, let $p_{ij}(z) := p_j - \sum_{k \neq i} a_{kj} z_k^*$ if $i \in I_j$ and $p_{ij}(z) = 0$ otherwise. We used a quantity similar to $p_{ij}(z)$ in Sect. 2.6.1. For each resource $i = 1, \ldots, m$, we can solve a single-resource revenue management problem. For the problem that takes place over resource i, the revenue associated with ODF j is $p_{ij}(z)$ and the probability of observing a request for ODF j at time period t is λ_{tj}. Notice that we no longer assume that the fares will arrive low-to-high, but a robust heuristic is to compute protection levels as if the arrivals are low-to-high and then use standard nesting. The low-to-high protection levels can be computed using dynamic programming or a heuristic such as EMSR-b. When a request for ODF j arrives at state (t, x), the request is accepted if the net fare is accepted for *each* resource $i \in I_j$.

The problem with a direct implementation of this approach is that there may be too many different ODF's mapped to resources. Some of these ODF's will have very similar net contributions, and some will have very small demands. A natural idea is to aggregate the ODF's into buckets of similar net contributions and aggregate the demands within each bucket. This exercise results in a more manageable single-resource allocation problem. The idea of aggregating ODF's into buckets was developed by the operations research group at American Airlines and it is known as displacement adjusted virtual nesting (DAVN). This method used in conjunction with the bid-prices obtained from the deterministic linear program combines the ideas of bid-prices to compute net fare contributions and the ideas of single-resource controls. Not surprisingly, the method performs quite well.

Example 2.19 Consider the data of Example 2.3. Suppose that three buckets are defined such that bucket 1 contains all net fares greater than \$120, bucket 2 all fares between \$60 and \$120, and bucket 3 all fares below \$60. The ODF's that utilize resource 1 are 1, 2, 5, and 6, and the net fares for resource 1 are $p_{11}(\bar{z}) = 150$, $p_{12}(\bar{z}) = 100$, $p_{15}(\bar{z}) = 170 = 250 - 80$, $p_{16}(\bar{z}) = 170 - 80 = 90$. Accordingly fares 1 and 5 belong to bucket 1, with corresponding expected demands 30 and 30. Fares 2 and 6 belong to bucket 2 with corresponding expected demands 60 and 40. We can aggregate fares 1 and 5 of bucket 1 into a single weighted average fare of 160 with aggregate demand 60. Similarly, for bucket 2 we can aggregate fares 2 and 6 into a single weighted average fare of 96 with aggregate demand 100. An EMSR-b calculation reveals that it is optimal to protect 58 units of capacity to bucket 1 and thereby authorize up to 32 units of capacity to book under bucket 2.

We can repeat the procedure for resource 2 and obtain net fares $p_{23}(\bar{z}) = 120$, $p_{24}(\bar{z}) = 80$, $p_{25}(\bar{z}) = 250 - 100 = 150$, and $p_{26}(\bar{z}) = 170 - 100 = 70$, with corresponding demands 20, 80, 30, and 40. Fares 3 and 5 map into bucket 1 while fares 4 and 6 map into bucket 2. We can aggregate fares 3 and 5 of bucket 1 into a single weighted average fare of 138 with aggregate demand 50. Similarly, for bucket 2 we can aggregate fares 4 and 6 into a single weighted average fare of 76.67 with aggregate demand 120. An EMSR-b calculation reveals that it is optimal to protect 48 units of capacity to bucket 1 and thereby authorize up to 42 units of capacity to book under bucket 2. The total expected revenue, estimated through simulation, is \$19,785 or 2.0% higher than the PAC heuristic applied once at the beginning of the sales horizon.

2.8 Approximate Dynamic Programming

In this section, we show that we can represent the dynamic programming formulation of the network revenue management problem given in (2.11) as a linear program. Although the numbers of decision variables and constraints in this linear programming representation grow exponentially fast with the number of resources, we can build on the linear programming representation to construct general value function approximation strategies. In particular, since the initial capacity of resource i is c_i, letting $C_i = \{0, 1, \dots, c_i\}$, the set of possible states in the dynamic program in (2.11) is $C = C_1 \times \dots \times C_m$. We can obtain the value functions $\{V(t, \cdot) : t = 1, \dots, T\}$ by solving the linear program

$$\min \quad \vartheta(T, c) \tag{2.19}$$

$$\text{s.t.} \quad \vartheta(t, x) \geq \sum_{j \in N} \lambda_{tj} \left\{ p_j u_j + \vartheta(t-1, x - u_j A_j) \right\}$$

$$\forall t = 1, \dots, T, \ x \in C, \ u \in \mathcal{U}(x),$$

where the decision variables are $\{\vartheta(t, x) : t = 1, \dots, T, \ x \in C\}$. In the linear program above, we assume that the value of the decision variable $\vartheta(0, x)$ is set to zero for all $x \in C$ by convention. It is a standard result in Markov decision processes that the optimal objective value of the linear program above is $V(T, c)$, providing the optimal total expected revenue for the network revenue management problem.

One difficulty with problem (2.19) is that it includes one decision variable for each possible state of the system and for each time period, which gets intractable for reasonably large problems. To get around this difficulty, we use $\{\phi_k(\cdot) : k = 1, \ldots, K\}$ to denote a set of fixed basis functions, where each basis function $\phi_k(\cdot) : C \to \Re$ maps the state space to the set of real numbers. Multiplying the basis functions with the scalars $\{r_{tk} : k = 1, \ldots, K\}$ and adding them up, we propose approximating the value function $V(t, \cdot)$ with $\sum_{k=1}^{K} r_{tk}\, \phi_k(\cdot)$. To choose a good set of multipliers $\{r_{tk} : t = 1, \ldots, T,\ k = 1, \ldots, K\}$, we can insert the value function approximation $\sum_{k=1}^{K} r_{tk}\, \phi_k(\cdot)$ into problem (2.19) to obtain the linear program

$$\min \quad \sum_{k=1}^{K} r_{Tk}\, \phi_k(c) \tag{2.20}$$

$$\text{s.t.} \quad \sum_{k=1}^{K} r_{tk}\, \phi_k(x) \geq \sum_{j \in N} \lambda_{tj} \left\{ p_j\, u_j + \sum_{k=1}^{K} r_{t-1,k}\, \phi_k(x - u_j\, A_j) \right\}$$

$$\forall t = 1, \ldots, T,\ x \in C,\ u \in \mathcal{U}(x),$$

where the decision variables are $\{r_{tk} : t = 1, \ldots, T,\ k = 1, \ldots, K\}$. If we use a manageable number of basis functions, then the number of decision variables in problem (2.20) is reasonable, but problem (2.20) has one constraint for each time period, for each possible state of the system, and for each possible decision, which can easily get too many. One can deal with problem by using column generation on the dual of problem (2.20).

A natural question is what basis functions are appropriate to use. One approach is to use linear value function approximations. In particular, one can use linear value function approximations of the form $\sum_{i \in M} r_{ti}\, x_i + r_{t,m+1}$, which corresponds to using $m + 1$ basis functions with $\phi_i(x) = x_i$ for all $i \in M$ and $\phi_{m+1}(x) = 1$. The multipliers $\{r_{ti} : i = 1, \ldots, m+1\}$ correspond to the slope and intercept parameters of the linear value function approximations of the form $\sum_{i \in M} r_{ti}\, x_i + r_{t,m+1}$. It turns out one can show that using linear value function approximations are equivalent to using the problem in (2.10). Thus, we can solve problem (2.10) directly rather than using linear value function approximations and solving problem (2.20) via column generation.

Another possible class of basis function is separable ones of the form $\sum_{i \in M} v_i(t, x_i)$. This value function approximation requires specifying the value of $v_i(t, x_i)$ for each $x_i \in C_i$, $i \in M$, and $t = 1, \ldots, T$, and it is equivalent to using $\sum_{i \in M}(1 + c_i)$ basis functions with $\phi_{i,\delta_i}(x) = \mathbf{1}(x_i = \delta_i)$ for all $i \in M$ and $\delta_i = 0, \ldots, c_i$. The multipliers $\{r_{ti,\delta_i} : i \in M,\ \delta_i = 1, \ldots, m\}$ correspond to the value of the approximation $v_i(t, x_i)$ at $x_i = \delta_i$. The value function approximation can be written in terms of the basis functions as $\sum_{i\in M}\sum_{\delta_i=0}^{c_i} r_{ti,\delta_i}\,\phi_{i,\delta_i}(x) = \sum_{i\in M}\sum_{\delta_i=0}^{c_i} r_{ti,\delta_i}\,\mathbf{1}(x_i = \delta_i)$. Interestingly enough, one can show that separable basis functions are equivalent to the Lagrangian relaxation strategy described earlier in this chapter.

As discussed above, linear or separable value function approximations reduce to other known value function approximation strategies. Nevertheless, there is still the opportunity to use value function approximations that are more sophisticated than linear and separable to obtain strong value function approximations from problem (2.20).

2.9 End of Chapter Problems

1. An airline operates a flight network among three locations A, B, and C. The first flight is from A to B, and the second flight is from B to C. There are three ODF's. The fares associated with the ODF's are given in Table 2.3.

   Assume that the time-to-go is t and the remaining capacities is $x = (1, 1)$. The value function at the next time period, $t - 1$, is given in Table 2.4. For example, if the remaining capacities at $t - 1$ is given by $x = (1, 0)$, then $V(t - 1, x) = 25$.

   (a) For each ODF, decide whether it is optimal to open or close it.
   (b) Compute $V(t, x)$ assuming that it is equally likely to observe a request for each of the three ODF's.

2. An airline operates two fights, one flight from San Francisco to Denver and another one from Denver to St. Louis. The capacity on the first flight is 100 and the capacity on the other is 120. The ODF's, fares and expected demands are given in Table 2.5.

   (a) Write down and solve the deterministic linear program corresponding to this network. Report the optimal primal and dual solutions.
   (b) Consider the bid-price policy discussed in Sect. 2.4. Suppose there is a request for a discount seat on the flight from San Francisco to St. Louis. Should we accept this request on the basis of the bid-prices?

Table 2.3 ODF's and fares

ODF	Fare
A–B, direct	50
B–C, direct	100
A–C, through B	200

Table 2.4 Value function at the next time period

x	$V(t-1,x)$
(1,1)	150
(1,0)	25
(0,1)	125

Table 2.5 ODF's, fares and expected demands

ODF		Fare	Demand
1	San Francisco to Denver, full fare	150	30
2	San Francisco to Denver, discount	100	60
3	Denver to St. Louis, full fare	120	20
4	Denver to St. Louis, discount	80	80
5	San Francisco to St. Louis, full fare	250	30
6	San Francisco to St. Louis, discount	225	20

Table 2.6 ODF's and fares

Itinerary	Origin-destination	Fare
1	A–B	$150
2	B–C	$50
3	A–C	$275

Table 2.7 Value function at the next time period

	B–C			
A–B	0	1	2	3
0	0	600	900	1000
1	300	700	1000	1300
2	400	800	1100	1300
3	500	850	1100	1300

(c) On the basis of the bid-prices, indicate for each ODF, whether the ODF would be open or closed.

3. An airline operates a flight network among three locations A, B, and C. There are two flight legs. One flight leg is from A to B. The other flight leg is from B to C. The capacity on both flight legs is 3. The ODF's and their fares are given in Table 2.6. Assume that we can have at most one itinerary request at each time period. The value function for time period $t-1$ for each inventory pair $x = (x_1, x_2)$ for integer values $0 \leq x_i \leq 3$, $i = 1, 2$, is given in Table 2.7. For example, if the remaining capacity on flight leg A–B is 2 and the remaining capacity on flight leg B–C is 1, then the value of the value function at time period $t-1$ is $V(t-1, (2, 1)) = 800$.

(a) Assume that the remaining capacity on flight leg A–B is 1 and the remaining capacity on flight leg B–C is 2 at t. Is it optimal to accept a request for itinerary 1 at time-to-go t?

(b) Assume that the remaining capacity on flight leg A–B is 2 and the remaining capacity on flight leg B–C is 3 t. Is it optimal to accept a request for itinerary 3 at time-to-go t?

4. Note that we can write the dynamic program in (2.11) equivalently as

$$\begin{aligned} V(t,x) \;=\; \max \quad & \sum_{j\in N} \lambda_{tj} \Big\{ p_j\, w_j + V(t-1, x - w_j \sum_{i\in M} a_{ij}\, e_i) \Big\} \\ \text{s.t.} \quad & a_{ij}\, w_j \le x_i \qquad \forall i \in M,\ j \in N \\ & w_j \in \{0,1\} \qquad \forall i \in M,\ j \in N. \end{aligned}$$

Relax the first set of constraints above by associating the Lagrange multipliers $\beta = \{\beta_{ti} : t = 1,\dots,T,\ i \in M\}$ with them. Let $\{V^\beta(t,\cdot) : t = 1,\dots,T\}$ be the value functions that we obtain after relaxing the first set of constraints in the dynamic program above.

(a) For any set of Lagrange multipliers β satisfying $\beta_{ti} \ge 0$ for all $t = 1,\dots,T$ and $i \in M$ show that $V(t,x) \le V^\beta(t,x)$ for all $x \in \mathbb{Z}_+^m$ and $t = 1,\dots,T$. (Hint: You can use an argument very similar to the one in the proof of Theorem 2.14.)

(b) Show that $V^\beta(t,x)$ has a closed-form expression and $V^\beta(t,x)$ is a linear function of the capacities x.

(c) Considering Part (a), to find the tightest possible upper bound on the optimal total expected revenue, we can solve the problem

$$\min_{\beta \in \Re_+^{Tmn}} V^\beta(T,c).$$

Give an equivalent linear programming formulation for this minimization problem.

(d) Write the dual of the linear program that you constructed in Part c. Can you give an intuitive interpretation for the dual?

5. Consider the Lagrangian relaxation strategy discussed in Sect. 2.6.2. In this problem, we will show two properties of the optimal Lagrange multipliers.

(a) Show that there exists an optimal solution α^* to problem (2.18) such that if $a_{ij} = 0$, then $\alpha_{tij} = 0$ for all $t = 1,\dots,T$.

(b) Show that there exists an optimal solution α^* to problem (2.18) such that $p_j = \sum_{i\in M} \alpha_{tji}$ for all $j \in N$ and $t = 1,\dots,T$.

6. Consider the Lagrangian relaxation strategy discussed in Sect. 2.6.2. Let ζ^* be the optimal objective value of problem (2.18). Thus, ζ^* corresponds to the tightest upper bound that we can obtain on the optimal total expected revenue by using the Lagrangian relaxation strategy. Recall that $\bar{V}(T,c)$, which is given by the optimal objective value of problem (2.6), corresponds to the upper bound

provided by the deterministic linear program. Show that

$$\zeta^* \leq \bar{V}(T, c).$$

In other words, the upper bound provided by the Lagrangian relaxation strategy is at least as tight as the one provided by the deterministic linear program. (Hint: Instead of using the decision variables $\{y_j : j \in N\}$ in problem (2.6), use the decision variables $\{y_{ij} : i \in M \cup \{\psi\}, j \in N\}$ to write the constraints of problem (2.6) equivalently as

$$\sum_{j \in N} a_{ij}\, y_{ij} \leq c_i \quad \forall i \in M$$

$$0 \leq y_{ij} \leq \Lambda_j \quad \forall i \in M \cup \{\psi\},\ j \in N$$

$$y_{ij} = y_{\psi j} \quad \forall i \in M,\ j \in N.$$

In this case, consider the dual of problem (2.6) and construct a feasible solution to problem (2.18) by using the optimal dual solution to problem (2.6).)

7. Consider the problem allocating m different types of capacity to n different types of demand over T time periods. At each time period, we observe one unit of demand of type j with probability p_j, with $\sum_{j=1}^{n} p_j = 1$. If we want to serve this demand, then we get to decide what type of capacity to use to serve the demand. In particular, if we use capacity of type i to serve one unit of demand of type j, then we generate a revenue of r_{ij} and consume one unit of capacity of type i. Initially, we have c_i units of capacity of type i available. The objective is to maximize the total expected revenue over T time periods.

   (a) Formulate the problem as a dynamic program.
   (b) Give a linear program whose optimal objective value provides an upper bound on the optimal total expected revenue. Show that the optimal objective value of your linear program is indeed an upper bound on the optimal total expected revenue.
   (c) Building on your linear program, give a dynamic programming decomposition approach similar to the one in Sect. 2.6.1 that provides an upper bound on the optimal total expected revenue that is tighter than the one provided by the linear program. Show that your dynamic programming decomposition approach indeed provides an upper bound that is tighter than the one provided by the linear program.

2.10 Bibliographical Remarks

Under the assumption that the fares associated with the ODF's are uniformly bounded continuous random variables, Talluri and van Ryzin (1998) show that the bid-price heuristic based on the deterministic linear program (2.6) is asymptotically optimal. An independent proof of a result similar to Theorem 2.7 can be found in Reiman and Wang (2008). Resolving a deterministic linear program at multiple reading dates has provable asymptotic optimality properties; see Maglaras and Meissner (2006). The benefits from frequent resolving can be substantial. Ciocan and Farias (2012b); Jasin and Kumar (2013); Jasin (2014) and Jasin (2015) give results showing that resolving results in a bounded optimality gap, even when demand process is not known with certainty. Resolving may also hurt performance if it is done towards the end of the sales horizon when the deterministic linear program suffers more significantly from ignoring randomness; see Cooper (2002).

Talluri and van Ryzin (1999) proposed the randomized linear program. Topaloglu (2009a) shows that the randomized linear program has asymptotic optimality properties. Problem (2.10) appears in Kunnumkal and Topaloglu (2010c) and Tong and Topaloglu (2014). The former paper derives problem (2.10) by using a Lagrangian relaxation strategy on the dynamic programming formulation of the network revenue management problem, whereas the latter paper derives problem (2.10) by using linear approximations to the value functions. To deal with the large number of constraints in problem (2.10), Vossen and Zhang (2015) propose solving a version of this problem where the constraints are aggregated. They iteratively solve an aggregated version, sequentially disaggregating the constraints until they reach the optimal solution. The observations in Example 7.8 regarding resolving the linear program are consistent with other reports; see de Boer et al. (2002) and Williamson (1992). The Lagrangian relaxation approach in Sect. 2.6.2 is due to Topaloglu (2009b). Bertsimas and Popescu (2003) use the optimal objective value of the deterministic linear program to approximate the value function and use the greedy policy with respect to the value function approximations. Kirshner and Nediak (2015) compute time-dependent bid-prices that are succinctly parameterized as a function of time.

Considering the approximate dynamic programming approach in Sect. 2.8, Adelman (2007) proposes using linear basis functions. A related linear program for tuning the coefficients of the basis functions appears in Farias and Van Roy (2007). Tong and Topaloglu (2014) show that using linear value function approximations is identical to using the more sophisticated version of the deterministic linear program given in (2.10). The latter result indicates that one can solve problem (2.10) rather than using column generation on the dual of problem (2.20). Meissner and Strauss (2012) use separable value function approximations of the form $\sum_{i \in M} v_i(t, x_i)$. Zhang and Vossen (2015) and Kunnumkal and Talluri (2016b) show that using separable basis functions is equivalent to the Lagrangian relaxation strategy described earlier in this chapter. Kunnumkal and Talluri (2016a) compare the tightness of the upper bounds on the optimal total expected revenue provided by different relaxation strategies. Barz et al. (2013) go one step beyond linear and separable value function approximations and use problem (2.20) with sophisticated basis functions to obtain strong value function approximations.

Cooper and Homem de Mello (2007) study structural properties of the value functions arising in the network revenue management setting. Akan and Ata (2009) and Ata and Akan (2015) show that the bid-prices form a martingale, remaining constant in expectation. Under slightly different modeling assumptions, Pang et al. (2014) show sub- or super-martingale properties of bid-prices. Amaruchkul et al. (2007, 2011); Levina et al. (2011) and Levin et al. (2012) consider cargo revenue management problems, where the space requirement of the cargo becomes known only at the departure time. In the same vein, Zhuang et al. (2012) study a model where the space requirement of the accepted requests is random. Feng et al. (2015) give an overview of cargo operations, including revenue management. Shumsky and Zhang (2009) and Yu et al. (2015) study a revenue management model with upgrades and characterize the structure of the optimal policy by using protection levels. Zhang et al. (2016) employ the regret minimization framework for allocating limited resources between two classes of customers.

Upgrades and upsells in the context of revenue management are treated in Gallego and Stefanescu (2009) and in Gallego and Stefanescu (2012). Bertsimas and Shioda (2003) construct revenue management models for restaurants, whereas Roels and Fridgeirsdottir (2009) construct revenue management models for online display advertising. The papers by Bertsimas and de Boer (2005); van Ryzin and Vulcano (2008a) and Topaloglu (2008) use stochastic approximation methods to tune succinctly parameterized policies for network revenue management. These policies are either characterized by protection levels or bid-prices. Perakis and Roels (2010) consider the case where the demand distribution is unknown and incorporate robust optimization methods into the mathematical programming-based network revenue management models. Levi and Radovanovic (2010); Chen et al. (2017c) and Lei and Jasin (2018) give policies with performance guarantees for revenue management problems with reusable resources. Bassamboo et al. (2009) also study systems with reusable products. Netessine and Shumsky (2005); Shumsky (2006); Wright et al. (2010) and Hu et al. (2013b) study revenue management problems in airline alliances. Wang et al. (2018); Stein et al. (2019) and Ma and Simchi-Levi (2017) study dynamic matching problems where the demand randomly arriving over time needs to matched to resources that are available at the beginning of the selling horizon. The authors give policies with performance guarantees. Calmon et al. (2019) study a network revenue management model where customers make repeated purchases and their available budget depends on their experience in previous purchases. Talluri et al. (2010) give an account of proving the practical improvements provided by a newly implemented revenue management system.

Appendix

Proof of Proposition 2.5 Consider an ODF $j \in F$, and notice that $\beta_j^* \geq p_j - \sum_{i \in M} a_{ij} z_i^* > 0$, where we use the fact that (z^*, β^*) is feasible to problem (2.8). Therefore, the second set of constraints for ODF j in problem (2.6) has a strictly positive dual variable, implying that its slack must be zero and we obtain $y_j^* = \Lambda_j$. The last equality implies that $j \in F'$. So, $F \subseteq F'$. Similarly, consider an ODF $j \in R$. Thus, we have $\sum_{i \in M} a_{ij} z_i^* + \beta_j^* - p_j > 0$, where we use the fact that $\sum_{i \in M} a_{ij} z_i^* > p_j$ and $\beta_j^* \geq 0$. Therefore, the first set of constraints for ODF j in problem (2.8) has a strictly positive slack, which implies that the dual variable associated with this constraint must be zero at the optimal solution and we obtain $y_j^* = 0$. The last equality implies that $j \in R'$. So, $R \subseteq R'$. Since $F \cup P \cup R = F' \cup P' \cup R' = N$, it also follows that $P' \subseteq P$. □

Proof of Theorem 2.7 Consider a variant of the PAC heuristic based on always admitting requests for product $j \in N$ with probability y_j^*/Λ_j. Notice that these probabilities are independent of the scaling parameter b, since the solution to the primal and dual linear programs are insensitive to the scaling parameter. Notice also that this variant of the PAC heuristic ignores inventory considerations, and it may end up overbooking the resources. Because the demand for product j is Poisson with parameter $b\Lambda_j$ and we admit demands with probability y^*/Λ_j, sales for product j form a thinned Poisson process with mean by_j^* for each $j \in N$. Consequently, the expected revenue from this variant of the PAC heuristic is $b \sum_{j \in N} p_j y_j^* = \bar{V}^b(T, c) = b\bar{V}(T, c)$. From this expected revenue, we need to deduct the cost for overbooking capacity. Let S_j denote the random sales for product j, from our earlier discussion this is a Poisson random variable with mean by_j^* and variance by_j^*. Suppose that we are charged an overbooking cost θ_i for each unit of capacity of resource i that we consume in excess of capacity. Then the overbooking costs are equal to

$$\sum_{i \in M} \theta_i [\sum_{j \in N} a_{ij} S_j - bc_i]^+.$$

Consequently, the expected revenue of this variant of the PAC heuristic, net of overbooking costs is of the form

$$\underline{\Pi}^b(T, c) = b\bar{V}(T, c) - \sum_{i \in M} \theta_i \, \mathbb{E}[\sum_{j \in N} a_{ij} S_j - bc_i]^+.$$

We now claim that if we select

$$\theta_i \geq \max\{p_j : a_{ij} = 1, j \in N\},$$

then

$$\underline{\Pi}^b(T, c) \leq \Pi^b(T, c) \leq V^b(T, c) \leq \bar{V}^b(T, c), \tag{2.21}$$

where $\bar{V}^b(T, c)$ is the optimal objective value of the deterministic linear program with a scaling factor of b. The first inequality above follows from a sample path argument. Notice that as long as the capacities are not violated, both the PAC heuristic and the alternative policy make the same decisions. If the alternative policy sells a ticket for an ODF and violates the capacity, then it incurs a penalty that is larger than the revenue from the sold ODF, losing revenue from the sale. Due to this decision, the alternative policy may also consume capacities of other available resources. Thus, the alternative policy not only loses money from the sale, but it is also left with even less capacity than the PAC heuristic. So, the revenue net of overbooking costs is always smaller than the revenue generated by the PAC heuristic. The second inequality follows because the PAC is a heuristic and its performance is bounded above by the expected revenue of the optimal policy. The last inequality follows because the optimal objective value of the deterministic linear program is an upper bound on the optimal total expected revenue.

Dividing the string of inequalities in (2.21) by $\bar{V}^b(T,c)$ results in the following string of inequalities:

$$1 - \frac{\sum_{i \in M} \theta_i \, \mathbb{E}[\sum_{j \in N} a_{ij} S_j - bc_i]^+}{\bar{V}^b(T,c)} = \frac{\Pi^b(T,c)}{\bar{V}^b(T,c)} \leq \frac{\Pi^b(T,c)}{\bar{V}^b(T,c)} \leq 1.$$

Consequently, if we can show that

$$\lim_{b \to \infty} \frac{\sum_{i \in M} \theta_i \, \mathbb{E}[\sum_{j \in N} a_{ij} S_j - bc_i]^+}{\bar{V}^b(T,c)} = 0,$$

then it would follow that

$$1 = \lim_{b \to \infty} \frac{\Pi^b(T,c)}{\bar{V}^b(T,c)} \leq \lim_{b \to \infty} \frac{\Pi^b(T,c)}{V^b(T,c)} \leq 1.$$

For each i, consider the random variable $Z_i = \sum_{j=1}^n a_{ij} S_j$ corresponding to the aggregate demand for resource i under the variant of the PAC heuristic that accepts request for product j with probability y_j^*/Λ_j regardless of capacity. Note that $\mathbb{E}[Z_i] = b \sum_{j \in N} a_{ij} y_j^* \leq bc_i$, where we use the fact that $\{y_j^* : j \in N\}$ is a feasible solution to problem (2.6). Also, the variance of Z_i satisfies $\text{Var}[Z_i] = b \sum_{j \in N} a_{ij} y_j^*$, where we have used the fact that $a_{ij}^2 = a_{ij}$ and that both the mean and the variance of S_j are equal to by_j^*. We now use the bound on partial expectations

$$\mathbb{E}[(Z-z)^+] \leq 0.5(\sqrt{\sigma^2 + (z-\mu)^2} - (z-\mu)) \leq \frac{1}{2}\sigma + \frac{1}{2}(|z-\mu| - (z-\mu)),$$

that holds for all random variables Z with mean μ and variance σ^2, and arbitrary constant z; see Gallego (1992). Notice that the last term vanishes when $z \geq \mu$. Applying the bound to the random variable Z_i and to the constant $z_i = bc_i \geq \mathbb{E}[Z_i]$, we obtain

$$\mathbb{E}[\sum_{j=1}^n a_{ij} S_j - c_i]^+ \leq \frac{1}{2}\sqrt{b \sum_{j \in N} a_{ij} y_j^*}.$$

Multiplying by the last expression by θ_i adding over i and dividing by $b\sum_{j=1}^{n} p_j y_j^*$, we see that

$$\frac{\sum_{i\in M}\theta_i\,\mathbb{E}[\sum_{j\in N} a_{ij} D_j - bc_i]^+}{\bar{V}^b(T,c)} \le \frac{\frac{1}{2}\sum_{i\in M}\theta_i\sqrt{\sum_{j\in N} a_{ij} y_j^*}}{\sqrt{b}\sum_{j\in N} p_j\, y_j^*}.$$

Notice that the ratio goes to zero at rate $1/\sqrt{b}$ as $b \to \infty$. □

Proof of Theorem 2.8 The proof of the first inequality is essentially identical to that of Theorem 2.2 and we omit it. To see the second inequality, let $\{y_{tj}^* : t = 1, \dots, T,\ j \in N\}$ and $\{x_{ti}^* : t = 1, \dots, T,\ i \in M\}$ be an optimal solution to problem (2.10). For each $i \in M$, adding the first two sets of constraints overall $t = 2, \dots, T$ yields $\sum_{i\in N} a_{ij} \sum_{t=2}^{T} y_{tj}^* + x_{1i}^* = c_i$. On the other hand, for each $i \in M$, adding the third set of constraints for $t = 1$ overall $j \in N$ yields $\sum_{i\in N} a_{ij}\, y_{1j}^* \le \sum_{j\in N} \lambda_{1j}\, x_{1i}^* \le x_{1i}^*$. Combining the inequalities, we obtain $\sum_{i\in N} a_{ij} \sum_{t=1}^{T} y_{tj}^* \le c_i$ for all $i = 1, \dots, n$, implying that the solution $\{\sum_{t=1}^{T} y_{tj}^* : j \in N\}$ satisfies the first set of constraints in problem (2.6). Furthermore, adding the fourth set of constraints in problem (2.10) overall $t = 1, \dots, T$, we obtain $\sum_{t=1}^{T} y_{tj}^* \le \sum_{t=1}^{T} \lambda_{tj}$ for all $j \in N$, so that the solution $\{\sum_{t=1}^{T} y_{tj}^* : j \in N\}$ satisfies the second set of constraints in problem (2.6) as well. Also, we have $\sum_{i\in N} p_j \sum_{t=1}^{T} y_{tj}^* = \tilde{V}(T,c)$ by the definition of $\{y_{tj}^* : t = 1, \dots, T,\ j \in N\}$. Therefore, $\{\sum_{t=1}^{T} y_{tj}^* : j \in N\}$ is a feasible solution to problem (2.6) and it provides an objective value of $\tilde{V}(T,c)$ for this problem, which imply that the optimal objective value of problem (2.6) can only be larger than $\tilde{V}(T,c)$, yielding $\bar{V}(T,c) \ge \tilde{V}(T,c)$. □

Proof of Lemma 2.15 The function $[p_j - \sum_{i\in M}\alpha_{\tau ij}]^+$ is convex in α. Noting (2.17), it is enough to show that $v_i^\alpha(t, x_i)$ is a convex function of α. The dynamic program in (2.16) characterizes $v_i^\alpha(t, x_i)$. Thus, by the discussion in Sect. 2.8, the value functions $\{v_i^\alpha(t, \cdot) : t = 1, \dots, T\}$ can be obtained by solving the linear program

$$\begin{aligned}
\min\quad & v_i(t, x_i) \\
\text{s.t.}\quad & v_i(\tau, x_i) \ge \sum_{j\in N} \lambda_{\tau j}\Big\{\alpha_{\tau ij}\, w_{ij} + v_i(\tau - 1, x_i - w_{ij}\, a_{ij})\Big\} \\
& \qquad\qquad \forall \tau = 1, \dots, T,\ x_i \in C_i,\ w_i \in \mathcal{U}_i(x_i),
\end{aligned}$$

where the decision variables are $\{v_i(\tau, x_i) : \tau = 1, \dots, T,\ x_i \in C_i\}$. The optimal objective value of the problem above provides $v_i^\alpha(t, x_i)$. The set of Lagrange multipliers α appear only on the right side of the constraints above. Thus, the optimal objective value of the problem above is convex in α by linear programming duality and the desired result follows. □

Chapter 3
Overbooking

3.1 Introduction

Early on, many airlines adopted the policy of not penalizing booked customers for canceling reservations at any time before departure. Some would not even penalize those that did not show up for booked flights. In essence, an airline ticket was "like money" since it could be used at full face value for a future flight or redeemed for cash at any future date. In the 1960s, no-shows were becoming a problem for airlines who found that flights that were fully booked were departing with many empty seats. In response, the airlines began to overbook as a means of hedging against no-shows. If a flight had more passengers show up than there were seats available, then the airlines would bump some passengers. The bumped passengers would be re-booked on a later flight. In addition, bumped passengers would be given other compensation, often a meal at the airport and a discount certificate applicable to future travel. The cost to the airline of bumping a passenger is called the denied boarding cost. The denied boarding cost would include the cost of putting a bumped passenger on another flight to her destination, the cost of any direct compensation to the bumped passenger, the cost of the meals or lodging that the airline provides to each bumped passenger, and the cost of "ill will" incurred by bumping the passenger. These costs can be different for each flight. For example, a passenger bumped from the last flight of the day will be provided with a hotel room at the airline's expense.

While unpopular with passengers, overbooking was effective at increasing load factors and revenues. This raised the issue of determining the right booking limit for a flight. When overbooking is allowed, the booking limit can exceed the capacity on the flight, allowing the airline to book more passengers than the capacity. If the booking limit is set too low, there will be lots of empty seats. On the other hand, if the booking limit is set too high, the benefits of filling the aircraft would be overwhelmed by the denied board costs paid. Determining the optimal booking

G. Gallego, H. Topaloglu, *Revenue Management and Pricing Analytics*,
International Series in Operations Research & Management Science 279,
https://doi.org/10.1007/978-1-4939-9606-3_3

limit was one of the first revenue management problem to be successfully analyzed utilizing the methods of operations research.

Airlines significantly changed their overbooking policies over the years. For example, airlines instituted auctions as a mechanism for identifying people who would be willing to forego their seat on a flight in return for compensation in the form of a future flight discount. This practice proved to be popular with passengers and dramatically reduced the number of involuntary denied boardings. Airlines now also sell non-refundable or partially refundable tickets, particularly at lower costs. Both of these developments have implications for the analysis of overbooking policies.

In this chapter, we study a variety of overbooking models. These models can be viewed as the extensions of the models considered throughout the book to deal with overbooking. In Sect. 3.2, we begin with a static, overbooking model with a single fare class. We characterize the optimal booking limit. In Sect. 3.3, we move on to an overbooking model with multiple fare classes over a single flight leg and characterize the structure of the optimal policy. In Sect. 3.4, we conclude the chapter with overbooking models over a network of flight legs.

3.2 Overbooking for a Single Fare Class

Suppose that a flight has capacity c, the unconstrained demand at a single fare $p > 0$ is D. We assume that passengers who do not show up are given a full refund, and that the unit cost for denied boarding is θ. Let b be the booking limit. Then $N := \min(D, b)$ is the number of bookings. The goal is to find a booking limit that maximizes the expected profit, which is given by the difference between the expected revenue from sold seats and the expected cost of denied boardings.

Let $Z(N) := Z(\min(D, b))$ denote the number of passengers that show up for the flight. We assume that each passenger shows up for the flight with probability q independent of everyone else, so $Z(N)$ is a conditional binomial random variable with parameters (N, q). We can express the expected profit as a function of the booking limit as

$$R(b) := p\,\mathbb{E}[Z(\min(D, b))] - \theta\,\mathbb{E}[Z(\min(D, b)) - c]^{+}]. \tag{3.1}$$

An optimal booking limit, say b^*, is the largest maximizer of $R(b)$. The next proposition provides a formula for $R(b + 1) - R(b)$ that shows that this quantity is always non-negative when $\theta < p$, and in this case, there is no need to impose a booking limit. When $\theta > p$, the quantity can change only from positive to negative, so b^* is the smallest b such that $R(b + 1) - R(b) < 0$.

Proposition 3.1

$$R(b+1) - R(b) = \mathbb{P}\{D \geq b+1\}\, q\, (p - \theta\, \mathbb{P}\{Z(b) \geq c\}).$$

$$b^* = \min\left\{b \geq 0 : \mathbb{P}\{Z(b) \geq c\} > \frac{p}{\theta}\right\}. \tag{3.2}$$

Formula (3.2) has some resemblance to Littlewood's rule derived in Chap. 1.

The optimal booking limit given above may yield high booking limits and result in large numbers of denied boardings. In fact, assuming that the demand quantities are large enough that we always have $\min\{D, b^*\} = b^*$, the formula for the optimal booking limit given above implies that the fraction of flights with shows exceeding capacity is roughly p/θ. When $\theta = 2\,p$, roughly half of the flights will have denied boardings. This observation motivates the adoption of frequency-based policies by many airlines, where airlines set a target frequency, say f, for the fraction of booked passengers that would be denied boarding. Under this policy, the airlines would set a booking limit as the largest integer b such that

$$\frac{\mathbb{E}\{[Z(\min(D, b)) - c]^+\}}{\mathbb{E}\{Z(\min(D, b))\}} \leq f.$$

In many cases, airlines use hybrid policies, where they calculate the booking limit that maximizes the expected profit and the booking limit that limits the fraction of passengers that are denied boarding, and use the smallest of the two booking limits.

3.3 Overbooking for Multiple Fare Classes

In this section, we present a model for a single flight with multiple fare classes and overbooking. There are n fare classes indexed by $1, \ldots, n$. We assume that the fare classes are ordered such that $p_1 \geq p_2 \geq \ldots \geq p_n$ and the demand from different fare classes arrive sequentially in the low-before-high order. Throughout this section, we make a number of simplifying assumptions to obtain a tractable model. First, we ignore the cancellations and assume that there are only no-shows. Furthermore, the no-show probability for all customers is the same and the no-show decisions of the different customers are independent of each other. The probability that a customer shows up for the flight does not depend on when she booked the ticket. Finally, the refunds and the denied-service costs are the same for all customers. These assumptions imply that the number of no-shows and the cost of no-shows are only a function of the total number of reservations on hand. As a result, we need to retain only a single state variable that keeps track of the total number of reservations, which helps keep the dynamic programming formulation tractable.

Among our assumptions, the most restrictive ones are perhaps the assumptions that the no-show probability, the refunds, and the denied-service costs are the same for all customers. In practice, cancellation options and penalties are often linked to a particular class, so no-show rates and costs can vary significantly from one class to the next. In certain cases, reservations from groups may be canceled simultaneously, which makes the assumption of independent show-up decisions somewhat unrealistic. There seems to be reasonable empirical evidence to support the assumption that the show-up probabilities of the customers do not depend on when they made their reservations.

As in previous section, the capacity on the flight is c. Each reservation shows up with probability q. We use the random variable $Z(y)$ to capture the number of passengers that show up for the flight given that we have y reservations just before the departure time. Thus, $Z(y)$ is binomially distributed with parameters (y, q). We ignore cancellations and assume that we do not give any refunds to the passengers who do not show up, but we will shortly discuss how to relax both of these assumptions. The cost of denying boarding to a reservation is θ. We use the random variable D_j to capture the demand from fare class j. Our goal is to find a policy to decide how much demand to accept from each class to maximize the total expected profit, where the total expected profit is given by the difference between the revenue from the accepted bookings and the penalty cost from the denied boardings.

For any j, we let $V_j(y)$ to denote the optimal total expected revenue that can be obtained from classes $j, \ldots, 1$, given that we have y reservations on hand at the beginning of stage j. Notice that instead of remaining capacity, we use the number of reservations on hand as the state variable. At the beginning of stage j, we observe the demand from fare class j. Knowing the number of reservations, we decide how many new requests to accept. After all of the n stages, a portion of the reservations show up. If the number of reservations that show up exceed the capacity available, then we incur the denied boarding cost. The sequence of events that we use here is different from the one in Chap. 1, where we first choose the booking limit, then accept as much demand as the booking limit allows. It turns out that both of these sequence of events give rise to the same policy, and our goal is to demonstrate an alternative dynamic programming formulation for the multiple class revenue management problems. Using u to denote the portion of the demand that we accept from a fare class and following the sequence of events that we just described, the dynamic programming formulation of the problem is given by

$$V_j(y) = \mathbb{E}\left\{ \max_{0 \le u \le D_j} p_j\, u + V_{j-1}(y+u) \right\}, \tag{3.3}$$

where we charge the denied boarding cost of the reservations that we cannot accommodate on the flight through the boundary condition

$$V_0(y) = -\theta\, \mathbb{E}\{[Z(y) - c]^+\}. \tag{3.4}$$

In this section, we will show that the optimal policy has the following structure. At each stage j, there exists a booking limit b_j^* such that it is optimal to bring the total number of accepted reservations as close as possible to b_j^* after making the decisions for class j. In the sequence of events for our dynamic program, we observe the demand from fare class j first, then decide what portion of this demand to accept. However, the structure of the optimal policy is such that we bring the total number of accepted reservations as close as possible to some fixed number b_j^* after making the decisions for fare class j. Thus, at the beginning of fare class j, we can set the booking limit to b_j^* before we even observe the demand from fare class j. This implies that assuming that we observe the demand before we decide what portion to accept or setting a booking limit before we observe the demand result in identical policies.

3.3.1 Optimal Booking Limits

Assume that the value functions $\{V_j(\cdot) : j = 1, \dots, n\}$ computed through the dynamic program in (3.3) are concave. This implies that $\Delta V_j(z) := V_j(z) - V_j(z - 1)$ is decreasing in z for all j.

Under this assumption, we show that the optimal policy can be characterized by a booking limit b_j^* for each stage j, such that it is optimal to bring the total number of reservations as close as possible to b_j^* after making the decisions for class j. Once we show this result, we will verify that concavity of the value function.

Theorem 3.2 *Assume that $V_{j-1}(\cdot)$ is concave, and let b_j^* be the maximizer of the concave function $p_j\, z + V_{j-1}(z)$ over* $[0, \infty]$. *Then,*

$$b_j^* = \min\{z \geq 0 : p_j + \Delta V_{j-1}(z + 1) < 0\}.$$

In this case, setting

$$u^*(y) = \begin{cases} 0 & \text{if } b_j^* < y \\ b_j^* - y & \text{if } y \leq b_j^* \leq y + D_j \\ D_j & \text{if } b_j^* > y + D_j. \end{cases} \tag{3.5}$$

solves problem (3.3).

The next result confirms the concavity of the value functions.

Theorem 3.3 *The value functions $\{V_j(\cdot) : j = 1, \dots, n\}$ computed through the dynamic program in* (3.3) *are concave.*

3.3.2 Class-Dependent No-Show Refunds

In the dynamic program in (3.3), we assume that if a passenger does not show up, then we do not give any refund and the probability of showing up for all passengers is the same. In practice, there are different restrictions that come along with different classes. As a result, passengers with tickets for different classes get different refunds when they do not show up and the probability of showing up is different for different classes. Allowing different show-up probabilities for different fare classes is difficult, because this extension requires using a high-dimensional state variable that keeps track of the reservations for each fare class separately. However, we can incorporate no-show refunds without too much difficulty and these no-show refunds could be different for different classes.

Assume that customers of class j who do not show up at the departure time of the flight are given a refund of h_j that is strictly less than the revenue p_j. We continue using all of the assumptions in our earlier model. Since whether a customer does not show up is completely independent of all other decisions and events in the system, we can charge the expected refund at the time the reservation is accepted, instead of the time of service. Thus, if we accept a reservation from a customer of class j, it yields an expected revenue of $p_j-(1-q)\,h_j$. In this case, we can use $p_j-(1-q)\,h_j$ in place of p_j in our earlier dynamic program.

3.3.3 Incorporating Cancellations

We can incorporate cancellations into our model, as long as the cancellation probabilities for the different fare classes are the same. We use ρ to denote the probability that a customer cancels her reservations at any stage. Given that we have y reservations on hand, we use $Z'(y)$ to denote the number of reservations that we still have on hand after observing the cancellations at the current stage. Thus, $Z'(y)$ is binomially distributed with parameters $(y,\, 1-\rho)$. In this case, the dynamic programming formulation of the problem is given by

$$V_j(y) = \mathbb{E}\left\{ \max_{0\le u\le D_j} p_j\, u + V_{j-1}(Z'(y+u)) \right\},$$

with the same boundary condition as in (3.4).

Using an induction argument that is very similar to the one used earlier in this section, we can show that the value functions $\{V_j(\cdot) : j = 1, \dots, n\}$ are concave. In this case, the optimal policy can be characterized by one booking limit b_j^* for each class j such that it is optimal to bring the number of reservations on hand as close as possible to b_j^* after making the decisions for fare class j. The optimal booking limit b_j^* for class j is the maximizer of the function $p_j\, y + V_{j-1}(Z'(y))$ over the interval $[0, \infty]$. Therefore, b_j^* can be computed as

$$b_j^* = \min\{y \ge 0 : p_j + \Delta V_{j-1}(Z'(y+1)) < 0\}.$$

3.4 Overbooking over a Flight Network

In this section, we give a dynamic programming formulation for the network model with overbooking. Following this formulation, we provide a deterministic linear programming approximation that is an upper bound on the optimal total expected revenue. Furthermore, this linear program can be used to extract control policies. Throughout this section, we use the independent demand model and adopt a discrete time formulation. There are m resources in the network indexed by $M := \{1, \dots, m\}$. We denote the vector of initial capacities by $c = (c_1, \dots, c_m) \in \mathcal{Z}^m$. There are T time periods in the selling horizon. We count the time periods backwards. In particular, time period T corresponds to the beginning of the selling horizon, whereas time period 1 is the last time period in the selling horizon. Time period 0 corresponds to the departure time of the flights. We use a single index to capture the ODF's. The set of ODF's in $N := \{1, \dots, n\}$. At time period t, we have a request for ODF j with probability λ_{tj}. The fare for ODF j is p_j. If we deny boarding to customer with a ticket for ODF j, then we incur a penalty of θ_j. Let $a_{ij} = 1$ if ODF j uses resource i, and $a_{ij} = 0$ if ODF j does not use resource i. We allow both cancellations and no-shows. The probability that a reservation for ODF j is retained from time period t to $t-1$ is q_{tj}. In other words, if we have a reservation for ODF j at time period t, this reservation cancels by time period $t-1$ with probability $1-q_{tj}$. Notice that q_{t1} is the probability that a reservation for ODF j is retained from period 1 to period 0, which corresponds to the show probability of a customer with a reservation for ODF j. The cancellation and no-show behavior of each customer is independent of the others. Furthermore, the cancellation decisions at different time periods are independent. Given that we have x_j reservations for itinerary j at time period t, we use $S_{tj}(x_j)$ to denote the number of reservations that we retain from time period t to $t-1$. Due to our assumptions, $S_{tj}(x_j)$ has a binomial distribution with parameters (x_j, q_{tj}). We use the vector $S_t(x) = (S_{tj}(x_j))_{j \in N}$ to capture the vector of retained reservations.

For any time to go t, we use (t, x) to represent the state of the system, where $x = (x_1, \dots, x_n)$ captures the number of reservations on hand for each ODF. To capture the decisions at any time period, we use the vector $u = (u_1, \dots, u_n)$, where $u_j = 1$ if accept a request for ODF j, and $u_j = 0$ otherwise. In this case, using $e_j \in \Re^n_+$ to denote the unit vector with a one in the j-th component, the dynamic programming formulation of the overbooking problem over a network is given by

$$V(t, x) = \max_{u \in \{0,1\}^n} \left\{ \sum_{j \in N} \lambda_{tj}\, u_j \left\{ p_j + \mathbb{E}\{V(t-1, S_t(x + e_j))\} \right\} \right.$$

$$\left. + \left\{ 1 - \sum_{j \in N} \lambda_{tj}\, u_j \right\} \mathbb{E}\{V(t, S_t(x))\} \right\},$$

where the expectations involve the random variables $S_t(x + e_j)$ and $S_t(x)$. Notice that the capacities of the resources do not play a role in the dynamic program above. Since we are allowed to overbook, the number of accepted reservations can exceed the available capacities on the resources. Thus, the capacities come into play when we compute the cost of denying boarding to the passengers that cannot be accommodated on the flights in the boundary condition of the dynamic program. For the boundary condition, we assume that the airline solves an optimization problem to decide which passengers should be allowed boarding so that the total penalty of denied boardings is minimized. (Our boundary condition is perhaps a bit optimistic in the sense that it would be difficult to solve a centralized optimization problem to decide which customers should be denied boarding.) Using the decision variable y_j to capture the number of reservations for ODF j that we deny booking, the boundary condition of our dynamic program is given by

$$\begin{aligned} V(0,x) = -\min \quad & \sum_{j \in N} \theta_j \, y_j \\ \text{s.t.} \quad & \sum_{j \in N} a_{ij} \, [x_j - y_j] \leq c_i \quad \forall i \in M \\ & y_j \leq x_j \;\; \forall j \in N \\ & y_j \in \mathcal{Z}_+ \;\; \forall j \in N. \end{aligned} \tag{3.6}$$

The objective function above minimizes the total cost of denied reservations. The first constraint ensures that the reservations that remain after denied boardings can be accommodated on the flights. The second constraint ensures that the number of denied bookings cannot exceed the number of reservations for each ODF.

Solving the dynamic program above is difficult because the state variable is a high-dimensional vector. Next, we give a tractable linear programming approximation that can be used to obtain an upper bound on the optimal total expected profit.

3.4.1 Linear Programming-Based Upper Bound on $V(T, 0)$

Since we start with no reservations on hand, the optimal total expected profit in our overbooking problem is given by $V(T, 0)$. We give a linear programming approximation that can be used to obtain an upper bound on $V(T, 0)$. We observe that a reservation booked at time period t is retained until the departure time with probability $Q_{tj} := q_{tj} \times q_{t-1,j} \times \ldots \times q_{1,j}$. Using the decision variable w_{tj} to capture the expected number of accepted reservations for ODF j at time period t and y_j to capture the number of reservations for ODF j that we deny boarding, we consider the linear program

$$\begin{aligned} \bar{V}(T,0) := \max \quad & \sum_{t=1}^{T} \sum_{j \in N} p_j\, w_{tj} - \sum_{j \in N} \theta_j\, y_j && && (3.7) \\ \text{s.t.} \quad & \sum_{t=1}^{T} \sum_{j \in N} a_{ij}\, Q_{tj}\, w_{tj} - \sum_{j \in N} a_{ij}\, y_j \le c_i && \forall i \in M \\ & \sum_{t=1}^{T} Q_{tj}\, w_{tj} - y_j \ge 0 && \forall j \in N \\ & w_{tj} \le \lambda_{tj} && \forall t = 1, \ldots, T,\ j \in N \\ & w_{tj},\ y_j \ge 0 && \forall t = 1, \ldots, T,\ j \in N. \end{aligned}$$

The objective function above accounts for the total expected profit, which is the difference between the revenue from the accepted reservations and the penalty cost of denied boardings. The expected number of accepted reservations for ODF j at time period t is w_{tj}. These reservations are retained until the end of the selling horizon with probability Q_{tj}. Therefore, $\sum_{t=1}^{T} Q_{tj}\, w_{tj}$ is the total expected number of reservations for ODF j retained until the departure time, which implies that $\sum_{t=1}^{T} \sum_{j \in N} a_{ij}\, Q_{tj}\, w_{tj}$ corresponds to the total expected capacity consumption of resource i by all the reservations that have been accepted over the selling horizon. On the other hand, $\sum_{j \in N} a_{ij}\, y_j$ gives the capacity of resource i released by the denied boardings. So, the first constraint ensures that the expected capacity consumption for each resource, after considering the capacity released by denied boardings, cannot exceed the capacity of the resource. The second constraint ensures that the number of denied boardings for passengers with a ticket for ODF j does not exceed the expected number of accepted reservations for ODF j. The third constraint is a demand constraint, ensuring that the expected number accepted reservations for ODF j at time period t does not exceed the expected demand for the same ODF at the same time period.

In the next theorem, we show that the optimal objective value of problem (3.7) is an upper bound on the optimal total expected profit. In contrast to our earlier linear programming-based upper bounds, our proof technique here does not use the Jensen's inequality, because owing to cancellations and no-shows, the random quantities would not appear on the right-hand side of a similar perfect hindsight linear program.

Theorem 3.4 $V(T, 0) \le \bar{V}(T, 0)$.

In the description of the dynamic program and the upper bound (3.7), we kept the fare p_j as time invariant. This was convenient to keep the exposition manageable. In practice, however, consumers who book in period t and either cancel or do not show may get a partial or full refund. Thus, it is more convenient to model the term $p_j w_{tj}$ as $p_{tj} w_{tj}$ where p_{tj} is the net revenue per booking after discounting refunds. As an example, if a consumer obtains a refund $r_{tj} < p_j$ if he cancels a time t booking for product j, then the average revenue per booking is $p_{tj} = p_j - r_{tj}(1 - Q_{tj})$. If $r_{tj} = p_j$, then $p_{tj} = p_j Q_{tj}$ models the case of full refunds, while the case $r_{tj} = 0$, results in $p_{tj} = p_j$ for the case with no refunds. We wrote the dynamic program and the upper bound as if $r_{tj} = 0$, but it is possible to modify both formulations to account for partial refunds so the upper bound remains valid. If we further set $p_{tj} \leftarrow p_{tj}/Q_{tj}$ for all t and all j, then we can modify the objective function in (3.7) to read

$$\sum_{t=1}^{T} \sum_{j \in N} p_{tj}\, Q_{tj}\, w_{tj} - \sum_{j \in N} \theta_j\, y_j.$$

This will be the objective function for (3.7) we will work from now on. In this version, the quantity p_{tj} is the net revenue per surviving booking, whereas $p_{tj} Q_{tj}$ is the net revenue per booking.

3.4.2 Book-and-Bump Strategy

A book-and-bump strategy occurs when an airline books passengers at a fare, say p_{tj}, and later bump them if needed by compensating them at level $\theta_j < p_{tj}$. To see how this may happen, suppose first that θ_j is sufficiently high so that $y_j^* = 0$ is optimal in (3.7), and suppose there is a period, say t, such that $w_{tj}^* < \lambda_{tj}$. Suppose now that we reduce θ_j so that now $\theta_j < p_{tj}$. We claim that it is now optimal to accept all requests in period t for product j. Indeed, suppose that we accept $\delta = \lambda_{tj} - w_{tj}^*$ additional requests of product j in period t, this brings additional profits $p_{tj} Q_{tj} \delta$, but we have to pay θ_j for each one of the $Q_{tj}\delta$ units we expect to survive. Thus, the change in profits is equal to $[p_{tj} - \theta_j] Q_{tj} \delta > 0$, showing that if $\theta_j < p_{tj}$, then it is optimal to set $w_{tj}^* = \lambda_{tj}$, and that this may involve booking some consumers with the idea of later bumping them later at a profit. On the other hand, if $p_{tj} < \theta_j$ for all t, then we claim that it is optimal to set $y_j^* = 0$. To see this, suppose for a contradiction that $y_j^* > 0$, so there must be a period t such that $w_{tj}^* > 0$. Reducing w_{tj}^* by ϵ and decreasing y_j^* by ϵ reduces revenues by $p_{tj} Q_{tj} \epsilon$ and costs by $\theta_j Q_{tj} \epsilon$ for a net savings of $-[p_{tj} - \theta_j] Q_{tj}\, \epsilon > 0$, contradicting the optimality of $y_j^* > 0$.

3.4.3 Upper Bound for High Overbooking Penalties

A book-and-bump strategy is unfair, unpopular, and illegal. Consequently, most airlines would plan their overbooking models by setting unit overbooking cost $\theta_j > p_{tj}$ for all t and for all j. In this case, $y_j^* = 0$ for all j. This means that in solving the linear program (3.7), we do not overbook beyond adjusting for the expected number of cancellations and no-shows. This implies that we can reformulate the problem ignoring the y_j variables keeping in mind that in the stochastic version of the problem we pay an overbooking cost θ_j for each unit of product j that is overbooked. The updated LP is

$$\bar{V}(T,0) := \max \quad \sum_{t=1}^{T} \sum_{j \in N} p_{tj}\, Q_{tj}\, w_{tj} \tag{3.8}$$

$$\text{s.t.} \quad \sum_{t=1}^{T} \sum_{j \in N} a_{ij}\, Q_{tj}\, w_{tj} \le c_i \quad \forall i \in M$$

$$0 \le w_{tj} \le \lambda_{tj} \qquad \forall t = 1, \ldots, T, \; j \in N$$

This LP is essentially of the same form as the one for the model without cancellations, so there should be hope for heuristics based on its solution. Indeed, if we make the transformation $x_{tj} = Q_{tj} w_{tj}$, the LP (3.8) is as a time-variant version of the model without cancellations except that $x_{tj} \le \lambda_{tj} Q_{tj}$. Notice that $\lambda_{tj} Q_{tj}$ represents the net demand for product j at time t after filtering the demand that cancels or does not show.

3.4.4 Heuristics Based on the Linear Program

We can derive the probabilistic acceptance control (PAC) heuristic from the linear programming upper bound (3.8) exactly as we did in the previous chapter. Let $\{w^*_{tj} : t = 1, \ldots, T, \; j \in N\}$ be the optimal solution to problem (3.8). In period t, a request for product j arrives with probability λ_{tj}, and the PAC heuristic accepts it with w^*_{tj}/λ_{tj} and rejects it with probability $1 - w^*_{tj}/\lambda_{tj}$. Consider now a system where the capacities and the arrival rates λ_{tj} are scaled by an integer factor b, so now the number of arrivals for product j in period t is a binomial with parameters b and λ_{tj}. The PAC heuristic would filter the arrivals by the factor w^*_{tj}/λ_{tj} so in the scaled model, the number of requests accepted by the PAC heuristic is binomial with parameter b and w^*_{tj}. For the upper bound, the expected demand is $b\lambda_{tj}$, and the solution is bw^*_{tj}. Let $\bar{V}^b(T,0)$, $V^b(T,0)$, and $V^b_h(T,0)$ denote the upper bound, the optimal expected revenue, and the expected revenue of the PAC heuristic for the scaled system. Clearly, $V^b(T,0) = b\bar{V}(T,0)$ as bw^*_{tj} is an optimal solution to the scaled LP. We will now show that the PAC heuristic is asymptotically optimal.

Theorem 3.5

$$\lim_{b\to\infty} \frac{V^b_h(T,0)}{V^b(T,0)} \ge \lim_{b\to\infty} \frac{V^b_h(T,0)}{\bar{V}^b(T,0)} \to 1,$$

We can also use a bid-price heuristic based on the solution to the dual problem:

$$\min_{z \geq 0} \left\{ c'z + \sum_{t=1}^{T} \sum_{j \in N} \lambda_{tj} Q_{tj} (p_{tj} - \sum_{i \in M} a_{ij} z_i)^+ \right\}.$$

The heuristic accepts a request at time t if $p_{tj} \geq \sum_{i \in M} a_{ij} z_i^*$. The heuristic is not asymptotically optimal, but as in the case without cancellations and no-shows, it performs very well if the system is resolved frequently during the sales horizon for moderately large problems as those found in practice.

3.4.5 Other Approximation Strategies

We demonstrated that the linear programming approach that we had developed for network revenue management without overbooking naturally extends to the case where overbooking is allowed. Unfortunately, other approaches that we developed for network revenue management without overbooking do not easily extend to the case where overbooking is allowed. In the chapter on network revenue management problems with independent demand, we discussed two ways of decomposing the dynamic programming formulation of the network revenue management problem by the resources. The first approach exploited the deterministic linear program, and the second approach used Lagrangian relaxation. Under overbooking, even if we can decompose the problem by the resources, the problem that takes place over each resource is intractable because solving the single-resource revenue management problem requires a high-dimensional state variable that keeps track of the numbers of reservations for each ODF. There is some work that is based on decomposing the network overbooking problem by the resources and approximating the single-resource revenue management problems. This work is discussed in the bibliographical remarks at the chapter. In Table 3.1, we compare the bid-price policy derived from the linear program in (3.7) with such a decomposition approach. There are four test problems in this table, encoded by the pair (q, ρ), where q is the probability that an accepted request shows up and ρ is the ratio between the total expected demand for the capacities and the total capacity. In particular, we have $Q_{tj} = q$ for all $t = 1, \ldots, T$ and $j \in N$ and $\rho = q \sum_{i \in M} \sum_{t=1}^{T} \sum_{j \in N} a_{ij} \lambda_{tj} / \sum_{i \in M} c_i$. In all of the test problems, the airline network has a hub and spoke structure. There is one hub and four spokes. There is a flight leg from each spoke to the hub and a flight leg from the hub to each spoke. There is a high-fare and a low-fare ODF connecting each origin-destination pair. The fare of a high-fare ODF is eight times the fare of the corresponding low-fare ODF. The arrival process for the requests is set up such that the requests for the low-fare ODF's tend to arrive earlier, whereas the requests for the high-fare ODF's tend to arrive later. The first column in the table shows the upper bound on the optimal total expected profit given by the optimal objective value of problem (3.7), whereas the second and third columns show the total expected revenues obtained by the bid-price heuristic and the decomposition approach. The

Table 3.1 Performance of the bid-price heuristic and the decomposition approach

Problem (q, ρ)	Upper bound	Bid-price heuristic	Decomp. approach	Gaps Bid-price	Decomp.
(0.9, 1.2)	\$30,754	\$29,286	\$29,514	4.77%	4.03%
(0.9, 1.6)	\$31,744	\$30,324	\$30,841	4.47%	2.84%
(0.95, 1.2)	\$28,983	\$27,386	\$27,676	5.51%	4.51%
(0.95, 1.6)	\$23,995	\$22,720	\$22,983	5.31%	4.22%

last two columns show the percent gap between the total expected revenues of the policies and the upper bound on the optimal total expected profit. The results indicate that the decomposition approach yields noticeable improvements over the bid-price heuristic, especially when the capacities are tight.

In the same chapter, we also discussed approximate dynamic programming methods to approximate the value functions. These approaches do not readily extend to the overbooking setting either. In particular, if overbooking is allowed, then the linear program that we used to calibrate the value function approximations includes one constraint for each possible state of the system and the right side of this constraint involves the bumping cost associated with each state. The presence of this constraint makes overbooking problems intractable.

3.5 End of Chapter Problems

1. Consider a flight with 100 seats and a passenger fare of \$130. The denied boarding cost is \$390 per denied boarding, and the no-show rate is 0.16 (assuming a binomial no-show model). Demand for this flight is extremely high; in fact, for any booking limit $b < 200$, bookings will always hit the booking limit.
   (a) Assume that only the passengers who show up for the flight pay the fare of \$130; others are fully refunded. What is the optimal booking limit in this case? What is the corresponding expected net profit? How much does the airline gain from overbooking in this case? (That is, compute the expected revenue under the assumption that the airline does not overbook at all and compare to the overbooking case.)
   (b) For this part, assume that all passengers pay the fare of \$130 at the time of the reservation, regardless of whether or not they show up for the flight. Determine the optimal booking limit, the corresponding expected profit, and the gain from overbooking. (Hint: For this problem, you will need to modify the profit function to account for the fact everyone pays the fare and derive an expression that needs to be satisfied by the optimal booking limit.)

Table 3.2 Complementary cumulative distribution function of $Y(n)$

n	$\mathbb{P}\{Y(n) \geq 50\}$
50	0.0002957647
51	0.0025139996
52	0.0109987484
53	0.0330590953
54	0.0769040347
55	0.1479328364
56	0.2455974389
57	0.3627949618
58	0.4880498145
59	0.6091295054
60	0.7162850318

2. Consider the following overbooking problem. We first choose a booking limit b. After this, a random demand D occurs. For each unit of demand that we accept, we generate a revenue of $\$p$. Assume that every accepted booking request will show up for the flight. The capacity of the plane is c.

   If the number of customers that show up at the departure time exceeds the capacity of the plane, then we offer every customer a voucher worth $\$f$ for use on future flights. The customers who accept the vouchers will voluntarily give up their reservations. We assume that each customer independently declines the voucher, and thus keeps his/her existing reservation with probability $\beta \in (0, 1)$.

   After offering the vouchers, if the number of remaining customers still exceeds the capacity c of the plane, then we begin an involuntary denied boarding process. For each booking that cannot be accommodated on the plane, we incur a penalty cost of $\$\theta$.

   (a) Let $R(b)$ denote the expected profit under the booking limit b. Provide an expression for $R(b)$.
   (b) Assuming that $p - f(1 - \beta) < \theta\beta$, determine the integer-valued optimal booking limit. Your answer should only involve probabilities that can be computed by simple table lookups and the problem parameters given above.
   (c) Suppose that $c = 50$, $p = 100$, $f = 200$, $\theta = 300$, and $\beta = 0.85$. Let $Y(n)$ denote a binomial random variable with parameters n and 0.85. Table 3.2 gives the value of $\mathbb{P}\{Y(n) \geq 50\}$ for different values of n. Using this table and the formula from Part (b), determine the optimal booking limit in this case.

3. Consider the model in Sect. 3.3. Assume that the fares satisfy $p_1 \leq p_2 \leq \ldots \leq p_n$. Show that the optimal booking limits $b_1^*, \ldots, b_n^*$ satisfy $b_1^* \leq b_2^* \leq \ldots \leq b_n^*$.
4. We are purchasing a certain product over the time periods $1, 2, \ldots, T$. The demand for the product occurs at the end of these T time periods, say time period $T + 1$. The price of the product fluctuates randomly over the time periods $1, 2, \ldots, T$ and we need to decide how many units of product we should purchase at each time period.

We use the random variable P_t to denote the price of the product at time period t. We use the random variable D to denote the demand for the product, which occurs at time period $T+1$. For each unit of demand that we cannot satisfy, we incur a shortage cost of \$$\theta$. We are interested in minimizing the total expected cost, which is the sum of the product purchasing cost and the shortage cost.

(a) Formulate the problem as a dynamic program. Clearly give your state and decision variables, and write down the boundary condition at the end of T time periods.
(b) By using backward induction over the time periods, show that the value function is convex.
(c) Assume that the price can take only three different values, a high, a medium, and a low value. Show that in order to be able to make the optimal purchasing decision at each time period, we only need to store $3T$ values. That is, we only need to store three values for each time period. Clearly indicate how each one of these $3T$ values should be computed.

5. Consider a single-flight overbooking problem without any cancellations, but with no-shows. The customers arrive over the time periods $1, 2, \ldots, T$. There are n possible price levels indexed by $1, 2, \ldots, n$. If we sell a ticket at price level j, then we generate a revenue of \$$p_j$. With probability λ_{jt}, a customer that is interested in price level j arrives into the system at time period t. We need to decide whether to accept or reject each customer request. For simplicity, assume that $\sum_{j=1}^{n} \lambda_{jt} = 1$ so that there is always one customer arrival at each time period.

   At the departure time of the flight, which we assume to happen at time period $T+1$, each reservation shows up with probability q. A no-show with a reservation at price level j is given a refund of \$$h_j$. The capacity of the flight is c and for each customer that we cannot board on the flight, we incur a cost of \$$\theta$.

   (a) Let x_{jt} be the number of accepted reservations that we have on hand for price level j at the beginning of time period t. Using the n-dimensional vector $x_t = (x_{1t}, x_{2t}, \ldots, x_{nt})$ as the state variable, formulate a dynamic program that maximizes the expected profit. In your dynamic program, make sure to charge the no-show refunds at the departure time of the flight. (Hint: Let e_j be the j-th unit vector in $\Re^n$. If you accept a request for price level j at time period t, then your state changes from x_t to $x_t + e_j$.)
   (b) We can charge the expected refund cost at the time of accepting a customer request. This amounts to assuming that the revenue associated with price level j is $p_j - (1-q)\,h_j$. Since we charge the expected refund cost at the time of accepting a customer request and each reservation shows up with the same probability q, we now need to keep track of only the total number of accepted requests.

      Let z_t be the total number of accepted reservations that we have on hand at the beginning of time period t. Using the scalar z_t as your state variable, formulate a dynamic program that maximizes the total expected profit.

(c) Denote the value function in Part a as $V_t(x_t)$ and the value function in Part b as $J_t(z_t)$. Use backward induction over the time periods to show that $V_t(x_t) = J_t(\sum_{j=1}^{n} x_{jt}) - \sum_{j=1}^{n} (1-q)\, h_j\, x_{jt}$. (Hint: Recall that if $B_1(n_1, q)$ is a binomial random variable with parameters n_1 and q, and $B_2(n_2, q)$ is a binomial random variable with parameters n_2 and q that is independent of $B_1(n_1, q)$, then $B_1(n_1, q) + B_2(n_2, q)$ is a binomial random variable with parameters $n_1 + n_2$ and q.)

6. Is the PAC heuristic asymptotically optimal if the condition $\theta_j > p_{tj}$ for all t and for all j fails to hold?

3.6 Bibliographical Notes

Simon (1968) proposes auctions as a possible way to handle involuntary denied boardings. Rothstein (1971) gives one of the first systematic treatments of the overbooking problem, where dynamic programming is used to develop an overbooking policy for American Airlines. Chatwin (1998, 1999) give a dynamic programming formulation of the overbooking problem with a single class and characterize the structure of the optimal policy. Lautenbacher and Stidham (1999) study overbooking problems with multiple fare classes over a single resource. The cancellation model in this paper assumes that there can be at most one cancellation at each time period, and the probability of having a cancellation increases as the number of reservations on hand increases. In contrast, we use a binomial cancellation model, which allows multiple cancellations at each time period.

Kleywegt (2001) and Dai et al. (2019) give deterministic approximations to overbooking problems to extract heuristic control policies, some of which have asymptotic optimality guarantees. Karaesmen and van Ryzin (2004a) consider an overbooking model with substitutable flights, where the passengers bumped from one flight can be accommodated on the next one. Karaesmen and van Ryzin (2004b) study various decomposition strategies for the overbooking problem over a flight network. Erdelyi and Topaloglu (2010) leverage the linear programming approximation given in this chapter to decompose the network overbooking problem by the resources. Solving the single-resource overbooking problems is still difficult when the cancellation and no-show probabilities are class-specific. The authors use approximations to the single-resource overbooking problems. The numerical example in Sect. 3.4.5 is taken from this paper. Erdelyi and Topaloglu (2009) use a separable approximation to the bumping cost. In this case, they show that the dynamic programming formulation of the network overbooking problem decomposes by the ODFs, and the single-ODF overbooking problem turns out to be completely tractable. Aydin et al. (2013) present an overbooking model over a single resource. Their cancellation model is similar to the one in our network overbooking model in this chapter, in the sense that the number of cancellations at each time period is binomially distributed. Kunnumkal and Topaloglu (2011b) use stochastic approximation methods to compute bid prices for overbooking over a network of flight legs. Kunnumkal et al. (2012) give a randomized version of the linear program in (3.7) to capture the randomness in the show-up decisions more accurately.

Appendix

Proof of Proposition 3.1 We can write $Z(N)$ as $Z(N) = \sum_{i=1}^{N} X_i$, where the X_i's are independent Bernoulli random variables with probability q. Clearly $Z(\min(D, b+1)) - Z(\min(D, b)) = X_{b+1} \times \mathbf{1}(D \geq b+1)$, where $\mathbf{1}(\cdot)$ is the indicator function. Consequently, we get

$$\mathbb{E}\{Z(\min(D, b+1)) - Z(\min(D, b))\} = q\,\mathbb{P}\{D \geq b+1\}, \tag{3.9}$$

Similarly, note that we always have $Z(\min(D, b+1)) \geq Z(\min(D, b))$. Furthermore, $Z(\min(D, b+1))$ and $Z(\min(D, b))$ can differ by at most 1. Thus, if $Z(\min(D, b)) < c$, then we have $Z(\min\{b+1, D\}) \leq c$. On the other hand, if $Z(\min(D, b)) \geq c$, then $Z(\min\{b+1, D\}) \geq c$. In this case, we obtain

$$\begin{aligned}
&[Z(\min(D, b+1)) - c]^+ - [Z(\min(D, b)) - c]^+ \\
&\qquad = \begin{cases} Z(\min(D, b+1)) - Z(\min(D, b)) & \text{if } Z(\min(D, b)) \geq c \\ 0 & \text{otherwise} \end{cases} \\
&\qquad = \begin{cases} X_{b+1} & \text{if } D \geq b+1 \text{ and } Z(\min(D, b)) \geq c \\ 0 & \text{otherwise} \end{cases} \\
&\qquad = \begin{cases} X_{b+1} & \text{if } D \geq b+1 \text{ and } Z(b) \geq c \\ 0 & \text{otherwise.} \end{cases}
\end{aligned}$$

Using the chain of equalities above, we get

$$\begin{aligned}
&\mathbb{E}\{[Z(\min(D, b+1)) - c]^+ - [Z(\min(D, b)) - c]^+\} \\
&\qquad\qquad = q\,\mathbb{P}\{D \geq b+1\}\,\mathbb{P}\{Z(b) \geq c\}.
\end{aligned} \tag{3.10}$$

Using (3.9) and (3.10) in (3.1), we obtain

$$R(b+1) - R(b) = \mathbb{P}\{D \geq b+1\}\, q\,(p - \theta\,\mathbb{P}\{Z(b) \geq c\}),$$

from which the formula for b^* follows. □

Proof of Theorem 3.2 In (3.3), we need to solve the problem

$$\max_{0 \le u \le D_j} \Big\{ p_j \, u + V_{j-1}(y+u) \Big\}.$$

We define a new decision variable z such that $z = y + u$. Since y is the number of reservations just before making the decisions for class j and u is the number of reservations we accept from class j, the decision variable z can be interpreted as the number of reservations after making the decisions for fare class j. After the change of variables, the problem is equivalent to

$$\max_{y \le z \le y+D_j} \Big\{ p_j \, z + V_{j-1}(z) \Big\} - p_j \, y. \tag{3.11}$$

Since the last term $p_j \, y$ does not affect the optimal solution, we can concentrate on the following problem

$$\max_{y \le z \le y+D_j} \Big\{ p_j \, z + V_{j-1}(z) \Big\}. \tag{3.12}$$

Since $V_{j-1}(\cdot)$ is concave, the objective function of problem (3.12) above is concave. Thus, the problem above maximizes a concave function subject to the constraint that the decision variable lies in the interval $[y, y + D_j]$.

Let b_j^* be the maximizer of the concave function $p_j \, z + V_{j-1}(z)$ over $[0, \infty]$. The maximizer can be computed as

$$b_j^* = \min\{z \ge 0 : p_j \, (z+1) + V_{j-1}(z+1) \le p_j \, z + V_{j-1}(z)\}.$$

which yields the desired result.

We can characterize the optimal solution to the constrained problem above depending on whether b_j^* is in the interval $[y, y + D_j]$ or lies to the left or the right side of this interval. In particular, using z^* to denote the solution to problem (3.12), we have

$$z^* = \begin{cases} y & \text{if } b_j^* < y \\ b_j^* & \text{if } y \le b_j^* \le y + D_j \\ y + D_j & \text{if } b_j^* > y + D_j. \end{cases} \tag{3.13}$$

We show the three cases above, along with the maximizer b_j^* of the function $p_j \, z + V_{j-1}(z)$ and the interval $[y, y + D_j]$ in Fig. 3.1. If $b_j^* < y$, then the number of reservations we have y is already larger than the optimal booking limit b_j^*. Thus, the only way to get as close as possible to b_j^* after making the decisions for class j is not to accept any reservations from class j. In other words, we keep the number of reservations on hand at y. This situation corresponds to the first case above. If $b_j^* < y$, then it is optimal to set $z^* = y$. If $y \le b_j^* \le y + D_j$, then $b_j^* - y \le D_j$. So, we can accept $b_j^* - y$ reservations from class j to bring the number of reservations on

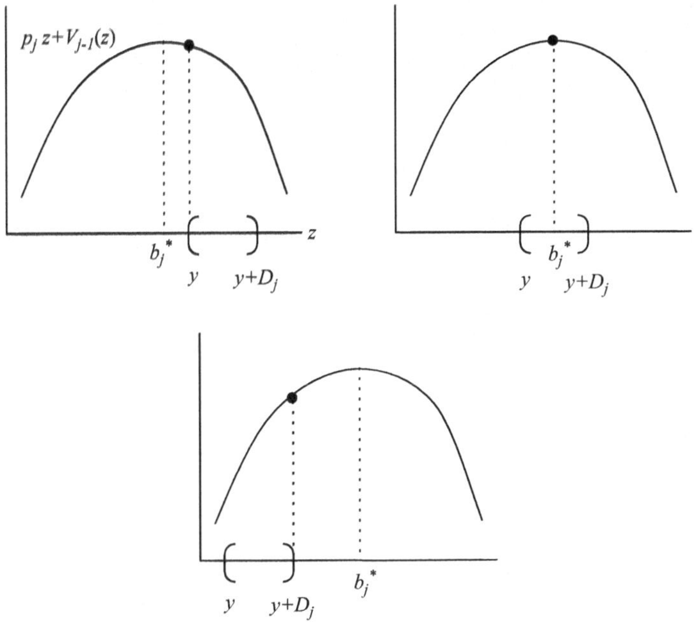

Fig. 3.1 Optimal decision for class j

hand to exactly b_j^* after making the decisions for class j. This situation corresponds to the second case above. If $y \le b_j^* \le y + D_j$, then it is optimal to set $z^* = b_j^*$. Lastly, if $b_j^* > y + D_j$, then $D_j < b_j^* - y$. Thus, the only way to get as close as possible to b_j^* after making the decisions for class j is to accept all of the demand from class j, in which case, the number of reservations that we have after making the decisions for class j goes up to $y + D_j$. This situation corresponds to the third case above. If $b_j^* > y + D_j$, then it is optimal to set $z^* = y + D_j$. Noting the change of variables $z = y + u$ and using (3.13), as a function of y, an optimal solution to problem (3.3) is given by the expression in the theorem. □

Proof of Theorem 3.3 We show the result by using induction over the classes in reverse order. Since $Z(y)$ is a binomial random variable with parameters (y, q), we can write $Z(y) = \sum_{i=1}^{y} X_i$, where $X_1, X_2, \dots$ are independent Bernoulli random variables with parameter q. In this case, we have

$$
\begin{aligned}
[Z(y+1) - c]^+ - [Z(y) - c]^+ &= \begin{cases} Z(y+1) - Z(y) & \text{if } Z(y) \ge c \\ 0 & \text{otherwise} \end{cases} \\
&= \begin{cases} X_{y+1} & \text{if } Z(y) \ge c \\ 0 & \text{otherwise,} \end{cases}
\end{aligned}
$$

which implies that $\mathbb{E}\{[Z(y+1)-c]^+ - [Z(y)-c]^+\} = q\,\mathbb{P}\{Z(y) \geq c\}$. Since $Z(y)$ is a binomial random variable with parameters (y, q), $\mathbb{P}\{Z(y) \geq c\}$ is increasing in y. Therefore, $\mathbb{E}\{[Z(y+1)-c]^+ - [Z(y)-c]^+\}$ is increasing in y. In this case, $\mathbb{E}\{[Z(y)-c]^+\}$ is convex in y, which implies that $V_0(y) = -\theta\,\mathbb{E}\{[Z(y)-c]^+\}$ is concave in y, as desired. This discussion establishes the base case for the induction argument. Next, we assume that the value function $V_{j-1}(\cdot)$ is concave and show that $V_j(\cdot)$ is also concave.

Assume that $V_{j-1}(\cdot)$ is concave. By using the same change of variables used to obtain problem (3.11), we can write the dynamic program in (3.3) as

$$V_j(y) = \mathbb{E}\left\{\max_{y \leq z \leq y+D_j} [p_j\, z + V_{j-1}(z)]\right\} - p_j\, y.$$

We define

$$W_j(y, D_j) = \max_{y \leq z \leq y+D_j} [p_j\, z + V_{j-1}(z)], \tag{3.14}$$

so that $V_j(y) = \mathbb{E}\{W_j(y, D_j)\} - p_j\, y$. If we can show that $W_j(y, D_j)$ is concave in y, then $\mathbb{E}\{W_j(y, D_j)\}$ is concave in y as well, in which case, it follows that $V_j(y) = \mathbb{E}\{W_j(y, D_j)\} - p_j\, y$ is concave, which is the result we are after. Thus, we proceed to showing that $W_j(y, D_j)$ is concave in y.

By the induction assumption $V_{j-1}(\cdot)$ is concave. We let b_j^* be the maximizer of the concave function $p_j\, z + V_{j-1}(z)$ over the interval $[0, \infty]$. Since $V_{j-1}(\cdot)$ is concave, the discussion that we used to obtain the three cases in (3.13) still holds. In this case, letting z^* be the optimal solution to problem (3.14), z^* is still given by the three cases in (3.13). Noting that the optimal objective function of problem (3.14) is $W_j(y, D_j)$, we have

$$W_j(y, D_j) = \begin{cases} p_j\, y + V_{j-1}(y) & \text{if } b_j^* < y \\ p_j\, b_j^* + V_{j-1}(b_j) & \text{if } y \leq b_j^* \leq y + D_j \\ p_j\,(y + D_j) + V_{j-1}(y + D_j) & \text{if } b_j^* > y + D_j \end{cases}$$

$$= \begin{cases} p_j\, y + V_{j-1}(y) & \text{if } b_j^* < y \\ p_j\, b_j^* + V_{j-1}(b_j^*) & \text{if } b_j^* - D_j \leq y \leq b_j^* \\ p_j\,(y + D_j) + V_{j-1}(y + D_j) & \text{if } y < b_j^* - D_j. \end{cases}$$

We plot the function $p_j\, y + V_{j-1}(y)$ as a function of y on the left side of Fig. 3.2. Notice that the maximizer of this function over $[0, \infty]$ is b_j^*. We plot the function $W_j(y, D_j)$ on the right side of Fig. 3.2. Notice that the functions $p_j\, y + V_{j-1}(y)$ and $W_j(y, D_j)$ are identical for y in $[b_j^*, \infty]$. For y in the interval $[b_j^* - D_j, b_j^*]$, the function $W_j(y, D_j)$ takes the constant value $b_j^* + V_{j-1}(b_j^*)$, which is the maximum value of $p_j\, y + V_{j-1}(y)$. Lastly, for y in the interval $[0, b_j^* - D_j]$, the

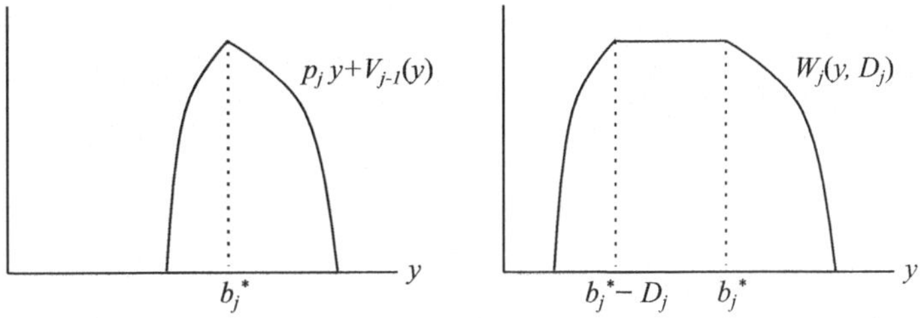

Fig. 3.2 Concavity of the value function for class j

function $W_j(y, D_j)$ takes the value of $p_j\,(y+D_j)+V_{j-1}(y+D_j)$. In other words, over the last interval, the function $W_j(y, D_j)$ is a shifted version of the function $p_j\,(y+D_j)+V_{j-1}(y+D_j)$. Thus, intuitively speaking, the function $W_j(y, D_j)$ is obtained by "cutting" the function $p_j\,y+V_{j-1}(y)$ in half at the point $y = b_j^*$, "shifting" the left portion of the function D_j units to the left, and "filling in" the middle with the constant value $b_j^* + V_{j-1}(b_j^*)$. Since $b_j^* + V_{j-1}(b_j^*)$ is the maximum value of the function $p_j\,y+V_{j-1}(y)$, it follows that $W_j(y, D_j)$ is concave, which is the desired result. □

Proof of Theorem 3.4 We let $D_{tj} = 1$ if there is a demand for ODF j at time period t, otherwise $D_{tj} = 0$. In this case, D_{tj} is a Bernoulli random variable with parameter λ_{tj} so that $\mathbb{E}\{D_{tj}\} = \lambda_{tj}$. We let the random variable W_{tj}^* be the number of accepted bookings for ODF j at time period t under the optimal policy and the random variable X_{tj}^* be the number of bookings for ODF j accepted at time period t that survive until the departure time. Thus, X_{tj}^* is a binomial random variable with parameters (W_{tj}^*, Q_{tj}). Thus, we have $\mathbb{E}\{X_{tj}^*\} = Q_{tj}\,\mathbb{E}\{W_{tj}^*\}$. Lastly, we let the random variable Y_j^* be the number of denied bookings for ODF j under the optimal policy. Under the optimal policy, we have the inequalities

$$\sum_{t=1}^{T}\sum_{j\in N} a_{ij}\,X_{tj}^* - \sum_{j\in N} a_{ij}\,Y_j^* \le c_i \quad \forall\, i \in M$$

$$Y_j^* \le \sum_{t=1}^{T}\sum_{j\in N} X_{jt}^* \quad \forall\, j \in N$$

$$W_{tj}^* \le D_{tj} \quad \forall\, t = 1, \ldots, T,\ j \in N.$$

The first inequality states that the capacity consumption of each resource, after accounting for the denied boardings, does not exceed the available capacity of the resource. The second inequality states that the number of denied boardings for each ODF cannot exceed the accepted bookings for the ODF. The third inequality states that the number of accepted bookings for each ODF at each time period cannot exceed the demand for the ODF. Taking expectations on both sides of the inequalities above and noting that $\mathbb{E}\{X^*_{tj}\} = Q_{tj}\,\mathbb{E}\{W^*_{tj}\}$, the inequalities above imply that setting $w_{tj} = \mathbb{E}\{W^*_{tj}\}$ and $z_j = \mathbb{E}\{Y^*_j\}$ for all $t = 1, \ldots, T$, $j \in N$ provides a feasible solution to problem (3.7). The total profit from the optimal policy is $\sum_{t\in T}\sum_{j\in N} p_j\, W^*_{tj} - \sum_{j\in N}\theta_j\, Y^*_j$, in which case, taking expectations, the total expected profit from the optimal policy is $V(T,0) = \sum_{t\in T}\sum_{j\in N} p_j\,\mathbb{E}\{W^*_{tj}\} - \sum_{j\in N}\theta_j\,\mathbb{E}\{Y^*_j\}$. Thus, setting $w_{tj} = \mathbb{E}\{W^*_{tj}\}$ and $y_j = \mathbb{E}\{Y^*_j\}$ for all $t = 1, \ldots, T$, $j \in N$ provides a feasible solution to problem (3.7) and the objective value provided by this solution is equal to $V(T,0)$. In this case, it follows that the optimal objective value of problem (3.7) is at least $V(T,0)$, so we obtain $\bar{V}(T,0) \geq V(T,0)$.

□

Proof of Theorem 3.5 Since the number of requests that arrive for product j in period t is a Bernoulli random variable with success probability λ_{tj}, the number admitted by the PAC heuristic is a thinned Bernoulli with probability w^*_{tj}. From this number, a fraction Q_{tj} will survive, so the number of bookings for period t that survive is also thinned Bernoulli with probability $Q_{tj}w^*_{tj}$. This shows that the expected revenues associated with the PAC heuristic, aggregating over all products, is equal to $\sum_{t=1}^T\sum_{j\in N} p_{tj}Q_{tj}w^*_{tj} = \bar{V}(T,0)$, where the equality uses the fact that $\{w^*_{tj} : t = 1, \ldots, T,\ j \in N\}$ is an optimal solution to problem (3.8).

Now, we consider the expected cost $\mathbb{E}[V(0,X)]$, where X is the vector of reservations on hand at the end of the horizon and $V(0,x)$ is the optimal objective value of problem (3.6). Clearly $X_j = \sum_{t=1}^T X_{tj}$, where X_{tj} is Bernoulli random variable with mean $Q_{tj}w_{tj}$. Since the X_{tj}'s are independent over t, it follows that X_j has mean $\sum_{t=1}^T Q_{tj}w^*_{tj}$ and variance $\sum_{t=1}^T Q_{tj}w^*_{tj}(1 - Q_{tj}w^*_{tj}) \leq \sum_{t=1}^T Q_{tj}w^*_{tj}$.

A feasible solution to program $V(0,X)$ in (3.6) is to pay the overbooking fee θ_j for each unit of product j booking in excess of the mean, yielding the feasible solution $y = \{y_j : j \in N\}$ with $y_j = (X_j - \mathbb{E}[X_j])^+$. Consequently, it follows that

$$\mathbb{E}[V(0,X)] \geq -\sum_{j\in N}\theta_j\,\mathbb{E}(X_j - \mathbb{E}[X_j]))^+ \geq -\frac{1}{2}\sum_{j\in N}\theta_j\sqrt{\sum_{t=1}^T Q_{tj}w^*_{tj}},$$

where we have used the fact that for any random variable with finite second moment $\mathbb{E}[(X - \mathbb{E}[X])^+] \leq 0.5\sqrt{\mathrm{Var}[X]}$. Thus, a lower bound on the expected revenue from the PAC heuristic is given by

$$V_h(T,0) = \bar{V}(T,0) + \mathbb{E}[V(0,X)] \geq \bar{V}(T,0) - \frac{1}{2}\sum_{j\in N}\theta_j\sqrt{\sum_{t=1}^T Q_{tj}w^*_{tj}}.$$

Clearly bw^*_{tj} is the solution to the linear program scaled by a factor b, so $\bar{V}^b(T,0) = b\bar{V}(T,0) \geq V^b(T,0) \geq V^b_h(T,0)$. From the bound on $\mathbb{E}[V(0,X)]$ we see that

$$V_h^b(T,0) = \bar{V}^b(T,0) - \mathbb{E}[V^b(0,X)] \geq \bar{V}^b(T,0) - \frac{1}{2}\sum_{j\in N}\theta_j \sqrt{b\sum_{t=1}^{T} Q_{tj} w_{tj}^*}.$$

Dividing by $\bar{V}^b(T,0)$ and letting $b \to \infty$, we find that

$$\lim_{b\to\infty} \frac{V_h^b(T,0)}{V^b(T,0)} \geq \lim_{b\to\infty} \frac{V_h^b(T,0)}{\bar{V}^b(T,0)} \to 1,$$

completing the proof. □

Part II
Revenue Management Under Customer Choice

Chapter 4
Introduction to Choice Modeling

4.1 Introduction

Revenue management models were originally developed under the assumption of stochastically independent demands. This assumption is untenable when products are close substitutes. In this case, the demand for a particular product may depend on the set of competing products that are available in the market. For example, when a product is removed from an assortment, its demand may be recaptured by another product in the assortment, or it may spill to competitors or the no-purchase alternative. Conversely, adding a product to the assortment may cannibalize the demand for other products in the assortment or may induce new demand. In this chapter, we study discrete choice models that help capture demand as a function of the offered products. In later chapters, we will use these discrete choice models to formulate optimization problems to choose profit or revenue maximizing assortments when the prices of the products are fixed, or to find the profit or revenue maximizing prices to charge for the offered products. Consequently, when discussing discrete choice models in this chapter, we pay particular attention to those discrete choice models that are rich enough to capture substitution effects and for which the corresponding optimization problems are computationally tractable.

In Sect. 4.2, we describe what we mean by a discrete choice model and formulate assortment and price optimization problems at a high level, but we defer the solution of such problems to later chapters. In Sect. 4.3, we discuss the maximum utility model, which provides a flexible framework to construct various choice models. In Sect. 4.4, we discuss the basic attraction model, of which the well-known multinomial logit model is a special case. In Sect. 4.5, we generalize this model to allow the attraction of the no-purchase alternative to depend on the products excluded from the assortment. A special case of this general attraction model is the independent demand model, where demand for each product is independent of other products offered. In this section, we also discuss the independence of

G. Gallego, H. Topaloglu, *Revenue Management and Pricing Analytics*, International Series in Operations Research & Management Science 279, https://doi.org/10.1007/978-1-4939-9606-3_4

irrelevant alternatives property, which is a shortcoming of the last two choice models mentioned. In Sect. 4.6, we explain the nested logit model, which alleviates this property, at least to a certain extent. In Sect. 4.7, we demonstrate how we can mix basic attraction models to generate richer choice models. In Sect. 4.8 we present the exponomial model based on reflected exponential utilities, while in Sect. 4.9 we study random consideration set models, where the preference order of the products is the same for all consumers, but different consumers drop different products from consideration. Finally, in Sect. 4.10 we discuss a general approach for choice modeling, where each consumer arrives with a particular ordering of products in mind, and she purchases the highest ranking available product. We then present the Markov chain, where consumers purchase their preferred product if available, and otherwise navigate according to a Markov chain until they find an available alternative, which may be the no-purchase alternative. Bounds and approximations are presented in Sect. 4.11 with an interpretation that may fit retail settings better than traditional choice models.

4.2 Discrete Choice Models

We will assume that the set of potential products that could be offered is $N := \{1, \ldots, n\}$. For any subset $S \subseteq N$, denote by $\pi_j(S)$ the probability that a consumer will select product $j \in S$, with $\pi_j(S) = 0$ if $j \notin S$. Let $\Pi(S) := \sum_{j \in S} \pi_j(S)$ denote the probability of a sale when subset S is offered. The complement $\pi_0(S) := 1 - \Pi(S)$ denotes the probability that the consumer selects an outside alternative. An outside alternative may mean either that the consumer does not purchase or that she purchases from another vendor. The outside alternative is always implicitly available. For this reason, it would be more appropriate to write $\pi_j(S_+)$ for $j \in S_+ := S \cup \{0\}$. However, we follow here the convention of writing $\pi_j(S)$ instead of the more cumbersome $\pi_j(S_+)$. Notice that under a discrete choice model a consumer will select exactly one product from the set S_+ when assortment S is offered. This is appropriate for transportation and lodging choices. Later we describe models that relax this assumption that may be more appropriate in a retail setting where consumers may buy more than one product.

A tedious way to describe any choice model is to list $\pi_j(S)$ for each $j \in S$ and for each $S \subseteq N$. This approach would require specifying $n\,2^{n-1}$ scalars which are far too many parameters to list or estimate. This is primarily why researchers have focused on parsimonious choice models that are based on a few parameters typically of the order $O(n)$ or $O(n^2)$. Most of the parametric choice models described in this chapter are parsimonious in nature.

In addition to our interest in finding realistic choice models to describe demand, we are also interested in finding a subset of products, say $S \subseteq N$, to offer to consumers with the objective of maximizing expected profits or revenues. Let p_j and z_j be, respectively, the price and the unit cost of product $j \in N$, and define the vectors $p := (p_1, \ldots, p_n)$ and $z := (z_1, \ldots, z_n)$. Then the expected profit obtained by offering the subset S is given by $R(S, z) := \sum_{j \in S}(p_j - z_j)\pi_j(S)$. The assortment optimization problem is that of finding a subset $S \subseteq N$ that maximizes $R(S, z)$, yielding

$$\mathcal{R}(z) := \max_{S \subseteq N} \; R(S, z). \tag{4.1}$$

The assortment optimization problem is combinatorial in nature and arises frequently in retailing and revenue management. We take on the question of formulating and solving such problems in the next chapter on assortment optimization.

If the prices are also decision variables, then the choice model needs to be price aware. Let $\pi_i(N, p)$ be the probability of selecting product i when the set N is offered. We are interested in maximizing $R(p, z) := \sum_{j \in N}(p_j - z_j)\pi_j(N, p)$ over all vectors $p \geq z$, yielding

$$\mathcal{R}(z) := \max_{p \geq 0} \; R(p, z).$$

For many choice models, if $p_j = \infty$, then $\pi_j(N, p) = 0$. Thus, setting $p_j = \infty$ is equivalent to not offering product j, indicating that optimizing over prices could implicitly select an assortment of products to offer. Problems of the form above are nonlinear optimization problems that arise in dynamic pricing and will also be encountered in a later chapter.

4.3 Maximum and Random Utility Models

Suppose there is a full preference ordering among the products. We assume without loss of generality that the products are labeled so the preferences are $1 \prec 2 \prec \ldots \prec n$. The maximum utility model (MUM) assigns $\pi_i(S) = 1$ for $i \in S$, if and only if $j \in S, j \neq i$ implies $j \prec i$. In other words, consumers select the highest ranked product in S, which corresponds to the product with the highest utility in the set S. The MUM is often presented by assigning cardinal utilities, say $\{u_i : i \in N\}$ to the products. If the u_i's are all different, then we can order them $u_1 < u_2 < \ldots < u_n$, and for any $S \subseteq N$ the maximum utility $\max_{i \in S} u_i$ is attained by the highest ranked product in S. If instead $u_1 \leq u_2 \leq \ldots \leq u_n$, then there may be more than one product in S attaining $\max_{i \in S} u_i$, in which case choice probabilities are assigned uniformly among such products.

Random utility models (RUM) add a random noise component to the utilities of the products, $U_i = u_i + \epsilon_i$, $i \in N$, where the ϵ_i's are mean zero, possibly dependent random variables. Another way to introduce randomness into the MUM is to assume a distribution of consumer types, each with a certain preference ordering. So, product $i \in S$ is selected by types of consumers for whom i is the highest ranked product in S.

All RUM's satisfy the regularity property that the probability of choosing a product does not increase as other products are added to the assortment. There are known cases where the regularity property does not hold, and readers are cautioned that in such cases there is a cost of approximating the choice model by a RUM. Nevertheless, RUM's are an important class to which we devote most of the chapter. We provide references to more general discrete choice models at the end of the chapter.

4.4 Basic Attraction and Multinomial Logit Models

The basic attraction model (BAM) is a discrete choice model where each product $j \in N$ has an attraction value $v_j > 0$, capturing the attractiveness of product j to a consumer. Similarly, the attraction value $v_0 > 0$ represents the attractiveness of the no-purchase alternative. The choice model is given by

$$\pi_j(S) = \frac{v_j}{v_0 + V(S)} \quad \forall j \in S, \tag{4.2}$$

where $V(S) := \sum_{j \in S} v_j$. Consequently, products with higher attraction values are more likely to be selected.

The introduction of BAM is based on postulating two choice axioms and demonstrating that a discrete choice model satisfies the axioms if and only if it is of the BAM form. To describe these axioms, we need additional notation. For any $S \subseteq T$, let $\pi_S(T) := \sum_{j \in S} \pi_j(T)$ denote the probability that a consumer selects a product in S when the set T is offered. Also, $\pi_{S_+}(T) := \pi_S(T) + \pi_0(T) = 1 - \pi_{T \setminus S}(T)$, where $T \setminus S$ is the set difference of T from S. The Luce axioms can be written as follows:

- Axiom 1: If $\pi_i(\{i\}) \in (0, 1)$ for all $i \in T$, then for any $Q \subseteq S_+$, $S \subseteq T$

$$\pi_Q(T) = \pi_Q(S)\, \pi_{S_+}(T).$$

- Axiom 2: If $\pi_i(\{i\}) = 0$ for some $i \in T$, then for any $S \subseteq T$ such that $i \in S$

$$\pi_S(T) = \pi_{S \setminus \{i\}}(T \setminus \{i\}).$$

Axiom 1 implies that the probability of selecting any set $Q \subseteq S_+$, when set T is offered, is equal to the probability of selecting Q when S is offered times the probability of selecting S_+ when T is offered assuming that $S \subseteq T$. Axiom 2 implies that if alternative i has no probability of being chosen, then it can be deleted without affecting the choice probabilities.

The celebrated multinomial logit (MNL) model is a special case of the BAM that arises from a RUM. Under a RUM, each product j has a random utility $U_j =$

$u_j + \epsilon_j$ for $j \in N_+$ and the probability that product $j \in S$ is selected is given by $\pi_j(S) = \mathbb{P}\{U_j \geq U_i \; \forall i \in S_+\}$. We can think of ϵ_j as an idiosyncratic variation on the mean utility or as errors in measuring the utility.

If ϵ_j is a normal random variable, then the resulting choice model is known as the Probit model. Unfortunately, there is no closed-form expression for the selection probabilities under the Probit model and that has discouraged its use to a certain extent. However, a closed-form expression for the selection probabilities can be obtained if $\{\epsilon_j : j \in N_+\}$ are modeled as independent and identically distributed Gumbel random variables all having the same scale parameter. The cumulative distribution function of a Gumbel random variable X with location and scale parameters ν and ϕ is given by

$$F(x : \nu, \phi) = \exp(-\exp(-\phi\,(x - \nu))). \tag{4.3}$$

The mode of this distribution is ν, while the median is $\nu - \ln(\ln 2)/\phi$. The mean is $\mathbb{E}[X] = \nu + \gamma/\phi$, where γ is the Euler constant. The variance is given by $\text{Var}[X] = \pi^2/6\phi^2$, where π is the ratio of a circle's circumference to its diameter. To obtain a mean zero random variable, we set $\nu = -\gamma/\phi$. Notice that the variance is inversely proportional to ϕ^2. If $\{\epsilon_j : j \in N_+\}$ are mean zero Gumbel random variables all with scale parameter ϕ and independent across the products, then

$$\pi_j(S) = \frac{e^{\phi u_j}}{1 + \sum_{k \in S} e^{\phi u_k}} \quad \forall j \in S,$$

and this is known as the MNL model, and a special case of the BAM.

As ϕ becomes large, the variance of ϵ_j becomes small and the choice probabilities concentrate on the product or products with the largest mean utility. Thus, only the products with the largest mean utilities are purchased resulting in the MUM. On the other hand, when ϕ becomes small, the probability of selecting any offered product converges to a uniform distribution, where each product is equally likely to be selected. This behavior arises because when the variance is much larger than the mean, the consumer loses the ability to reliably select products with higher mean utility.

4.5 Generalized Attraction Model

One of the shortcomings of the BAM is that the attraction value v_0 of the no-purchase option is a fixed parameter and does not depend on the subset of offered products. There is considerable empirical evidence that the BAM may be too optimistic in estimating demand recapture probabilities when the first choice of a consumer is not part of the offer set S. In particular, the BAM assumes that even if a consumer prefers some product $j \in \bar{S} := N \setminus S$, she must select a product in S_+. This approach ignores the possibility that the consumer may look for the products in $\bar{S}$ elsewhere or at a later time.

As an example, suppose that a consumer prefers a certain wine, and the store does not have it. The consumer may then either buy one of the wines in the store, go home without purchasing, or drive to another store and look for the specific wine she wants. The BAM precludes the last possibility; it implicitly assumes that the search cost for an alternative source of product $j \in \bar{S}$ is infinity, or equivalently that there is no competition. As an illustration, suppose that the consideration set is $N = \{1, 2\}$ and that $v_0 = v_1 = v_2 = 1$, so $\pi_j(\{1, 2\}) = 33.3\%$ for $j = 0, 1, 2$. Under the BAM, eliminating choice 2 results in $\pi_j(\{1\}) = 50\%$ for $j = 0, 1$. Suppose, however, that product 2 is available across town and the attraction value for product 2 from the alternative source is $w_2 = 0.5$. Then the choice set of a consumer, when product 2 is not offered, is in reality $\{1, 2'\}$ with $2'$ representing product 2 in the alternative location with shadow attraction w_2. Under this model,

$$\pi_0(\{1\}) = \frac{1.5}{2.5} = 60\%, \quad \pi_1(\{1\}) = \frac{1}{2.5} = 40\%.$$

To formally define the generalized attraction model (GAM), we assume that in addition to the attraction values $\{v_j : j \in N\}$, there are shadow attraction values $\{w_j : j \in N\}$ with $w_j \in [0, v_j]$ for all $j \in N$. Under the GAM, for any subset $S \subseteq N$, the selection probabilities are given by

$$\pi_j(S) = \frac{v_j}{v_0 + W(\bar{S}) + V(S)} \qquad \forall j \in S, \tag{4.4}$$

where $W(R) := \sum_{j \in R} w_j$ for any $R \subseteq N$. Thus, the GAM can be viewed as a version of the BAM, where the attractiveness of the no-purchase alternative is inflated to $v_0 + \sum_{j \in \bar{S}} w_j$ as a function of the shadow attraction values of the alternatives that are not offered. The no-purchase probability is given by

$$\pi_0(S) = \frac{v_0 + W(\bar{S})}{v_0 + W(\bar{S}) + V(S)}.$$

The GAM can be obtained axiomatically by modifying one of the Luce axioms. The special case of the GAM with $w_j = 0$ for all $j \in N$ recovers the BAM. As with the BAM, it is possible to normalize the parameters so that $v_0 = 1$ when $v_0 > 0$.

The parsimonious GAM (p-GAM) is given by $w_j = \theta\, v_j$ for all $j \in N$ for some $\theta \in [0, 1]$. In this model, the shadow attraction values are a constant factor of the original attraction values. The special case $\theta = 0$ recovers the BAM.

At the other extreme, the case $\theta = 1$ results in the independent demand model (IDM). Under the IDM, the probability $\pi_j(S)$ of selecting product $j \in S$ is independent of the offer set S, as long as S includes product j. If product j is removed, then all of its demand is lost to the no-purchase alternative. This implies that there are nonnegative constants, say v_0 and $\{v_j : j \in N\}$, such that

$$\pi_j(S) = v_j \quad \forall j \in S,$$

and $v_0 + V(N) = 1$. The IDM model may reflect what happens in very competitive situations where consumers can readily find another vendor offering products not in S. In most practical situations, however, it is reasonable to expect that some of the demand for product $j \notin S$ may be recaptured by other products in S.

The p-GAM can serve to test the competitiveness of the market, by testing the hypothesis $H_0 : \theta = 0$ or $H_0 : \theta = 1$, to see whether the BAM applies or the independent demand model applies. To test these hypotheses, we can use offer sets S_t and realized sales s_t over a certain period of time $t = 1, \ldots, T$ and obtain the likelihood functions $L(v, \theta)$. We would reject $H_0 : \theta = 0$ in favor of $H_1 : \theta = \theta_1 > 0$ if $\max_v L(v, 0)/\max_v L(v, \theta_1)$ is sufficiently small. Likewise, we would reject $H_0 : \theta = 1$ against $H_1 : \theta = \theta_1 < 1$ if $\max_v L(v, 1)/\max_v L(v, \theta_1)$ is sufficiently small. If these tests fail, then a GAM maybe a better fit to the data.

There is an alternative, perhaps simpler, way of presenting the GAM by using the following transformation. For all $j \in N$, we let $\tilde{v}_j = v_j - w_j$ and $\tilde{v}_0 = v_0 + W(N)$. For $S \subseteq N$, let $\tilde{V}(S) = \sum_{j \in S} \tilde{v}_j$. With this notation, the selection probabilities under the GAM given in (4.4) can equivalently be written as

$$\pi_j(S) = \frac{v_j}{\tilde{v}_0 + \tilde{V}(S)} \quad \forall\, j \in S \quad \text{and} \quad \pi_0(S) = 1 - \sum_{j \in S} \pi_j(S).$$

The form of the selection probability above is similar to the one under the BAM given in (4.2). This similarity becomes useful when extending assortment optimization results for the BAM to the GAM.

4.5.1 Independence of Irrelevant Alternatives

A close inspection of the selection probabilities under the BAM and GAM reveals that if we add a new product to an offered subset, then the purchase probability of all offered products decreases by the same relative amount. In particular, for any set $S \subseteq N$ and product $j, i \in S$ and $k \in \bar{S}$, the selection probabilities under the BAM and GAM satisfy

$$\frac{\pi_j(S)}{\pi_j(S \cup \{k\})} = \frac{\pi_i(S)}{\pi_i(S \cup \{k\})}.$$

The expression on the left can be interpreted as the relative decrease in the purchase probability of product j when we add product k to the offered set, whereas the expression on the right can be interpreted as the relative decrease in the purchase probability of product i when we add product k to the offered set. This situation is referred to as the independence of irrelevant alternatives (IIA) property. It can equivalently be stated as the ratio $\pi_i(S)/\pi_j(S)$ is independent of the set S containing both products i and j. For the BAM and the GAM, this ratio is v_i/v_j.

Intuitively, the IIA property should not hold when k cannibalizes the demand for products j and i to different extents. To see the negative consequences that the

IIA can have on the BAM, we use the well-known red bus, blue bus paradox. In this paradox, a person has a choice between driving a car and taking either a red or a blue bus. It is implicitly assumed that both buses have ample capacity and depart at the same time. Let v_c represent the attraction value of driving a car and let v_b represent the attraction value of riding the red bus. Under the BAM with $v_0 = 0$, the probability of driving, when the choice set is between driving and riding the red bus, is $v_c/(v_c + v_b)$. Adding a blue bus with attraction value v'_b drops the probability of driving to $v_c/(v_c + v_b + v'_b)$. This drop has been widely criticized because the addition of an equally attractive blue bus in addition to the red bus should not influence the probability of driving to the extent it does under the BAM.

While the GAM also suffers from the IIA property, it can better handle the blue bus, red bus paradox. Indeed, by setting $w_b = v'_b$, the probability of driving remains unchanged by the introduction of the blue bus under the GAM as long as $v'_b \leq v_b$. More generally, the probability of driving can be modeled as $v_c/(v_c + \max(v_b, v'_b))$ as driving competes with the more attractive of the two buses. We can fit this into the GAM framework by setting $v_b \leftarrow \max(v_b, v'_b)$ and $w_b \leftarrow \min(v_b, v'_b)$. A more common fix for the IIA property is to use the nested logit model, which we describe next.

4.6 Nested Logit Model

In the nested logit (NL) model, the products are organized into nests such that the products in the same nest are regarded as closer substitutes of each other relative to the products in different nests. Under the NL model, the selection process of a consumer proceeds in two stages. First, the consumer selects either one of the nests or decides to leave without making a purchase. Second, if the consumer selects one of the nests, then the consumer chooses one of the products offered in this nest. To formulate the NL model, we use $M := \{1, \dots, m\}$ to denote the set of nests. For notational brevity, we assume that the number of products in each nest is the same and we use N to denote the set of products available in each nest. It is straightforward to generalize our formulation to the case where different nests have different numbers of products. We use $S_i \subseteq N$ to denote the set of products offered in nest i. Therefore, the sets of products offered over all nests are given by $\{S_1, \dots, S_m\}$. The attraction value of product j in nest i is given by v_{ij}. Under the NL model, if a consumer has already decided to make a purchase in nest i and the set $S_i \subseteq N$ of products is offered in this nest, then the consumer selects product $j \in S_i$ with probability

$$q_{j|i}(S_i) := \frac{v_{ij}}{V_i(S_i)},$$

where $V_i(S_i) := \sum_{j \in S_i} v_{ij}$. On the other hand, each nest i has a dissimilarity parameter $\gamma_i \in [0, 1]$. The parameter γ_i is a measure of how easily the products in nest i substitute for each other. In this case, if the sets of products offered over all nests are given by $\{S_1, \ldots, S_m\}$, then a consumer chooses nest i with probability

$$Q_i(S_1, \ldots, S_m) := \frac{V_i(S_i)^{\gamma_i}}{v_0 + \sum_{l \in M} V_l(S_l)^{\gamma_l}},$$

where v_0 denotes the attraction value of the no-purchase alternative. Thus, if we offer the sets of products $\{S_1, \ldots, S_m\}$, then the selection probability of product j in nest i is given by

$$Q_i(S_1, \ldots, S_m)\, q_{j|i}(S_i) = \frac{V_i(S_i)^{\gamma_i}}{v_0 + \sum_{l \in M} V_l(S_l)^{\gamma_l}} \frac{v_{ij}}{V_i(S_i)}.$$

In Sect. 4.4, we discuss that the MNL model can be interpreted as a RUM, where a consumer associates random utilities with the options and chooses the option providing thc largest utility. The NL model has the same kind of a random utility interpretation. In particular, assume that a consumer associates the utility $U_{ij} = u_{ij} + \epsilon_{ij}$ with product j in nest i, where u_{ij} and ϵ_{ij} are respectively the deterministic and random utility components. We assume that $\epsilon = \{\epsilon_{ij} : i \in M,\ j \in N\}$ has a type of generalized extreme value distribution and the joint distribution function for ϵ is given by

$$F(x; \gamma) = \exp\left(-\sum_{i \in M} \left(\sum_{j \in N} \exp(-x_{ij}/\gamma_i) \right)^{\gamma_i} \right),$$

where we use x and γ to denote the vectors $\{x_{ij} : i \in M,\ j \in N\}$ and $\{\gamma_i : i \in M\}$. With the generalized extreme value distribution above, the marginal distribution of ϵ_{ij} is Gumbel and has the form (4.3). For two distinct nests $l, k \in M$, the random utilities ϵ_{ij} and ϵ_{lk} are independent of each other, but for a given nest i, the random utilities ϵ_{ij} and ϵ_{ik} are positively correlated. The parameter $1 - \gamma_i$ measures the degree of correlation between the utilities in nest i. If the random utilities have this form of correlated generalized extreme value distribution and consumers choose the product that provides the largest utility, then the selection probabilities under the corresponding RUM have the same form as the selection probabilities under the nested logit model, when the attraction values are $v_{ij} = e^{u_{ij}/\gamma_i}$ for all $i \in M$, $j \in N$.

The RUM interpretation of the NL model explains why products in the same nest are closer substitutes of each other. If two products are in the same nest, then their utilities are positively correlated. Thus, if a consumer associates a high utility with product j in nest i, then this consumer is also likely to associate a high utility with product k in nest i. In this case, if product j is not available but product k is available, then the consumer is likely to substitute product k for product j. As γ_i approaches to zero, the utilities of the products in the same nest become more positively correlated, so they become closer substitutes. When $\gamma_i = 1$ for all $i \in M$, the utilities of the products are uncorrelated and the NL model reduces to the MNL model. In some settings, the NL model has been used with dissimilarity parameters

$\{\gamma_i : i \in M\}$ exceeding one. This form of the NL model does not necessarily have a random utility interpretation. However, when one estimates the parameters of the NL model from the data, it is conceivable to end up with estimates of the dissimilarity parameters exceeding one that perform well both with training and testing data. Consequently, models with $\gamma_i > 1$ are often fit in practice even if they lose their RUM interpretation.

One of the attractive features of the BAM, GAM, and NL model is that the assortment optimization problem formulated in (4.1) is tractable when consumers choose according to one of these choice models. In the later chapters, we show how to solve problem (4.1) under these choice models.

4.7 Mixtures of Basic Attraction Models

One option to add richness to the BAM is to consider a version of BAM with multiple consumer segments. This adds heterogeneity in tastes. In particular, we assume that there are multiple consumer types and we use G to denote the set of all possible consumer types. Customers of type g associate the attraction value v_j^g with product j and the attraction value v_0^g with the option of not making a purchase. An arriving consumer is of type g with probability α^g. In this case, if we offer the set S of products, then the selection probability for product $j \in S$ is

$$\pi_j(S) = \sum_{g \in G} \alpha^g \frac{v_j^g}{v_0^g + \sum_{i \in S} v_i^g} \qquad \forall\, j \in S.$$

The choice model above is referred to as a mixture of BAMs. It can be shown that any discrete choice model that arises from a RUM can be approximated to any degree of accuracy by a mixture of BAM's. This result indicates that the mixture of BAM's can be quite flexible in modeling a variety of consumer choice patterns. However, as we will see in the next chapter, the mixture of BAMs leads to difficulties in assortment optimization when we try to find a revenue maximizing set of products to offer to consumers. In Sect. 4.10 we describe an alternative choice model that can provide a good approximation to the mixture of BAM's, while keeping the corresponding assortment optimization problem tractable.

4.8 The Exponomial Model

In the exponomial model, the utility of alternative i is $U_i = u_i + \epsilon_i$ for $i \in N_+$, where the ϵ_i's are independent, mean zero random variables of the form $\epsilon_i = \theta - \tau_i$ and τ_i is an exponential random variable with mean θ. Suppose that a subset $S \subseteq N$ is offered, the cardinality of S_+ is m, and that the products in S_+ are sorted in increasing order of u_i. Let $\{1, \ldots, m\}$ be the label of the products in S_+ under this sorting. Let $G(0) = 0$, and

$$G(i) := \frac{\exp(-\sum_{j=i}^{m}(u_j - u_i)/\theta)}{m - i + 1} \quad \text{for} \quad i \in \{1, \ldots, m\}.$$

Then, the choice probabilities are given as follows:

$$\pi_i(S_+) = G(i) - \sum_{j=1}^{i-1} \frac{G(j)}{m - i} \quad i \in \{1, \ldots, m\},$$

where sums over empty sets are assumed to be zero. Notice that in this model, the no-purchase alternative is one of the elements in $\{1, \ldots, m\}$. On the surface, the exponomial model seems difficult to work with, but on the positive side the parameters are easy to estimate.

4.9 Random Consideration Set Model

The random consideration set (RCS) model is characterized by a strict preference order $\prec$ on N, and by a vector λ of positive attention probabilities. We assume without loss of generality that the products are labeled so that $1 \prec 2 \prec \ldots \prec n$. Then for any subset $S \subseteq N$, the choice probability of product i is

$$\pi_i(S) = \lambda_i \prod_{j>i, j\in S} (1 - \lambda_j) \quad \forall i \in S, \tag{4.5}$$

with $\pi_i(S) = 0$ if $i \notin S$. The interpretation of the choice probability above is that a consumer forms a consideration set $C(S)$ by independently including each product i in her consideration set with probability λ_i. She then selects the most preferable available product in her consideration set. Thus, for a consumer to purchase product i, product i should be in her consideration set and all offered products that are preferred to product i should not be in her consideration set. In other words, $i \in S$ is chosen if and only if i is the highest ranked product in the random consideration set $C(S)$. The extreme case $\lambda_j = 1$ for all $j \in N$ corresponds to the MUM under a strict preference order. This case corresponds to having $\pi_i(S) = 1$ if and only if $i > j$ for all $j \in S$.

The RCS model assumes implicitly that $\lambda_0 = 1$ and that consumers pay no attention to products that are ranked below the no-purchase alternative, so consumers select the no-purchase alternative among assortments $S \subseteq N$ only from inattention to the alternatives in the assortment. Indeed,

$$\pi_0(S) = \lambda_0 \prod_{j\in S}(1 - \lambda_j) = \prod_{j\in S}(1 - \lambda_j) > 0,$$

if and only if all the products $j \in S$ have inattention probabilities $1 - \lambda_j > 0$.

Product cannibalization is asymmetric in this model, as the introduction of product i to an assortment S cannibalizes the demand of products $j \in S$ only if $j < i$. This can be seeing by noting that $\pi_j(S \cup \{i\}) = \pi_j(S)$ if $i < j$, while $\pi_j(S \cup \{i\}) = \pi_j(S)(1 - \lambda_i)$ if $i > j$. In words, demand for a product can only be cannibalized by a product higher up in the preference ordering. The cannibalization asymmetry exhibited by the RCS model is an important feature that is difficult to capture by other choice models. Thus, the RCS model is particularly useful when there is evidence of cannibalization asymmetry. It turns out that the RCS model is a special case of the Markov chain choice model presented next.

4.10 Markov Chain Choice Model

A random utility model can be viewed as a probability distribution over all preference lists, or permutations, over $N_+ = N \cup \{0\}$. A permutation σ is a bijection from $N_+ \to N_+$ resulting in the preference ordering $\sigma(1) > \sigma(2) > \ldots > \sigma(n+1)$. Each permutation σ of N_+ has a certain probability, say $\mathbb{P}(\sigma) \geq 0$, with $\sum_\sigma \mathbb{P}(\sigma) = 1$. The probability that consumers prefer product i when the full set of options N is offered is given by

$$\lambda_i := \pi_i(N) = \sum_\sigma \mathbb{P}\{\sigma(1) = i\}, \quad \forall i \in N_+, \tag{4.6}$$

where $\sigma(\ell)$ is the product with rank ℓ in the permutation σ. In words, the choice probability $\pi_i(N)$ is obtained by adding over all of those permutations that have product i in the first position. We refer to the λ_i as the first choice probability of product $i \in N$, as it is the probability of selecting product i when the entire set of products N is offered. Clearly, $\sum_{i \in N_+} \lambda_i = 1$.

If product i is not available, then consumers whose first choice is i substitute to product $j \neq i$ with probability

$$\rho_{ij} = \sum_\sigma \mathbb{P}\{\sigma(2) = j \mid \sigma(1) = i\} \quad \forall i \neq j, \ i \in N, \ j \in N_+.$$

In words, ρ_{ij} is the conditional probability that a consumer whose first choice is $i \in N$ will have $j \in N_+$ as its second choice. Because the no-purchase alternative is always available, we set $\rho_{0,j} = 0$ for all $j \in N$, and $\rho_{0,0} = 1$. Notice that $\sum_{j \in N_+} \rho_{ij} = 1$ for all $i \in N_+$.

Assuming that we know $\lambda_i = \pi_i(N)$ and $\pi_i(N \setminus \{j\})$ for all $i, j \in N$, we can compute the substitution probabilities using the formula

$$\rho_{ij} = \frac{\pi_j(N \setminus \{i\}) - \pi_j(N)}{\pi_i(N)} \quad \forall i \neq j, \ i \in N, \ j \in N_+.$$

If we knew the distribution $\mathbb{P}(\sigma)$ over all permutations, we could compute $\pi_i(S)$ by modifying (4.6) so that the sum includes all permutations where i appears before any other element in $S_+ = S \cup \{0\}$. This is a lot of work, even if we knew $\mathbb{P}(\sigma)$ for all σ. Suppose instead, that we try to approximate $\pi_i(S)$ based only on the information contained in $\lambda = \{\lambda_i : i \in N\}$ and $\rho = \{\rho_{ij} : i, j \in N\}$. To do this we need to make an assumption about what happens if a consumer does not find her first choice, say $i \notin S$, and transitions to some $j \in N_+$. If the transition is to 0, then the consumer will leave the system without purchase. If the transition is to $j \in S$, then the consumer will purchase product j as this is her second choice after i. When $j \in \bar{S}$ we make the Markovian assumption that the consumer will continue evolving according to a Markov chain as if her first choice was product j. She would then transition to product k with probability ρ_{jk} until she either transitions to a product in S or to the no-purchase alternative. This choice process corresponds to the one under the Markov chain (MC) choice model. Thus, the MC choice model can be interpreted as a permutation-based choice model, but once a consumer considers a particular product in her preference list, she loses track of the earlier products in the preference list.

Example 4.1 If $n = 2$, then there are six permutations of $N_+ = \{0, 1, 2\}$. If the probabilities of observing these permutations are given by $\mathbb{P}(0, 1, 2) = 0.2$, $\mathbb{P}(0, 2, 1) = 0.15$, $\mathbb{P}(1, 2, 0) = 0.2$, $\mathbb{P}(1, 0, 2) = 0.1$, $\mathbb{P}(2, 1, 0) = 0.15$, and $\mathbb{P}(2, 0, 1) = 0.2$, then $\lambda_1 = \mathbb{P}(1, 2, 0) + \mathbb{P}(1, 0, 2) = 0.3$, $\lambda_2 = \mathbb{P}(2, 1, 0) + \mathbb{P}(2, 0, 1) = 0.35$, and $\lambda_0 = \mathbb{P}(0, 1, 2) + \mathbb{P}(0, 2, 1) = 0.35$. Moreover, we have $\rho_{1,2} = \mathbb{P}\{\sigma(2) = 2 | \sigma(1) = 1\} = \mathbb{P}(\sigma(1, 2, 0))/\mathbb{P}\{\sigma(1) = 1\} = 2/3$ and consequently $\rho_{1,0} = 1/3$. Similarly, $\rho_{2,1} = \mathbb{P}\{\sigma(2) = 1 | \sigma(1) = 2\} = \mathbb{P}(\sigma(2, 1, 0))/\mathbb{P}\{\sigma(1) = 2\} = 0.15/0.3 = 1/2$, so $\rho_{2,0} = 1/2$. The full specification of the MC choice model is given by $\lambda = (0.3, 0.35, 0.35)$ and

$$\rho = \begin{pmatrix} 1 & 0 & 0 \\ 1/3 & 0 & 2/3 \\ 1/2 & 1/2 & 0 \end{pmatrix}.$$

Clearly $\pi_2(\{2\}) = \mathbb{P}(1, 2, 0) + \mathbb{P}(2, 1, 0) + \mathbb{P}(2, 0, 1) = 0.2 + 0.15 + 0.2 = 0.55$ as these are the three permutations where 2 is preferred to 0. However, this calculation requires the knowledge of the distribution among permutations. We can estimate $\pi_2(\{2\})$ from λ and ρ as follows. With probability $\lambda_1 = 0.3$ consumers prefer product 1 and among those 2/3 will transition to product 2. In addition, there is a $\lambda_2 = 0.35$ probability that consumers directly prefer product 2. Then, our estimate of $\pi_2(\{2\})$ is given by $\hat{\pi}_2(\{2\}) = 0.3 \times (2/3) + 0.35 = 0.55$ which is exactly $\pi_2(\{2\})$. Similarly, we can compute $\pi_1(\{1\}) = \mathbb{P}(1, 2, 0) + \mathbb{P}(1, 0, 2) + \mathbb{P}(2, 1, 0) = 0.2 + 0.1 + 0.15 = 0.45$, while the MC approximation results in $\hat{\pi}_1(\{1\}) = 0.3 + 0.35 \times (1/2) = 0.475$.

It is possible to write a system of equations to describe the selection probabilities under the MC choice model. We will denote by $\phi_j(S)$ the probability that a consumer considers product $j \notin S$ during the course of her choice process but does not purchase because $j \notin S$. By definition, $\phi_j(S) = 0$ for all $j \in S$. The quantities $\pi_j(S)$ for $j \in S$ and $\phi_j(S)$ for $j \in \bar{S}$ are related by the system of equations

$$\pi_j(S) = \lambda_j + \sum_{i \in \bar{S}} \phi_i(S)\, \rho_{ij} \quad \forall\, j \in S \tag{4.7}$$

$$\phi_j(S) = \lambda_j + \sum_{i \in \bar{S}} \phi_i(S)\, \rho_{ij} \quad \forall\, j \in \bar{S}. \tag{4.8}$$

The system of equations in (4.7) and (4.8) above can be interpreted as balance equations and are simply a reinterpretation of the steady-state probabilities. Considering the first set of equations, a consumer ends up purchasing product $j \in S$ either as a first choice with probability λ_j, or by visiting some product $i \in \bar{S}$ during the course of her choice process, not purchasing this product because it is not available, and by transitioning to product j. Similarly, for $j \notin S$, a consumer may consider product j as she enters the system, with probability λ_j, or after transitioning to j from other product $i \in \bar{S}$. The sets of equations in (4.7) and (4.8) consist of $|N|$ equations, which we can solve to obtain the $|N|$ probabilities $\{\pi_j(S) : j \in S\}$ and $\{\phi_j(S) : j \in \bar{S}\}$, with $\pi_0(S) = 1 - \sum_{j \in S} \pi_j(S)$.

Consider the BAM with parameters $v_0, v_1, \ldots, v_n$, normalized so that $\sum_{i \in N_+} v_i = 1$, then

$$\lambda_i = v_i, \quad \forall i \in N_+, \quad \text{and} \quad \rho_{ij} = v_j/(1 - v_i) \;\; \forall j \neq i, i \in N, j \in N_+. \tag{4.9}$$

In this case, the solution of the system of equations in (4.7)–(4.8) is $\pi_i(S) = v_i/(v_0 + V(S))$ for all $i \in S$ and $\phi_i(S) = v_i(1 - v_i)/(v_0 + V(S))$ for all $i \notin S$. This solution for $\pi_i(S)$ is exactly what the BAM would predict. Consequently, in the case of the BAM, all of the information about the choice model is contained in (λ, ρ). It can be shown that the GAM is also a special case of a MC choice model with a rank-one transition matrix, and the RCS model is a special case of the MC choice model with a rank-one triangular transition matrix. In a later chapter we will argue that the linear demand model $d(p) = a - Bp$, with $d(p) \geq 0$ can also be viewed as MC choice model when only products in S are allowed.

The MC choice model can also be used to approximate the mixture of BAM's, where consumers of type $g \in G$ choose according to a BAM with nonnegative attraction values $\{v_j^g : j \in N_+\}$ normalized so $v_0^g + \sum_{j \in N} v_j^g = 1$ for all $g \in G$. Consequently, the first choice and substitution probabilities λ_i^g and ρ_{ij}^g for type g consumers are of the form (4.9). Assume that a consumer is of type g with probability $\alpha^g > 0$, where we have $\sum_{g \in G} \alpha^g = 1$. The formulas for the first choice and substitution probabilities λ_i and ρ_{ij} for the mixed model are given by

$$\lambda_i = \sum_{g \in G} \alpha^g\, v_i^g, \; i \in N_+ \;\; \text{and} \;\; \rho_{ij} = \sum_{g \in G} \alpha^{g|i} \rho_{ij}^g \; j \neq i, i \in N, j \in N_+,$$

where

$$\alpha^{g|i} := \frac{\alpha^g v_i^g}{\sum_{c\in G} \alpha^c v_i^c} \quad \forall g \in G, \ \forall i \in N$$

is the conditional probability that a consumer whose first choice is i is of type g.

On the surface, the MC choice model seems difficult to work with because even computing the selection probabilities requires solving a system of equations. In the next chapter, we will demonstrate that the revenue-maximizing set of products to offer to consumers under this choice model can actually be computed in an efficient manner. Moreover, the MC choice model can be used to approximate any discrete choice model. For example, the model can be fitted by obtaining unbiased guess-timates of the transition probabilities $\{\rho_{ij} : i, j \in N\}$ from a group of managers, with those estimates perhaps averaged with weights that reflect experience in doing such estimations. Alternatively, if there is a metric for the distance between products, then the transition probabilities can be fitted as a decaying function of that metric. More precisely, if δ_{ij} is the distance between products i and j, then ρ_{ij} can be modeled as $\rho_{ij} = e^{-\beta\,\delta_{ij}}$ for some $\beta > 0$, which can be calibrated by making sure that $1 - \sum_{j\in N\setminus\{i\}} e^{-\beta\,\delta_{ij}}$ matches the probability of losing the consumer to the no-purchase alternative. The flexibility of the MC choice model together with the ease with which it is possible to find a revenue- or profit-maximizing assortment makes it a useful tool both as a choice model and as a mechanism to find optimal or near-optimal assortments.

Empirical evidence suggests that different choice models can be effective in fitting data. A parsimonious model such as the BAM can be too inflexible to capture choice behavior accurately, while a mixture of BAM's may be too flexible and may suffer from large errors on test data. The MC choice model, particularly the rank-one versions like the GAM, and triangular versions such as the RCS model have a good mixture of flexibility while still being fairly parsimonious.

4.11 Bounds and Approximate Choice Probabilities

Consider a random utility model where $U_i = u_i + \epsilon_i$ for all $i \in N_+$, $\mathbb{E}[\epsilon_i] = 0$ for all $i \in N_+$, and $\epsilon = (\epsilon_0, \epsilon_1, \ldots, \epsilon_n)$ follows an absolutely continuous joint distribution that is independent of u_i's. For any $S \subseteq N$, let

$$G(u, S) := \mathbb{E}[\max_{i\in S_+} U_i]$$

be the expected surplus for the consumer when the offered assortment is S. The Williams-Daly-Zachary theorem

$$\pi_i(S) = \frac{\partial G(u, S)}{\partial u_i} = \mathbb{P}(U_i \geq U_j, \forall j \neq i, j \in S_+)$$

gives the choice probabilities for al $i \in S_+$.

With the exception of a few models like the ones discussed in this chapter, computing $G(u, S)$ and $\pi_i(S)$ $i \in S_+$ can be quite difficult. In what follows, we obtain a bound on $G(u, S)$ and an approximation to $\pi_i(S)$ that is easy to compute even if the ϵ_i's are not independent. These approximations are fairly accurate and quite useful for cases where there are no closed-form solutions. The approximations themselves can be considered bona fide choice models as the examples below will show.

Let z be any constant and define

$$G(u, z, S) := z + \sum_{i \in S_+} \mathbb{E}[(U_i - z)^+].$$

It is easy to see that $G(u, z, S)$ is an upper bound on $G(u, S)$ for all z. Let

$$\bar{G}(u, S) := \min_z G(u, z, S).$$

Since $G(u, z, S)$ is convex in z, the first order optimality condition is also sufficient and is given by:

$$\sum_{i \in S_+} \mathbb{P}(U_i \geq z) = 1.$$

The root of this equation, say $z^*(S)$, exists given our continuity assumptions. Consequently, $\bar{G}(u, S) = G(u, z^*(S), S)$.

We can approximate $\pi_i(S)$ by the gradient of $\bar{G}(u, S)$, which we denote by $\bar{\pi}$. Clearly

$$\bar{\pi}_i(S) = \frac{\partial \bar{G}(u, S)}{\partial u_i} = \mathbb{P}(U_i \geq z^*(S)) \quad i \in S_+.$$

Notice that $\bar{\pi}_i(S)$ depends only on the marginal distribution of ϵ_i and $z^*(S)$, which makes $\bar{\pi}_i(S)$ much easier to compute than $\pi_i(S)$.

Example 4.2 Suppose $U_i = u_i + \epsilon_i$ and that $\epsilon_i = \tau_i - \theta$ for all $i \in N_+$, where the τ_i's are exponential random variables with mean θ. Let $v_i := \exp(u_i/\theta)$ for all $i \in N_+$. It is easy to see that in this setting,

$$\bar{\pi}_i(S) = \frac{v_i}{v_0 + V(S)} \quad \forall\ i \in S_+,$$

resulting in a BAM. Notice that the independence of the ϵ_i's was not required.

Example 4.3 Let $U_i := u_i + \epsilon_i$ and $\epsilon_i := \theta - \tau_i$, where the τ_i's are exponential with mean θ. Let $v_i := \exp(-u_i/\theta)$ for all $i \in N_+$ (notice the negative sign in the exponent). Suppose $S \subseteq N$ has the following property

$$|S| \max_{i \in S_+} v_i \leq v_0 + V(S) \quad \forall i \in S_+, \tag{4.10}$$

then a closed-form approximation for the exponomial model described in Sect. 4.8 is given by:

$$\bar{\pi}_i(S) = \frac{v_0 + \sum_{j \in S}(v_j - v_i)}{v_0 + V(S)} \quad \forall i \in S_+, \tag{4.11}$$

where again the independence of the ϵ_is was not required.

Condition (4.10) guarantees that all of the probabilities are non-negative. Otherwise the product with the largest $v_i, i \in S_+$ needs to be dropped. By repeatedly removing the product with the largest v_i from S_+, we eventually obtain a new set S satisfying condition (4.10) so that (4.11) is a closed-form approximation to the exponomial model. Notice that in the process of removing products, it is possible that product 0 is removed and in this case the term v_0 must be dropped from (4.11).

4.12 Choice Models and Retailing

Most of the applications of discrete choice models are to situations where the consumer would normally select a single alternative. Examples include transportation and lodging. As choice modeling and assortment optimization are finding their way into the retailing and e-commerce, the assumption of purchasing at most one product needs to be revisited as people may end up buying more than one pair of shoes even they go shopping with the expectation of buying a single pair. A reasonable model in the retail setting may be such that a customer selects non-negative thresholds $z_i, i \in S$, with the idea of buying all products whose utilities exceed their corresponding thresholds. Under this setting, the consumer may wish to maximize $\sum_{i \in S} \mathbb{E}[U_i | U_i \geq z_i] \mathbb{P}(U_i \geq z_i)$ subject to a bound, say $\sum_{i \in S} \mathbb{P}(U_i \geq z_i) \leq \kappa$, on the expected number of products to be purchased. We call this the threshold utility model (TUM). The upper bound $\bar{G}(u, S)$ on the consumer surplus $G(u, S)$, corresponds to the optimal selection of a uniform threshold ($z_i = z$ for all $i \in S$) for the case $\kappa = 1$ when the non-negativity of the thresholds is relaxed.

We can interpret the model in two different ways. First, after the consumer observes the utilities of the products, the consumer purchases all the products whose utility exceeds $z^*(S)$. The number of purchased products is random, and its expectation is κ. In this case $\bar{G}(u, S)$ measures directly the expected surplus. Alternatively, we can view the TUM as a consideration set model, where the consumer first observes the products $i \in S$ with $U_i \geq z^*(S)$, and then selects the one with the largest utility. If the set is empty, then the consumer selects the no-purchase alternative. It can be shown that under mild conditions the expected consumer surplus under this consideration set model is at least $e^{-\kappa}$ of $G(u, S)$.

4.13 End of Chapter Questions

1. Show that the IDM is a RUM by modeling it as a mixture of BAM's.
2. Show that the BAM and the GAM can be represented as rank-one MC choice models.
3. Give an interpretation of the general rank-one MC choice model.
4. Show that the random consideration set model can be represented as a MC choice model with a rank-one triangular transition matrix.
5. Consider a nested logit model with two nests. The assortment for nest 1, say S_1, must be selected from the set $N_1 = \{1, 2\}$ and the assortment for nest two, say S_2, must be selected from the set $N_2 = \{3, 4\}$. Suppose that dissimilarity parameters of the nests are $\gamma_1 = 0.8$ and $\gamma_2 = 0.5$. Finally, assume that $v_0 = 1, v_1 = 0.5, v_2 = 1.2, v_3 = 0.8$, and $v_4 = 0.5$.

   (a) Compute the first choice probabilities for all $j = 0, 1 \ldots, 4$.
   (b) Compute the transition probabilities ρ_{ij} for all $i \neq j, i \in N = N_1 \cup N_2$ and $j \in N_+ = N \cup \{0\}$.
   (c) Use the MC choice model to compute the probability that product 1 is selected if $S_1 = \{1\}$ and $S_2 = \{3, 4\}$ and compare this to the actual probabilities from the nested logit model.
   (d) Repeat Part c for $S_2 = \{3\}$ and for $S_2 = \{4\}$.

6. Compute $G(u, S)$ for the MNL model and verify that $\pi(S)$ is the gradient of $G(u, S)$ with respect to u.
7. Consider a Probit model where $U_i = u_i + \epsilon_i$ where the ϵ_i's are independent, mean zero normal random variables. Suppose that $N = \{1, 2\}$, $u = (0, 1)$ and the ϵ_i's have standard deviation equal to 2. Use simulation to compute $\pi_i(N)$ for $i \in N_+$ and then use the approximation of Sect. 4.11 to compute $\bar{\pi}_i(N)$ for $i \in N+$. Compare your results.
8. Let $U_i = u_i + \epsilon_i$ for all $i \in N_+$. Suppose there is a partition of the products $N = \cup_{j \in M} N_j$ so that for each $i \in N_j$, ϵ_i is an independent exponential with parameter θ_j. In addition, suppose that ϵ_0 is exponential with parameter 1. Let $v_i = \exp(u_i/\theta_j)$ for all $i \in N_j$, and $v_0 = \exp(u_0)$. Suppose that an assortment $S = \cup_{j \in M} S_j$ is offered where $S_j \subseteq N_j$ for each j. Show that there is a z^* such that

$$\sum_{j \in M} V_j(S_j) \exp(-z/\theta_j) + v_0 \exp(-z) = 1$$

where $V_j(S_j) = \sum_{i \in S_j} v_i$. Let $\gamma_j = \exp(1 - z^*/\theta_j)$, and show that

$$\bar{\pi}_i(S) = \frac{v_i}{V_i(S_j)} \frac{\gamma_j V_i(S_j)}{v_0 + \sum_{k \in M} \gamma_j V_k(S_k)} \quad \forall i \in S_j, \ \forall j \in M.$$

9. Consider the choice model (4.11) under assumption (4.10). Assume that the v's are normalized so that $v_0 + V(N) = 1$. Show that the MC approximation is given by $\lambda_i = \pi_i(N) = 1 - nv_i$ for all $i \in N_+$ and $\rho_{ij} = v_j/(1 - v_i)$ for all $i \in N$, $j \in N_+$, with $\rho_{0,0} = 1$. Show that the approximation is exact and therefore (4.11) is a special case of the MC choice model.

4.14 Bibliographic Remarks

The BAM was first proposed by Luce (1959), where the author postulated the two choice axioms discussed in this chapter. Luce (1977) also provides a discussion of the same axioms. The classical reference for the MNL model is McFadden (1980). The equivalence between random utility models and choice models that result from probability distributions over preference lists is due to Block (1974). The GAM was introduced by Gallego et al. (2015) by slightly modifying the Luce axioms. Early references on the NL model date back to Domencich and McFadden (1975) and McFadden (1978). Huh and Li (2015) study a multi-stage version of the NL model, where the products are divided into nests, the nests are divided into subsets and so on. Koppelman and Wen (2000) examine the properties of the paired combinatorial logit (PCL) model, which is a version of the NL model. In the PCL model, each nest has at most two products, but a product can occur in multiple nests. Wen and Koppelman (2001) study the generalized NL model, which includes the NL and PCL models. McFadden and Train (2000) show that any RUM model can be approximated to any degree of accuracy by using a mixture of BAM's. The generalized NL model is a special case of a substantially broader class of choice models, called the generalized extreme value models, which are discussed in Train (2002). Wang (2018a) studies a variant of the BAM, where the attraction value of a product can depend on the size of the market it garners.

Manzini and Mariotti (2014) discuss maximum random consideration set models. Blanchet et al. (2016) introduce the MC choice model. The MC choice model has been shown to be a special case of the RUM by Berbeglia (2016). The exponomial model was first introduced by Daganzo (1979) and further analyzed by Alptekinoglu and Semple (2016). Economists have been active developing rational inattention models, where consumers have a prior on the utility of each product and can reduce their uncertainty at a cost that is proportional to the reduction in entropy between the prior and the posterior. The resulting model has some semblance to the mixture of BAM's. The reader is referred to Caplin and Dean (2014) and Matejka and McKay (2015) for further information about rational inattention models. Ben-Akiva and Lerman (1985) give a thorough discussion of classical discrete choice models. The paper by van Ryzin (2005) discusses how discrete choice models can replace relying on the assumption that each consumer arrives with a fixed product in mind.

Huettner et al. (2018) study choice models where the consumers make a rational choice on which products to focus. Jagabathula and Rusmevichientong (2019) develop approaches to understand to what extent the choices of the consumers in the data can be explained by using the random utility maximization principle. Natarajan et al. (2009), Mishra et al. (2012, 2014), and Ahipasaoglu et al. (2019) build choice models by assuming that moment or marginal distribution information about the utilities are available and they estimate the joint utility distribution as the one that maximizes the expected utility of the consumer from her most preferred option. The authors show that this estimation problem can be solved efficiently. Feng et al. (2017, 2018) show that other arguments can be used to construct choice models and some of these arguments can yield broader classes of choice models. Berbeglia (2018) presents the generalized stochastic preference model that contains interesting examples of discrete choice models that are not RUMs. The upper bound $\bar{G}(u, S)$ is based on the well-known Lai and Robbins (1976) bound for maximally dependent random variables, while the interpretation of the bound in the retail setting is due to Gallego and Wang (2019).

While estimation is not covered in detailed in this book, it is of crucial importance to the implementation of good revenue management solutions. Important papers on estimation of choice models include Berbeglia et al. (2018), Farias et al. (2013), Jagabathula and Vulcano (2018), Jagabathula et al. (2018), Kok and Fisher (2007), Martinez-de-Albeniz and Saez-de-Tejada (2014), Phillips et al. (2015, 2016), Simsek and Topaloglu (2018), van Ryzin and Vulcano (2015), and Vulcano and van Ryzin (2012). Some of these papers are based on adaptations of the EM method for parametric models, while others attempt to fit non-parametric models which have important advantages in capturing complex behavior that may be at odds with the parametric models covered in this book.

Chapter 5
Assortment Optimization

5.1 Introduction

A fundamental question in revenue management involves deciding which fares to offer in response to a request from an origin to a destination. The solution depends both on available capacity, the time of departure, and how consumers makes choices. This problem needs to be solved in real time as travel requests arrive. The assortment optimization problem is crucial in other RM applications such as hotels and car rentals, and is becoming more important in retailing and e-commerce. The fundamental tradeoff in assortment optimization is that broad assortments result in demand cannibalization and spoilage, while narrow assortments result in disappointed consumers that may walk away without purchasing. The formulation can be interpreted broadly to include more strategic decisions such as the location of stores within a city. The profitability of an assortment can be best captured through a choice model that provides sale probabilities as a function of the set of products contained in the assortment. In this chapter, we formulate and solve assortment optimization problems for many of the choice models presented in the previous chapter.

We consider a problem setting where the firm has access to a finite set of products, among which management needs to select an assortment or subset to offer to its consumers. Consumers decide either to purchase one of the products in the assortment or leave without purchasing. We use discrete choice models to capture the selection probabilities among the products in the assortment and the no-purchase alternative. The goal of the firm is to find a set of products to offer to maximize the expected revenue. In this chapter we limit our attention to the static assortment optimization problem, but readers should be aware that an optimal assortment may change dramatically with the state of the system. As an example, a luxury watch maker that normally offers gold watches may add stainless steel watches to his offerings during hard times and remove them during good times. We will consider

G. Gallego, H. Topaloglu, *Revenue Management and Pricing Analytics*, International Series in Operations Research & Management Science 279, https://doi.org/10.1007/978-1-4939-9606-3_5

the case of time-varying discrete choice models with inventory considerations in the next chapter.

Assortment optimization problems are combinatorial in nature as the set of offered products determines the choice probabilities. In many cases the problems are also non-linear, so not surprisingly the general assortment optimization problem is NP-hard even to approximate it by a constant factor. Fortunately, however, there are choice models of practical and theoretical importance that admit a polynomial time solution. This is true, in particular, for many of the choice models covered in the previous chapter. For certain choice models, we are also able to solve *constrained* assortment optimization problems to deal, for example, with limits on the number of products that can be offered or with precedence constraints.

In Sect. 5.2, we formulate the assortment optimization problem for a generic choice model and draw attention to costs that may arise from the utilization of resources by the offered products. In Sect. 5.3, we demonstrate that the assortment problem under the maximum utility model (MUM) has a simple solution, which involves simply offering the highest revenue product. In Sect. 5.4, we see that the independent demand model (IDM) also has a simple solution that offers all products. In Sect. 5.5, we show that the assortment problem under the basic attraction model (BAM) is a linear fractional program, and that this program can be solved efficiently. In Sect. 5.6, we show that our observations for the assortment problem under the BAM also apply to the generalized attraction model (GAM). In Sect. 5.7, we discuss discrete choice models that are mixtures of BAM's. For this case, the resulting assortment optimization problem is NP-hard. In Sect. 5.8, we focus on the nested logit (NL) model and show that this can be solved efficiently. In Sect. 5.9, we provide an $O(n)$ algorithm for the random consideration set (RCS) model. In Sect. 5.10, we focus on the assortment problem under the Markov chain (MC) choice model and show that an optimal assortment can be solved via linear programming. The MC choice model can be used to approximate models for which the assortment optimization problem is NP-hard. In Sect. 5.11, we present constrained assortment problems under the BAM and the NL models. In Sect. 5.12, we present the notion of efficient sets and its applications to assortment optimization.

5.2 The Assortment Optimization Problem

Let $N := \{1, \dots, n\}$ denote the set of potential products that can be offered, and let $M := \{1, \dots, m\}$ be the set of resources utilized by the products. Let A be an $m \times n$ matrix where $A_{ij} \geq 0$ is the number of units of resource $i \in M$ consumed by product $j \in \{1, \dots, n\}$. For example, $A_{ij} = 1$ if product j uses one unit of resource i and $A_{ij} = 0$ otherwise. Let $z \in \Re^m_+$ be a vector of marginal cost of the resources, then $A'z$ is a vector in $\Re^n_+$ representing the marginal cost of the products. If we sell one unit of product j, then we collect a revenue of p_j and incur a marginal cost $A'_j z$ where A_j is the j-th column vector of A. If we offer assortment $S \subseteq N$, then a consumer purchases product $j \in S$ with probability $\pi_j(S)$. The probabilities $\{\pi_j(S) : j \in S,\ S \subseteq N\}$ can be governed by any choice model.

For any $S \subseteq N$, let

$$R(S, A'z) := \sum_{j \in S} (p_j - A'_j z)\, \pi_j(S)$$

denote the expected profit when the vector of marginal cost for the resources is z. The goal is to find an assortment that maximizes the expected profit, resulting in the assortment optimization problem

$$\mathcal{R}(A'z) := \max_{S \subseteq N} R(S, A'z) = \max_{S \subseteq N} \left\{ \sum_{j \in S} (p_j - A'_j z)\, \pi_j(S) \right\}. \tag{5.1}$$

It is possible to show that $\mathcal{R}(A'z)$ is decreasing convex in z. A brute force way of solving (5.1) is to compute $R(S, A'z)$ for the 2^n subsets of N and to pick an assortment with the highest expected profit. Unfortunately, this approach is practical only if n is reasonably small or the problem needs to be solved only sporadically.

The problem of maximizing $R(S, A'z)$ arises in situations where z is a vector of dual variables associated with capacity constraints on the resources used by the products. In revenue management, the dual variable of the capacity constraints changes due to the randomness of the sale process. Consequently, problem (5.1) has to be solved very frequently for different values of z and in an efficient way. This highlights the importance of finding efficient algorithms to solve (5.1). Fortunately, we will be able to deliver procedures and algorithms for most of the discrete choice models studied in the previous chapter. For brevity of notation, we will make the transformation $p_j \leftarrow p_j - A'_j z$ for all $j \in N$ and will write $R(S)$ as a short-hand for $R(S, 0)$ and $\mathcal{R}^*$ as a short-hand for $\mathcal{R}(0)$. The marginal costs will resurface once we obtain the pertinent results for each choice model and we will indicate what can be done to obtain optimal assortments for different values of z.

5.3 Maximum Utility Model

Given a full preference ordering $1 \prec 2 \prec \ldots \prec n$, and a product $i \in S \subseteq N$, the MUM assigns $\pi_i(S) = 1$ if and only if $j \in S, j \neq i$ implies $j \prec i$. An optimal assortment for the MUM is any assortment S with the following two properties. First, S contains the product with the largest revenue, say product k. Second, S may contain other products $i \in N$ but only if $i \prec k$. Notice that such products would not be purchased anyway. This implies that

$$\mathcal{R}^* = \max_{j \in N} p_j = p_k = R(\{k\}).$$

5.4 Independent Demand Model

In this section, we discuss the assortment optimization problem when consumers choose according to the IDM. Suppose there exist values $v_j \geq 0,\ j \in N$ and $v_0 \geq 0$ such that $V(N) + v_0 = 1$, where we set $V(S) := \sum_{j \in S} v_j$ for all $S \subseteq N$, and v_0 is the no-purchase probability. Under this parameterization, the IDM is given by

$$\pi_j(S) = v_j \quad \forall\, j \in S_+,\, S \subseteq N,$$

with $\pi_j(S) = 0$ for all $j \notin S$, $S_+ := S \cup \{0\}$, and 0 is the no-purchase alternative. It is easy to see for this model that if $p_j \geq 0$ for all $j \in N$, then

$$\mathcal{R}^* = R(N),$$

and more generally $S^* := \{j \in N : p_j \geq 0\}$ is an optimal assortment. Consequently, for the IDM it is optimal to offer all products with non-negative revenues, even those with small revenues p_j without fear that their presence may cannibalize the demand of higher revenue products. In practice, this is rarely true as the removal of a product may result in both spilled and recaptured demand.

5.5 Basic Attraction Model

We now discuss the assortment optimization problem when consumers choose according to the basic attraction model (BAM). Like in the IDM, each product $j \in N_+$ has an attraction, and the attraction values can be normalized so that $V(N) + v_0 = 1$. Recall that for the BAM,

$$\pi_j(S) = \frac{v_j}{v_0 + \sum_{i \in S} v_i} = \frac{v_j}{v_0 + V(S)} \quad \forall\, j \in S_+,\ S \subseteq N,$$

with $\pi_j(S) = 0$ for all $j \notin S$. If we offer the set $S \subseteq N$ of products, then the expected revenue obtained from a consumer is

$$R(S) = \sum_{j \in S} p_j\, \pi_j(S) = \frac{\sum_{j \in S} p_j\, v_j}{v_0 + V(S)}.$$

Our goal is to find a set of products to maximize the expected revenue from each consumer. This can be done by solving the problem

$$\mathcal{R}^* = \max_{S \subseteq N} R(S) = \max_{S \subseteq N} \left\{ \frac{\sum_{j \in S} p_j\, v_j}{v_0 + V(S)} \right\}. \tag{5.2}$$

Table 5.1 Expected revenues provided by nested-by-revenue assortments

Assort.	E_0	E_1	E_2	E_3	E_4	E_5
Exp. rev.	\$0	\$1.63	\$2.83	\$3.13	\$3.10	\$2.94

We assume here that the products are indexed such that $p_1 \geq p_2 \geq \ldots \geq p_n$. The class of nested-by-revenue assortments is $\{E_0, E_1, \ldots, E_n\}$, where $E_0 = \emptyset$ and $E_j := \{1, \ldots, j\}$, $j \in N$. In other words, the nested-by-revenue assortment E_j includes the j products with the largest revenues. We now show that problem (5.2) has an optimal solution in this class.

Theorem 5.1 *An optimal assortment for problem* (5.2) *is in the class of nested-by-revenue assortments* $\{E_0, E_1, \ldots, E_n\}$.

Knowing that a nested-by-revenue assortment is optimal for problem (5.2) is quite valuable as this reduces the number of assortments to search from 2^n to $n+1$. Thus, we can solve problem (5.2) in polynomial time. In the proof of Theorem 5.1, we show that an optimal assortment under the BAM is characterized by $S^* = \{j \in N : p_j \geq \mathcal{R}^*\}$. This characterization shows that it is optimal to offer the products whose revenues exceed $\mathcal{R}^*$, hence a nested-by-revenue assortment is optimal, but it does not allow us to solve problem (5.2) directly because we do not know the value of $\mathcal{R}^*$ before solving the assortment problem. Exercise 2 shows that the search can stop at E_j if j is the first index such that $R(E_j) > p_{j+1}$, and in this case $\mathcal{R}^* = R(E_j)$.

Example 5.2 Consider a problem instance with five products, where the revenues associated with the products are $(p_1, \ldots, p_5) = (7, 6, 4, 3, 2)$ and the attraction values are $(v_1, \ldots, v_5) = (3, 5, 6, 4, 5)$. The attraction value of the no-purchase alternative is 10. Table 5.1 shows the expected revenues associated with each one of the nested-by-revenue assortments. We observe that the largest expected revenue is $\mathcal{R}^* = 3.125$ provided by the assortment $S^* = \{1, 2, 3\}$. Thus, the optimal assortment to offer is $\{1, 2, 3\}$. Furthermore, we note that $p_1 \geq p_2 \geq p_3 \geq \mathcal{R}^*$, but $\mathcal{R}^* > p_4 \geq p_5$, validating the characterization of the optimal assortment $S^* = \{j \in N : p_j \geq \mathcal{R}^*\}$.

5.6 Generalized Attraction Model

We can extend the results presented here to the GAM. Under the GAM, in addition to the attraction values, we have a shadow attraction value w_j for each product j resulting in the selection probability

$$\pi_j(S) = \frac{v_j}{v_0 + \sum_{j \in \bar{S}} w_j + V(S)} = \frac{v_j}{v_0 + W(\bar{S}) + V(S)},$$

where $W(R) := \sum_{j\in R} w_j$ for any $R \subseteq N$ and $\bar{S} := N \setminus S$. With some minor algebraic manipulations we can write $R(S)$ in a form that is consistent with a modified version of the BAM:

$$R(S) = \frac{\sum_{j\in N} \frac{p_j v_j}{v_j - w_j}(v_j - w_j)}{v_0 + \sum_{j\in N} w_j + \sum_{j\in S}(v_j - w_j)}.$$

Notice that this expression is identical to the expected revenue function under the BAM, where the revenue associated with product j is $p_j v_j/(v_j - w_j)$, the attraction value of product j is $(v_j - w_j)$, and the attraction value of the no-purchase alternative is $v_0 + \sum_{j\in N} w_j$. Thus, exploiting our earlier result, we can reindex the products such that $p_1 v_1/(v_1 - w_1) \geq p_2 v_2/(v_2 - w_2) \geq \ldots \geq p_n v_n/(v_n - w_n)$. By our earlier discussion, an optimal solution is within the class $\{E_0, E_1, \ldots, E_n\}$. Under the GAM, the ordering of the products depends on both the revenues of the products and the attraction values. One can think of $p_j v_j/(v_j - w_j)$ as $p_j/(1-\theta_j)$, where $\theta_j := w_j/v_j$ is a measure of how competitive product j is. Consequently, products with high prices p_j and high competitive values θ_j are more likely to be in the optimal assortment. For the parsimonious GAM, $w_j = \theta v_j$, the ordering of the products is $p_1 \geq p_2 \geq \ldots \geq p_n$, just as in the original BAM.

5.7 Mixtures of Basic Attraction Models

Under a mixture of BAM's, there are multiple consumer types and consumers of each type select among the offered products according to a different BAM, whose parameters depend on the type of the consumer. We use G to denote the set of consumer types. An arriving consumer is of type g with probability α^g. A consumer of type g associates the attraction value v_j^g with product j and the attraction value v_0^g with the no-purchase alternative. If we offer the set S of products to consumers, then a consumer of type g selects product $j \in S$ with probability $v_j^g/(v_0^g + \sum_{i\in S} v_i^g)$. Thus, if we offer the set S of products, then we obtain an expected revenue of $\sum_{j\in S} p_j v_j^g/(v_0^g + \sum_{i\in S} v_i^g)$ from a consumer of type g. Therefore, by offering the set S of products, the expected revenue obtained from a consumer is given by

$$R(S) = \sum_{g\in G} \alpha^g \frac{\sum_{j\in S} p_j v_j^g}{v_0^g + V(S)^g}.$$

Unfortunately, the problem of finding a set that maximizes the expected revenue for the mixture of BAM's is NP-complete. On the positive side, nested-by-revenue assortments provide a revenue guarantee of at least $\max\{1/n, 1/(1 + \log(p_1/p_n))\}$ relative to the optimal revenue. The following example illustrates the difficulties associated with assortment optimization problems under a mixture of BAM's.

Example 5.3 Consider a problem instance with two consumer types and three products. The probability of observing each consumer type is $(\alpha^1, \alpha^2) = (0.5, 0.5)$. The revenues associated with the products are $(p_1, p_2, p_3) = (8, 4, 3)$. A consumer of type 1 associates the preference weights $(v_1^1, v_2^1, v_3^1) = (5, 20, 1)$ with the products and a consumer of type 2 associates the preference weights $(v_1^2, v_2^2, v_3^2) = (1/5, 10, 10)$. The preference weight of the no-purchase alternative is 1. The optimal assortment for type 1 consumers is $\{1\}$, while the optimal assortment for type 2 consumers is $\{1, 2\}$. The assortment that maximizes the expected revenue over the two consumer types is $\{1, 3\}$ providing an expected revenue of 4.48. This indicates that the solution to the mixture is non-nested. Furthermore, product 3 does not appear in the optimal assortment when we want to maximize the expected revenue from each of the two customer types separately, but this product appears in the optimal assortment when we want to maximize the expected revenue over the two customer types. The best revenue order assortment, $\{1, 2\}$, yields an expected revenue equal to 4.16 which is about 8% lower than the optimal expected revenue.

5.8 Nested Logit Model

Under the NL model, the products are grouped into nests. The choice process of a consumer can be modeled as a two-stage process. First, the consumer chooses one of the nests or decides not to purchase anything. Second, if the consumer chooses one of the nests, then the consumer selects one of the products in the chosen nest. We use $M := \{1, \ldots, m\}$ to index the set of nests. For notational brevity, we assume that each nest includes n products and we use $N := \{1, \ldots, n\}$ to denote the set of products that can be offered in each nest. Generalization to the case where different nests include different numbers of products is straightforward. We let p_{ij} be the revenue associated with product j in nest i. The attraction value of product j in nest i is v_{ij}. We use $S_i \subseteq N$ to denote the set of products offered in nest i. Under the NL model, if a consumer has already decided to make a purchase in nest i, then this consumer selects product $j \in S_i$ in nest i with probability $v_{ij}/V_i(S_i)$, where $V_i(S_i) := \sum_{j \in S_i} v_{ij}$. Consequently, the expected revenue from a consumer that has already decided to make a purchase in nest i is

$$R_i(S_i) := \frac{\sum_{j \in N} p_{ij}\, v_{ij}}{V_i(S_i)}.$$

The dissimilarity parameter of nest i is given by $\gamma_i \in [0, 1]$. As discussed in the previous chapter, having dissimilarity values in the interval [0, 1] ensures that the NL model is compatible with a random utility model. If the sets of products offered over all nests are given by $(S_1, \ldots, S_m)$, then a consumer chooses nest i with probability

$$Q_i(S_1, \ldots, S_m) := \frac{V_i(S_i)^{\gamma_i}}{v_0 + \sum_{l \in M} V_l(S_l)^{\gamma_l}},$$

where v_0 is the attraction value of the no-purchase alternative. If we offer the sets of products $(S_1, \ldots, S_m)$ over all nests, then the expected revenue per consumer is given by

$$R(S_1, \ldots, S_m) := \sum_{i \in M} R_i(S_i)\, Q_i(S_1, \ldots, S_m) = \frac{\sum_{i \in M} R_i(S_i)\, V_i(S_i)^{\gamma_i}}{v_0 + \sum_{i \in M} V_i(S_i)^{\gamma_i}}.$$

We can find a set of products to offer that maximizes the expected revenue obtained from a consumer by solving the problem

$$\mathcal{R}^* = \max_{\substack{(S_1, \ldots, S_m): \\ S_i \subseteq N \, \forall i \in M}} R(S_1, \ldots, S_m). \tag{5.3}$$

Problem (5.3) can be viewed as a binary fractional program if we identify with each subset S_i an incidence vector $x_i = (x_{i1}, \ldots, x_{in}) \in \{0, 1\}^n$ where $x_{ij} = 1$ if $j \in S_i$ and $x_{ij} = 0$ if $j \notin S_i$. That is, $x_{ij} = 1$ if we offer product j in nest i, otherwise $x_{ij} = 0$. Let $(x_1, \ldots, x_n)$ be any feasible solution to the assortment problem where x_i represents the assortment offered in nest $i \in M$. The fractional program in (5.3) is of the form

$$\mathcal{R}^* = \max_{\substack{(x_1, \ldots, x_m): \\ x_i \in \{0,1\}^n \, \forall i \in M}} \frac{\sum_{i \in M} f_i(x_i)}{v_0 + \sum_{i \in M} g_i(x_i)}, \tag{5.4}$$

where, with slight abuse of notation, we let $V_i(x_i) := \sum_{j \in N} v_{ij}\, x_{ij}$, $f_i(x_i) := R_i(x_i) V_i(x_i)^{\gamma_i}$, and $g_i(x_i) := V_i(x_i)^{\gamma_i}$ for all $i \in M$. Then, we have $\mathcal{R}^* \geq \sum_{i \in M} f_i(x_i)/(v_0 + \sum_{i \in M} g_i(x_i))$ for all $(x_1, \ldots, x_m)$ with $x_i \in \{0, 1\}^n$ for all $i \in M$ and the inequality is tight at the optimal solution $(x_1^*, \ldots, x_m^*)$ to problem (5.4). Rewriting the last inequality, we have

$$\sum_{i \in M} f_i(x_i) - \mathcal{R}^* \sum_{i \in M} g_i(x_i) \leq \mathcal{R}^* v_0, \quad \forall x_i \in \{0, 1\}^n, \; i \in M, \tag{5.5}$$

and once more, the inequality above is tight at the optimal solution $(x_1^*, \ldots, x_m^*)$ to problem (5.4). Consider the parametric program:

$$\Gamma(z) := \max_{\substack{(x_1, \ldots, x_m): \\ x_i \in \{0,1\}^n \, \forall i \in M}} \left\{ \sum_{i \in M} f_i(x_i) - z \sum_{i \in M} g_i(x_i) \right\}.$$

(Notice that z here is just a parameter, not the vector of marginal costs for the products.) It is easy to see that $\Gamma(z)$ is continuous, convex, and strictly decreasing in z. Furthermore, noting the inequality in (5.5) and observing that this inequality is tight at the solution $(x_1^*, \ldots, x_m^*)$, we have $\Gamma(\mathcal{R}^*) = v_0\,\mathcal{R}^*$. So, it follows that $\mathcal{R}^*$ corresponds to the smallest value of z that satisfies $\Gamma(z) \le v_0\, z$. Letting $x^*(z)$ be the solution to the problem above, then $x^* = x^*(\mathcal{R}^*)$ is an optimal solution to the fractional program (5.4).

Solving the fractional program (5.4) is therefore equivalent to finding the smallest value of z such that $\Gamma(z) \le v_0\, z^*$. Fortunately, the program to solve for $\Gamma(z)$ decomposes by the nests, so

$$\Gamma(z) = \sum_{i \in M} \max_{x_i \in \{0,1\}^n} \left\{ f_i(x_i) - z\, g_i(x_i) \right\}.$$

Thus, we can find the smallest value of z such that $\Gamma(z) \le v_0\, z^*$ by solving the linear program

$$\begin{aligned} \mathcal{R}^* = \min \quad & z \\ \text{s.t.} \quad & \sum_{i \in M} y_i \le v_0 z \\ & f_i(x_i) - z g_i(x_i) \le y_i \quad \forall x_i \in \{0,1\}^n, \ i \in M, \end{aligned} \tag{5.6}$$

where the decision variables are z and $y = (y_1, \ldots, y_m)$. This linear program has $1 + m$ decision variables and exponentially many constraints for each nest $i \in M$. However, this does not pose a difficulty as we can show that without loss of optimality we can limit our attention to nested-by-revenue assortments for each nest. In particular, we argue that it is enough to impose the second set of constraints only for the nested-by-revenue assortments within each nest. This result, in turn, implies that it is optimal to consider offering a nested-by-revenue assortment in each nest.

To state the result more precisely, we index the products in each nest so that the revenues associated with the products in a nest satisfy $p_{i1} \ge p_{i2} \ge \ldots \ge p_{in}$ for all $i \in M$. For a fixed value of $z \in \Re_+$, we will show that an optimal solution $x_i^* = (x_{i1}^*, \ldots, x_{in}^*)$ to the problem

$$\max_{x_i \in \{0,1\}^n} \left\{ f_i(x_i) - z\, g_i(x_i) \right\} = \max_{x_i \in \{0,1\}^n} \left\{ V_i(x_i)^{\gamma_i} R_i(x_i) - z\, V_i(x_i)^{\gamma_i} \right\} \tag{5.7}$$

is given by

$$x_{ij}^* = \begin{cases} 1 & \text{if } j \le k^* \\ 0 & \text{if } j > k^* \end{cases} \tag{5.8}$$

for some $k^* = 1, \ldots, n$. Thus, noting that the products are indexed such that $p_{i1} \geq p_{i2} \geq \ldots \geq p_{in}$, a nested-by-revenue assortment solves problem (5.7) for any $z \in \Re_+$. This result implies that it is enough to impose the second set of constraints in problem (5.6) only for the nested-by-revenue assortments. The following result is useful in proving the nested-by-revenue structure.

Theorem 5.4 *Fix $z \in \Re_+$ and let x_i^* be an optimal solution to problem* (5.7). *Define the scalar $u_i^* := z + (1 - \gamma_i)\,[R_i(x_i^*) - z]^+$, and let $\hat{x}_i$ be an optimal solution to the problem*

$$\max_{x_i \in \{0,1\}^n} \Big\{ V_i(x_i)\, R_i(x_i) - u_i^*\, V_i(x_i) \Big\}. \tag{5.9}$$

Then, $\hat{x}_i$ is also an optimal solution to problem (5.7).

Theorem 5.4 shows that we can obtain an optimal solution to problem (5.7) alternatively by solving problem (5.9). The advantage of working with problem (5.9) is that we can use the definition of $V_i(x_i)$ and $R_i(x_i)$ to write this problem equivalently as

$$\max_{x_i \in \{0,1\}^n} \Big\{ \sum_{j \in N} v_{ij}\,(p_{ij} - u_i^*)\, x_{ij} \Big\}.$$

An optimal solution x_i^* to the problem above is given by setting $x_{ij}^* = 1$ if $p_{ij} \geq u_i^*$, and $x_{ij}^* = 0$ if $p_{ij} < u_i^*$. Thus, letting k^* be such that $p_{i1} \geq p_{i2} \geq \ldots \geq p_{ik^*} \geq u_i^* > p_{i,k^*+1} \geq \ldots \geq p_{in}$, an optimal solution to problem (5.9) is of the form (5.8), as desired.

The preceding discussion shows that we can impose the second set of constraints in problem (5.6) only for the nested-by-revenue assortments. Thus, letting $e_j \in \Re^n_+$ be the j-th unit vector, we use $\sum_{j=1}^k e_j$ to denote the nested-by-revenue assortment that includes the first k products. Let $\mathcal{A}_i := \{x \in \{0,1\}^n : x = \sum_{j=1}^k e_j,\ k = 0, \ldots, n\}$, denote the class of nested-by-revenue assortments for nest $i \in M$. We interpret the sum over an empty index set as zero. In this case, problem (5.6) is equivalent to the linear program

$$\begin{aligned} \mathcal{R}^* = \min_{z,y}\ & z \\ \text{s.t.}\ & \sum_{i \in M} y_i \leq v_0\, z \\ & f_i(x_i) - z\, g_i(x_i) \leq y_i \quad \forall\, x_i \in \mathcal{A}_i,\ i \in M. \end{aligned} \tag{5.10}$$

Since there are only $1 + n$ nested-by-revenue assortments in each nest, the problem above is a linear program with $1+m$ decision variables and $1+m(1+n)$ constraints. Thus, the problem above finds the smallest value of such that $\Gamma(z) \leq v_0\, z$ and this value of z corresponds to the optimal objective value $\mathcal{R}^*$ of problem (5.3). Furthermore, if we solve the problem $\max_{x_i \in \mathcal{A}_i}\{f_i(x_i) - \mathcal{R}^*\, g_i(x_i)\}$, then we obtain the nested-by-revenue assortment that is optimal to offer in nest i.

Example 5.5 Consider a problem instance with two nests with dissimilarity parameters $\gamma_1 = 0.2$ and $\gamma_2 = 0.9$. There are four possible products that can be offered in each nest. The revenues associated with the products in the two nests are $(p_{11}, \ldots, p_{14}) = (9, 8, 7, 6)$ and $(p_{21}, \ldots, p_{24}) = (6, 5.5, 4, 3)$. The attraction values for all of the products are equal to 1. For this problem, solving the linear program in (5.10) yields $\mathcal{R}^* = 5.104$, so $u_1^* = 7.821$ and $u_2^* = 5.104$. Consequently, we can recover the optimal assortment in each nest as $S_1^* = \{j : p_{1j} \geq 7.821\} = \{1, 2\}$ and $S_2^* = \{j : p_{2j} \geq 5.104\} = \{1, 2\}$. The optimal assortments are nested-by-revenue in each nest. However, if we consider the optimal assortment $S_1^* \cup S_2^*$ over all nests, then this assortment is not nested-by-revenue. In particular, the products that we offer in the first nest have revenues 9 and 8, whereas the products that we offer in the second nest have revenues 6 and 5.5. This solution does not offer the product with revenue 7 in the first nest, but it offers the product that has revenue 6 in the second nest.

5.9 Random Consideration Set Model

The RCS model arises as an extension of the MUM applied to a full order $1 \prec 2 \prec \ldots \prec n$, where product $j \in N$ has attention probability $\lambda_j \in (0, 1]$. Under the RCS model,

$$\pi_j(S) = \lambda_j \prod_{k \in S: j \prec k} (1 - \lambda_k)$$

for all $j \in S$ with $\pi_0(S) = \prod_{k \in S}(1 - \lambda_k)$, and $\pi_j(S) = 0$ for all $j \notin S$. The following result provides the key for assortment optimization.

Lemma 5.6 *Let $S \subseteq N$, and let $k \in N$ be such that $i \prec k$ for all $i \in S$. Then,*

$$R(S \cup \{k\}) = (1 - \lambda_k) R(S) + \lambda_k p_k, \tag{5.11}$$

and consequently $R(S \cup \{k\}) > R(S)$ if and only if $p_k > R(S)$.

Proof For $i \in S, i \prec k$, $\pi_i(S \cup \{k\}) = \pi_i(S)(1 - \lambda_k)$, while $\pi_k(S \cup \{k\}) = \lambda_k$, from which (5.11) follows directly. Moreover $R(S \cup \{k\}) = R(S) + \lambda_k(p_k - R(S)) > R(S)$ if and only if $p_k > R(S)$. □

We will use $E_0 = \emptyset$ and $E_j = \{1, \ldots, j\}$ $j \in N$ to denote consecutive sets in the full ranking, but remark that the ordering here is *not* the revenue ordering discussed earlier. We are now ready to present an algorithm to solve the assortment optimization problem under the RCS model.

Algorithm Set $H(E_0) := 0$ and compute $H(E_j)$, $j \in N$ recursively as

$$H(E_j) := H(E_{j-1}) + \lambda_j (p_j - H(E_{j-1}))^+. \tag{5.12}$$

Also, let

$$\tilde{E}_j := \{i \in E_j : p_i > H(E_{i-1})\}, \quad j \in N.$$

The sequence $H(E_j)$, $j \in N$ considers whether or not to add product j to $\tilde{E}_{j-1}$ to form $\tilde{E}_j$ and does this if and only if $p_j > H(E_{j-1})$. For each $j \in N$, the algorithm requires one comparison, and three arithmetic operations, so it runs in $O(n)$ time. Next, we show that this algorithm yields an optimal assortment.

Theorem 5.7 *The sequence $H(E_j)$ is increasing in $j \in N$, with*

$$H(E_j) = R(\tilde{E}_j) \geq R(S) \;\; \forall\, S \subseteq E_j, \;\; j \in N,$$

so $S^* = \tilde{E}_n$ is an optimal assortment.

5.10 Markov Chain Choice Model

The MC choice model is characterized by the parameters $\{\lambda_i : i \in N\}$ and $\{\rho_{ij} : i, j \in N\}$, where λ_i is the probability that a customer arriving into the system has product i as her first choice, and ρ_{ij} is the probability that a customer transitions from product i to product j when she finds product i unavailable. In the previous chapter, we showed that a system of linear equations needs to be solved to compute the selection probabilities under the MC choice model. This suggests that we may need to solve up to exponentially many linear systems to solve the assortment problem.

Fortunately, there is an efficient way to solve the assortment optimization problem under the MC choice model. Let g_i be the optimal expected revenue that can be obtained from a consumer that is currently considering product i during the course of her choice process. If $i \in S$, then we obtain revenue $g_i = p_i$. Otherwise, the customer transitions to some product j with probability ρ_{ij}. Thus, if $i \notin S$, the expected revenue is equal to $\sum_{j \in N} \rho_{ij} g_j$. Therefore, $g = \{g_i : i \in N\}$ satisfy

$$g_i = \max\left\{p_i, \sum_{i \in N} \rho_{ij} g_j\right\} \quad \forall i \in N. \tag{5.13}$$

One way to find the values of $\{g_i : i \in N\}$ is to use value iteration. In particular, we can generate a sequence of values $g^t = \{g_i^t : i \in N\}$ for $t = 0, 1, \ldots$, where g^{t+1} is obtained from g^t by $g_i^{t+1} := \max\{p_i, \sum_{j \in N} \rho_{ij} g_j^t\}$ for all $i \in N$. The initial values g^0 can be chosen arbitrarily. It can be shown that $\lim_{t\to\infty} g^t = g$, where $g = \{g_i : i \in N\}$ satisfy (5.13). In addition to the value iteration approach, another approach for obtaining a solution to (5.13) is to solve the linear program

$$\begin{aligned} \mathcal{R}^* = \min \quad & \sum_{i\in N} \lambda_i\, g_i && (5.14) \\ \text{s.t.} \quad & g_i \geq p_i && \forall i \in N \\ & g_i \geq \sum_{i\in N} \rho_{ij}\, g_j && \forall i \in N, \end{aligned}$$

where the decision variables are $g = \{g_i : i \in N\}$. Letting g^* be an optimal solution to the linear program above, we can see that g^* corresponds to the value of g that satisfies (5.13). In particular, for each $i \in N$, either the first or the second constraint should be tight at the optimal solution because if one of these constraints is loose, then we can decrease the value of the decision variable g_i^* by a small amount without changing the feasibility of the solution g^*. In this way, we obtain a feasible solution to problem (5.14) that provides an objective value that is strictly better than the objective value provided by the optimal solution, which is a contradiction. Since for each $i \in N$, either the first or the second constraint should be tight at the optimal solution, we must have $g_i^* = \max\{p_i, \sum_{j\in N} \rho_{ij}\, g_j^*\}$, as desired. It is also worthwhile to note that the optimal objective value of problem (5.14) corresponds to the optimal expected revenue obtained from a consumer. Indeed, the optimal value of the decision variable g_i is the optimal expected revenue obtained from a consumer that is visiting product i, whereas λ_i is the probability that product i is the first choice. Thus, at the optimal solution, $\sum_{i\in N} \lambda_i\, g_i^*$ is the optimal expected revenue per consumer.

Example 5.8 Consider the data of Example 5.3. For this example, the revenues associated with the products are $(p_1, p_2, p_3) = (8, 4, 3)$. In the previous chapter, we discuss how we can estimate the first choice and transition probabilities λ and ρ for the MC choice model so as to approximate any choice model. Using that approach, we approximate the mixture of BAM's model in Example 5.3 by using the MC choice model with parameters

$$\lambda = (10.1\%, 62.0\%, 23.6\%)$$

and

$$[\rho_{ij}]_{i,j\in N} = \begin{pmatrix} 0.0\% & 93.0\% & 2.2\% \\ 52.3\% & 0.0\% & 33.9\% \\ 1.8\% & 89.3\% & 0.0\% \end{pmatrix}.$$

For this problem instance, the optimal solution to the linear program in (5.14) is $(g_1^*, \ldots, g_3^*) = (\$8.00, \$6.08, \$5.57)$, so the optimal offer set is $\{1\}$. This assortment provides an expected revenue of \$4.00 under the actual mixture of BAM's we are approximating. This captures about 90% of the optimal revenue attained by the offer set $\{1, 3\}$. Notice that computing the optimal assortment under the mixture of BAM's is hard, but we can efficiently compute an optimal assortment under the MC choice model.

Although the linear program (5.14) provides a fairly intuitive way of thinking about the assortment optimization problem, its dual program below may be closer in spirit to other models that seek to maximize expected revenues. Associating the dual variables $\{x_i : i \in N\}$ and $\{y_i : i \in N\}$ with the two sets of constraints, the dual of problem (5.14) is

$$\begin{aligned} \mathcal{R}^* = \max \quad & \sum_{i \in N} p_i\, x_i && (5.15) \\ & x_i + y_i - \sum_{j \in N} \rho_{ji}\, y_j = \lambda_i && \forall i \in N \\ & x_i \geq 0, \; y_i \geq 0 && \forall i \in N. \end{aligned}$$

This formulation can be used to obtain linear programming formulations for the assortment optimization problem for any discrete choice model with an exact MC representation. Discrete choice models with exact MC representations include the BAM, GAM, and RCS model.

5.11 Constrained Assortment Optimization

We study assortment problems with constraints on the products that can be part of the assortment. Cardinality constraints limit the number of products that can be offered. Business rules may impose precedence relationships on the products such that a product can only be offered if some other products are offered. In some instances, we may want to select both the assortment and the store or web-page location of each product in the assortment. Constrained assortment optimization problems are generally more difficult than their unconstrained counterparts. In this section, we focus on constrained assortment problems under the BAM and the NL model as these are the models that are known to be polynomially solvable.

5.11.1 Basic Attraction Model

Similar to our approach for the NL model, we represent assortments by their incidence vectors. In particular, we use the vector $x = \{x_j : j \in N\} \in \{0, 1\}^n$ to capture the assortment $S = \{j \in N : x_j = 1\}$, and denote by $\pi_j(x)$ the probability that a consumer selects product j when the assortment with incidence vector x is offered. Under the BAM, we have $\pi_j(x) = v_j\, x_j/(v_0+\sum_{i\in N} v_i\, x_i)$. We generically capture the constraints on the offered products by $\sum_{j\in N} a_{ij}\, x_j \le b_i$ for all $i \in L$ for some $|L| \times |M|$ dimensional constraint matrix $\{a_{ij} : i \in L,\ j \in N\}$ and $|L|$ dimensional right side vector $\{b_i : i \in L\}$. Throughout this section, we assume that the constraint matrix $\{a_{ij} : i \in L,\ j \in N\}$ is totally unimodular (TU). A TU matrix is a matrix for which every square non-singular submatrix is integer and has determinant $+1$ or -1. TU matrices are important in combinatorial optimization since they give a quick way to verify that a linear program has integral solutions. Later in this section, we give some practical cases where the constraints imposed on the set of offered products yield a TU constraint matrix. The problem we want to solve is of the form

$$\begin{aligned} \mathcal{R}^* = \max\quad & \frac{\sum_{j\in N} p_j\, v_j\, x_j}{v_0 + \sum_{j\in N} v_j\, x_j} \\ \text{s.t.}\quad & \sum_{j\in N} a_{ij}\, x_j \le b_i \qquad \forall\, i \in L \\ & x_j \in \{0, 1\} \qquad \forall\, j \in N, \end{aligned} \tag{5.16}$$

which maximizes the expected revenue obtained from a consumer under the BAM subject to some constraints on the offered set of products. The objective function of the problem above is a constrained binary problem with a fractional objective function. However, if the constraint matrix in the problem above is TU, then it is possible to show that this problem is equivalent to the linear program

$$\begin{aligned} \mathcal{R}^* = \max\quad & \sum_{j\in N} p_j\, y_j \\ \text{s.t.}\quad & \sum_{j\in N} y_j + y_0 = 1 \\ & \sum_{j\in N} \frac{a_{ij}}{v_j}\, y_j \le \frac{b_i}{v_0}\, y_0 \qquad \forall\, i \in L \\ & 0 \le \frac{y_j}{v_j} \le \frac{y_0}{v_0} \qquad \forall\, j \in N \end{aligned} \tag{5.17}$$

In the problem above, we interpret the decision variable y_j as the probability that a consumer purchases product j, whereas we interpret the decision variable y_0 as the probability that a consumer selects the no-purchase option. The objective function accounts for the expected revenue from a consumer. The first constraint ensures that a consumer either purchases a product or selects the no-purchase alternative. The second set of constraints ensure that the purchase probabilities are consistent with a feasible assortment, whereas the third set of constraints ensure that the purchase probabilities are consistent with the BAM. In the next theorem below, we will show that problems (5.16) and (5.17) are equivalent to each other.

Consider, for the moment, the formulation in (5.17) in the absence of the second set of constraints. The resulting linear program solves the unconstrained assortment problem. From the objective function it is clear that we want the y_j's to be large, and for this, we need y_0 to be small according to the first constraint, and large according to the third set of constraints. The tension is solved at $y_0 = v_0/(v_o + V(S^*))$ and $y_j = v_j y_0/v_0$ for $j \in S^*$ and $y_j = 0$ for $j \notin S^*$. The second set of constraints represent TU constraints on the assortments and may tilt the optimal y_0. The following theorem formally shows that problems (5.16) and (5.17) are equivalent.

Theorem 5.9 *Problems (5.16) and (5.17) have the same optimal objective value. Furthermore, if y^* is an optimal solution to problem (5.17), then setting $x_j^* = 1$ if and only if $y_j^* > 0$ gives an optimal solution to problem (5.16).*

Theorem 5.9 indicates that we can solve the assortment optimization problem under the BAM subject to a TU constraint structure by solving a linear program. There are several uses of this theorem, as discussed next.

Applications

Consider the case where we can offer at most C products in an assortment. In this case, the constraint on the assortment can be written as $\sum_{j \in N} x_j \leq C$ and the constraint matrix is a single row of consecutive ones. This constraint matrix is TU. Thus, we can use a linear program to solve the assortment problem under the BAM subject to a cardinality constraint. Similarly, the products may have been grouped into several categories, say $S_1, S_2, \ldots, S_L$ with $S_l \cap S_{l'} = \emptyset$ and we may be allowed to offer C_l products in category l. In this case, the constraints on the offered assortment can be written as $\sum_{j \in S_l} x_j \leq C_l$ for all $l = 1, \ldots, L$. This constraint matrix has L rows and the products can be ordered such that each row of the matrix includes consecutive ones. Such a matrix is known as an interval matrix and it is TU.

Consider now the joint pricing and assortment problem. The set of products is $P := \{1, \ldots, p\}$. We need to decide which set of products to offer and choose the prices for the offered products. The price of a product affects the revenue that we obtain from this product and the attraction value of the product. We assume that there are b distinct possible prices for each product and we let $B := \{1, \ldots, b\}$ be the index set of possible price levels. Let v_{kl} be the attraction value of product k if we charge the price level l for this product and ρ_{kl} be the revenue that we obtain from product k if we charge the price level l. To formulate this joint pricing and assortment problem as a pure assortment problem, we create b copies of each product, corresponding to different price levels that can be charged for the product. We refer to the multiple copies of a product as a virtual product. Thus, we have a total of $n = p \times b$ virtual products. We denote the set of all virtual products by $N := \{1, \ldots, n\}$ and use N_k to denote the set of virtual products that correspond to product k.

Thus, the virtual products in N_k correspond to charging different price levels for product k. We let v_j and p_j, respectively, be the attraction value and the revenue associated with virtual product j. In particular, if virtual product j corresponds to offering product k and price level l, then $v_j = \nu_{kl}$ and $p_j = \rho_{kl}$. In this case, the pricing and assortment problem can be cast as a pure assortment problem in terms of virtual products. We want to find a set of virtual products to offer, while ensuring that we offer at most one virtual product corresponding to a product. The latter condition ensures that we charge at most one price level for each product. Therefore, the constraints on the offered assortment can be written as $\sum_{j \in N_k} x_k \leq 1$ for all $k \in P$. Since $N_k \cap N_{k'} = \emptyset$, the products can be ordered such that each row of this constraint matrix includes consecutive ones to obtain an interval matrix. This implies that constraint matrix is TU, so under the BAM, the joint pricing and assortment problem with discrete price levels can be formulated as a linear program.

5.11.2 Nested Logit Model

We consider constrained assortment problems under the NL model. Solving constrained assortment problems under the NL model is somewhat more difficult than solving such problems under the BAM. Here, we consider the case where we have a cardinality constraint for the assortment offered in each nest. In particular, using $S_i \subseteq N$ to denote the set of products offered in nest i, the set of feasible assortments that can be offered in nest i is given by $\mathcal{F}_i := \{S_i \subseteq N : |S_i| \leq c_i\}$, which limits the cardinality of the assortment offered in nest i to c_i. Thus, the problem we want to solve is

$$\mathcal{R}^* = \max_{\substack{(S_1,\ldots,S_m):\\ S_i \in \mathcal{F}_i \, \forall i \in M}} \left\{ \frac{\sum_{i \in M} R_i(S_i)\, V_i(S_i)^{\gamma_i}}{v_0 + \sum_{i \in M} V_i(S_i)^{\gamma_i}} \right\}. \tag{5.18}$$

The objective function is the same as the expected revenue function under the NL model used in problem (5.3), but we impose the constraint that $S_i \in \mathcal{F}_i$ for the set offered in each nest i.

Our approach in this section closely parallels the one that we used for solving the unconstrained assortment problem under the NL model using fractional programming. In particular, we use the vector $x_i = (x_{i1}, \ldots, x_{in}) \in \{0, 1\}^n$ to describe the assortment that we offer in nest i, where $x_{ij} = 1$ if we offer product j in nest i, otherwise $x_{ij} = 0$. The set of feasible assortments that can be offered in nest i is given by $\mathcal{F}_i := \{x_i \in \{0, 1\}^n : \sum_{j \in N} x_{ij} \leq c_i\}$. As before, we let $V_i(x_i) := \sum_{j \in N} v_{ij}\, x_{ij}$, $R_i(x_i) := \sum_{j \in N} p_{ij}\, v_{ij}\, x_{ij} / V_i(x_i)$, $f_i(x_i) := R_i(x_i) V_i(x_i)^{\gamma_i}$ and $g_i(x_i) := V_i(x_i)^{\gamma_i}$. Then, we have $\mathcal{R}^* \geq \sum_{i \in M} f_i(x_i)/(v_0 + \sum_{i \in M} g_i(x_i))$ for all $(x_1, \ldots, x_m)$ with $x_i \in \mathcal{F}_i$ for all $i \in M$ and the equality is tight at the optimal solution $(x_1^*, \ldots, x_m^*)$ to problem (5.18). Thus, we have

$$\sum_{i\in M} f_i(x_i) - \mathcal{R}^* \sum_{i\in M} g_i(x_i) \le \mathcal{R}^* v_0, \quad \forall\, x_i \in \mathcal{F}_i,\ i \in M,$$

and once more, the inequality above is tight at the optimal solution $(x_1^*, \ldots, x_m^*)$ to problem (5.18). We consider the parametric program

$$\Gamma(z) := \max_{\substack{(x_1,\ldots,x_m):\\ x_i \in \mathcal{F}_i\, \forall\, i \in M}} \left\{ \sum_{i\in M} f_i(x_i) - z \sum_{i\in M} g_i(x_i) \right\}.$$

Similar to our earlier development, we observe that $\Gamma(z)$ is continuous, convex, and strictly decreasing in z and we have $\Gamma(\mathcal{R}^*) = v_0\, \mathcal{R}^*$. Thus, $\mathcal{R}^*$ corresponds to the smallest value of z that satisfies $\Gamma(z) \le v_0\, z$. To find the smallest value of z such that $\Gamma(z) \le v_0\, z$, we can solve the linear program

$$\begin{aligned} \mathcal{R}^* = \min\ & z \\ \text{s.t.}\ & \sum_{i\in M} y_i \le v_0 z \\ & f_i(x_i) - z g_i(x_i) \le y_i \quad \forall\, x_i \in \mathcal{F}_i,\ i \in M \end{aligned} \tag{5.19}$$

with the decision variables z and $y = (y_1, \ldots, y_m)$. The problem above has a large number of constraints. In the rest of this section, we argue that we can impose the second set of constraints only for a small collection of assortments in each nest and this result makes the linear program in (5.19) tractable. To that end, we consider the problem

$$\max_{x_i\in\mathcal{F}_i} \left\{ f_i(x_i) - z\, g_i(x_i) \right\} = \max_{x_i\in\mathcal{F}_i} \left\{ V_i(x_i)^{\gamma_i} R_i(x_i) - z\, V_i(x_i)^{\gamma_i} \right\} \tag{5.20}$$

If we can show that there exists a small collection of assortments that includes an optimal solution to problem (5.20) for any value of $z \in \Re_+$, then we can impose the second set of constraints in problem (5.19) for only the assortments in this collection. In the next theorem, we give a simple characterization of an optimal solution to problem (5.20) and this characterization becomes useful to limit the number of constraints in the linear program in (5.19). This theorem is a direct extension of Theorem 5.4 to the constrained case. Its proof is identical to that of Theorem 5.4 and we skip the proof.

Theorem 5.10 *Fix $z \in \Re_+$ and let x_i^* be an optimal solution to problem* (5.20). *Define the scalar $u_i^* := z + (1-\gamma_i)\,[R_i(x_i^*) - z]^+$, and let $\hat{x}_i$ be an optimal solution to the problem*

$$\max_{x_i\in\mathcal{F}_i} \left\{ V_i(x_i)\, R_i(x_i) - u_i^*\, V_i(x_i) \right\}. \tag{5.21}$$

Then, $\hat{x}_i$ is also an optimal solution to problem (5.20).

Theorem 5.10 implies that we can recover an optimal solution to problem (5.20) for a fixed value of $z \in \Re_+$ by solving problem (5.21). We are interested in finding an optimal solution to (5.20) for any value of $z \in \Re_+$. A particular value of z in problem (5.20) translates into an optimal solution x_i^*, which translates into the scalar $u_i^* = z + (1-\gamma_i)\,[R_i(x_i^*) - z]^+$. This observation implies that by solving problem (5.21) with this value of u_i^*, we can obtain an optimal solution to problem (5.20). Thus, to obtain the optimal solutions to problem (5.20) for any value of $z \in \Re_+$, it is enough to obtain the optimal solutions to the problem

$$\max_{x_i \in \mathcal{F}_i} \left\{ V_i(S_i)\, R_i(S_i) - u_i\, V_i(x_i) \right\} \tag{5.22}$$

for all values of $u_i \in \Re_+$. As a function of u_i, we use $x_i^*(u_i)$ to denote an optimal solution to problem (5.22). Next, we argue that the collection of assortments $\{x_i^*(u_i) : u_i \in \Re_+\}$ includes only $O(n^2)$ assortments in it and these assortments can be generated in a tractable fashion. To see this result, using the definitions of $V_i(x_i)$ and $R_i(x_i)$, we write problem (5.22) as

$$\max_{x_i \in \{0,1\}^n} \left\{ \sum_{j \in N} (p_{ij} - u_i)\, v_{ij}\, x_{ij} : \sum_{j \in N} x_{ij} \le c_i \right\}. \tag{5.23}$$

The problem above is a knapsack problem. The capacity of the knapsack is c_i. Each item occupies one unit of space in the knapsack and the utility from item j is $(p_{ij} - u_i)\, v_{ij}$. This knapsack problem can be solved by ordering the items according to their utilities and filling the knapsack starting from the item with the largest utility, as long as the utility of the item is positive. Thus, the solution to this knapsack problem depends only on the ordering of the utilities of the items, but not the exact values of the utilities.

We let $f_{ij}(u_i) := (p_{ij} - u_i)\, v_{ij}$ for all $j \in N$ to capture the utility of item j and let $f_{i0}(u_i) := 0$. For all $j \in N \cup \{0\}$, we note that $f_{ij}(u_i)$ is a linear function of u_i. The $n+1$ lines in $\{f_{ij}(\cdot) : j \in N \cup \{0\}\}$ intersect at $O(n^2)$ points. These points of intersection can be found by solving the equation $f_{ij}(u_i) = f_{ik}(u_i)$ for u_i for all distinct item pairs j and k; see Fig. 5.1. We use $\{\bar{u}_i^t : t \in \mathcal{G}_i\}$ to denote these intersection points. The important observation is that the ordering between the lines $\{f_{ij}(\cdot) : j \in N \cup \{0\}\}$ does not change over the intervals in between the intersection points and these intersection points correspond to the only places where the ordering between the lines $\{f_{ij}(\cdot) : j \in N \cup \{0\}\}$ can possibly change. For example, Fig. 5.1 shows the lines $\{f_{ij}(\cdot) : j \in N \cup \{0\}\}$ for a possible case with $N = \{1, 2, 3\}$. In this figure, we have $f_{i2}(u_i) \ge f_{i3}(u_i) \ge f_{i0}(u_i) \ge f_{i1}(u_i)$ for all $u_i \in [\bar{u}_i^3, \bar{u}_i^4]$. Thus, as u_i takes values over $\Re_+$, there are at most $O(n^2)$ different orderings between the lines $\{f_{ij}(\cdot) : j \in N \cup \{0\}\}$ and these orderings can be found by finding the intersection points between the $n+1$ lines. Since each ordering of the lines $\{f_{ij}(\cdot) : j \in N \cup \{0\}\}$ corresponds to an optimal solution to problem (5.23) for a particular value of u_i, there are at most $O(n^2)$ possible optimal solutions to problem (5.23) as u_i takes values over $\Re_+$.

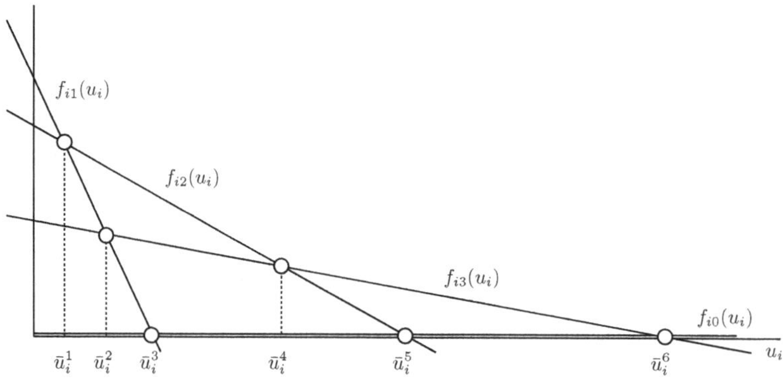

Fig. 5.1 The lines $\{f_{ij}(\cdot) : j \in N \cup \{0\}\}$ for a possible case with $N = \{1, 2, 3\}$

The discussion in the paragraph above shows that we can construct a collection of candidate solutions $\mathcal{A}_i := \{\hat{x}_i^t : t \in \mathcal{T}_i\}$ with $|\mathcal{T}_i| = O(n^2)$ such that $\mathcal{A}_i$ always includes an optimal solution to problem (5.23) for any value of u_i. Thus, by Theorem 5.10 and the discussion that follows this theorem, it is enough to impose the second set of constraints in problem (5.19) only for the solutions in the collection $\mathcal{A}_i$. So, we can solve the linear program

$$\begin{aligned} \mathcal{R}^* = \min_{z,y} \quad & z \\ \text{s.t.} \quad & \sum_{i \in M} y_i \le v_0\, z \\ & f_i(x_i) - z\, g_i(x_i) \le y_i \quad \forall\, x_i \in \mathcal{A}_i, \ \ i \in M \end{aligned}$$

to obtain the optimal objective value of problem (5.18). Since there are $O(n^2)$ assortments in each one of the collections $\mathcal{A}_1, \ldots, \mathcal{A}_m$, we can obtain the optimal assortment with a cardinality constraint in each nest by solving a linear program with $1 + m$ decision variables and $O(mn^2)$ constraints.

5.12 Convexity and Efficient Sets

Let $A_j \in \Re_+^m$ be the vector of resources utilized by one unit of product j and let A be an $m \times n$ matrix whose j-th column in A_j. For any $S \subseteq N$, let $A(S) := \sum_{j \in S} A_j\, \pi_j(S)$ be the expected resource utilization when assortment $S \subseteq N$ is offered. For any $z \in \Re_+^m$, let

$$R(S, A'z) := \sum_{j \in S} (p_j - A_j' z)\, \pi_j(S) = R(S) - A(S)'z,$$

and

$$\mathcal{R}(A'z) := \max_{S \subseteq N} R(S, A'z). \tag{5.24}$$

In revenue management, problem (5.24) needs to be solved frequently for different values of $z \in \Re^m_+$ which are often dual variables of capacity constraints. The following two results are useful for this purpose.

Theorem 5.11 *The function $\mathcal{R}(A'z)$ is convex in z. Moreover, if assortment S is optimal for each of the vectors $\{z^k : k = 1, \ldots, K\}$, then S is optimal for any z in the convex hull of $\{z^k : k = 1, \ldots, K\}$.*

The question then arises as which assortments may be optimal for different values of z. To answer this question we introduce the notion of efficient sets. For any $\rho \in \Re^m_+$, let

$$\begin{aligned} Q(\rho) := \max \quad & \sum_{S \subseteq N} R(S)\, t(S) \\ \text{s.t.} \quad & \sum_{S \subseteq N} A(S)\, t(S) \leq \rho \\ & \sum_{S \subseteq N} t(S) = 1 \\ & t(S) \geq 0 \qquad \forall S \subseteq N. \end{aligned} \tag{5.25}$$

Notice that the linear program (8.9) selects a convex combination of all possible subsets of N to maximize the expected revenue that can be obtained subject to the bound ρ on resource utilization. A set $E \subseteq N$ is said to be A-efficient if and only if $t(E) = 1$ is a solution to the linear program (8.9) for $\rho = \Pi(E)$. Equivalently, a set E is A-efficient if and only if $R(E) = Q(A(E))$. Let $\mathcal{E} := \{E \subseteq N : R(E) = Q(\Pi(E))\}$ be the collection of efficient sets. Notice that $\emptyset \in \mathcal{E}$ as $t(\emptyset) = 1$ is an optimal solution to the linear program above with $\rho = 0 \in \Re^m$. Computing the collection $\mathcal{E}$ may be a daunting task, but in situations where it is possible to pre-compute $\mathcal{E}$ and the cardinality of $\mathcal{E}$ is not large, the knowledge of $\mathcal{E}$ can greatly simplify problem (5.24) as explained in our next result.

Theorem 5.12 *For any $z \in \Re^m_+$,*

$$\mathcal{R}(A'z) = \max_{E \in \mathcal{E}} R(E, A'z). \tag{5.26}$$

This result implies that the optimization does not need to be over all the 2^n subsets, but merely over the efficient sets in $\mathcal{E}$.

Given $E \in \mathcal{E}$, let

$$O(E) := \{z \in \Re^m_+ : \mathcal{R}(A'z) = R(E, A'z)\}$$

be the set of vectors z for which E is optimal. Clearly $O(E)$ is a convex set for each E. Moreover, since for each $z \in \Re^m$, there is an efficient set $E \in \mathcal{E}$ that solves problem (5.26), it follows that

$$\cup_{E \in \mathcal{E}} O(E) = \Re^m_+.$$

The reader may wonder whether the collection $O(E)$ is mutually exclusive, so that a unique E is optimal for every $z \in \Re^m_+$. This is not quite the case, but with a slight tightening of the definition of efficient sets, it is possible to show that, for any two distinct efficient sets E and E', the interiors of $O(E)$ and $O(E')$ are disjoint, so E and E' can be both optimal for z only if z is in the common boundary of $O(E)$ and $O(E')$.

The value of this is that the optimization problem (5.26) is reduced to identifying a set $O(E)$ that contains z and then offering the corresponding efficient set knowing that this is an optimal assortment for this problem. One implication of these results is that for the single resource, multi-product problem the sets in $\mathcal{E}$ are non-overlapping intervals that cover $\Re_+$. To explicitly construct these intervals, assume that the efficient sets are $E_0, E_1, \ldots, E_k$ for some k and that the efficient sets are sorted so that $\Pi_j := \sum_{i \in E_j} \pi_i(E_j)$ is increasing in $j \in K = \{0, 1 \ldots, k\}$ with $E_0 = \emptyset$ and $\Pi_0 = 0$. Let $R_j := R(E_j)$ for all $j \in K$. Define $u_j := (R_j - R_{j-1})/(\Pi_j - \Pi_{j-1})$ for $j = 1, \ldots, k$. The concavity of $Q(\rho)$ implies that $u_0 > u_1 > \ldots > u_k > u_{k+1} = 0$, where for convenience we set $u_0 = \infty$ and $u_{k+1} = 0$. Then E_j is optimal for all $z \in O(E_j) = [u_{j+1}, u_j]$ for $j = 0, 1, \ldots, k$. At boundary points u_j, both E_j and E_{j-1} are optimal.

As can be seen, the value of identifying efficient sets can be helpful in speeding up the solution of assortment optimization problems where p is the vector of original prices and the costs of the products are in the cone $\{A'z : z \in \Re^m_+\}$. However, the notion of efficient sets is less useful when m is large. To see this consider the case $m = n$, and $A = I$, then $p - I'z = p - z, z \geq 0$ spans the entire space of profit contribution vectors that are less than or equal to p. If $p_i > 0$ for all $i \in N$, then for any $S \subseteq N$, we can construct a vector $z \in \Re^n$ such that $\mathcal{R}(z) = R(S, z)$. This suggest that the notion of efficient sets is not useful when the dimension of m is large because the class $\mathcal{E}$ can be as large as the class of all subsets of N.

For this reason it is practical to turn the question around and instead of asking what are the efficient sets for a given p and z, we fix S and ask what are the vector of profit contributions for which S is efficient. To this end, let

$$M(S) := \{p \in \Re^n_+ : R(S) \geq R(T) \ \forall\, T \subseteq N\}.$$

In words, $M(S)$ is the set of profit contribution vectors p for which S is an optimal assortment.[1] The following theorem asserts that $M(S)$ is a nonempty convex cone.

Theorem 5.13 *$M(S)$ is a nonempty convex cone in $\Re^n_+$.*

We will now give a full characterization of $M(S)$. Let $\pi(S) = (\pi_1(S), \ldots, \pi_n(S))$, so that $R(S) = \pi(S)'p$. Therefore, $R(S) \geq R(T)$ if and only if $(\pi(S) - \pi(T))'p \geq 0$, thus

$$M(S) = \{p \in \Re^n_+ : (\pi(S) - \pi(T))'p \geq 0 \quad \forall T \subseteq\}.$$

This tells us that $M(S)$ is the intersection of 2^n half-planes going through the origin. A cone in $\Re^n_+$ can be characterized by its extreme rays, so there exist vectors $r^1, \ldots, r^k$, so that

$$M(S) = \left\{p = \sum_{i=1}^{k} \alpha_i r^i : \alpha_i \geq 0 \ \forall i \in N\right\}.$$

This is a better characterization because there are at most $k \leq n$ extreme rays defining a cone in $\Re^n$.

We can define fatter convex cones, say $F(S)$ that contain $M(S)$ by insisting that S is only locally optimal. Indeed, for any set S, define the immediate neighbors of S, say $N(S)$ as the collection of subsets generated by either excluding a product in S or including a product not in S. Notice that there are only n assortments in $N(S)$. Then

$$F(S) = \{p \in \Re^n_+ : (\pi(S) - \pi(T))'p \geq 0 \quad \forall T \in N(S)\}.$$

Of course, $p \in F(S)$ does not guarantee that $p \in M(S)$, but in many cases S is either optimal or close to optimal. The idea can be used as a heuristic. Indeed, if S is currently optimal for p and the net fare changes to p' due to changes in capacity, we can check whether $p' \in F(S)$, and keep offering S if the answer is yes, and move to an improving $T \in N(S)$ otherwise.

5.13 End of Chapter Problems

1. Consider a flight with three fares $p_1 = 1150$, $p_2 = 950$, $p_3 = 650$, quality attributes $q_1 = 1000$, $q_2 = 850$, $q_3 = 750$, price sensitivity $\beta_p = -1$ and quality sensitivity $\beta_q = 1.25$. Suppose that the utility of fare i is $U_i = \mu_i + \epsilon_i$ where $\mu_i = \beta_p p_i + \beta_q q_i, i = 1, 2, 3$ and the ϵ_i's are independent Gumbel random variables with parameter $\phi = .01$.

[1] We use the term profit contribution vector to distinguish p from the price vector that can influence the choice probabilities.

(a) Compute the expected utilities $\mu_i, i = 1, 2, 3$.
(b) Compute the attraction values $v_i = \exp(\phi\mu_i), i = 1, 2, 3$.
(c) Let $v[0, j] = v_0 + v_1 + \ldots + v_j$. Assume there is an outside alternative with attractiveness $v_0 = 3$ compute $\pi_0(E_j) = v_0/v[0, j]$ and $\pi_k(E_j) = v_k/v[0, j], k \in E_j = \{1, \ldots, j\}$ for $j = 0, 1, 2, 3$, where $E_0 = \emptyset$.
(d) What can you say about the choice model when ϕ is very small, say $\phi = .1$? What if $\phi = 1$?
(e) Solve the assortment problem with cardinality constraint 1, 2, and 3.

2. For the BAM model, let $\Pi_j = \sum_{k \in E_j} \pi_k(E_j)$ and $R_j = \sum_{k \in E_j} p_k\pi_k(E_j)$ for $j = 0, \ldots, n$.

(a) Show algebraically that

$$\Pi_j = \Pi_{j-1}\frac{v[0, j-1]}{v[0, j]} + 1\frac{v_j}{v[0, j]}, \quad j = 1, \ldots, n$$

$$R_j = R_{j-1}\frac{v[0, j-1]}{v[0, j]} + p_j\frac{v_j}{v[0, j]} \quad j = 1, \ldots, n.$$

(b) Notice that $\Pi_j > \Pi_{j-1}$ always, but $R_j \leq R_{j-1}$ whenever $p_j \leq R_{j-1}$. Thus $p_j \leq R_{j-1}$ renders set E_j inefficient since it consumes more capacity and produces lower expected revenues. Show that if E_j is inefficient then so are sets $E_{j+1}, \ldots, E_n$.
(c) Consider Problem 1 with fares $p_1 = 1000$, $p_2 = 500$, $p_3 = 475$ and determine which of the sets E_1, E_2, E_3 are efficient.

3. Use the LP formulation (5.15) to find an exact LP formulation for the BAM. Repeat this for the GAM and for the RCS model.
4. Show that the assortment optimization problem with cardinality constraints can be solved in $O(n^2)$ time for the RCS model.
5. Assume that consumers choose under the BAM among three products. The revenues and the attraction values associated with the products are, respectively, $(p_1, p_2, p_3) = (1000, 850, 650)$ and $(v_1, v_2, v_3) = (1, 1.65, 0.22)$. The attraction value of the no-purchase alternative is $v_0 = 1$. Let e be the vector of ones. Show that the e'-efficient sets are $E_0 = \emptyset$, $E_1 = \{1\}$ and $E_2 = \{1, 2\}$.
6. Consider the GAM with the same data as the previous problem with shadow attraction values given by $w_1 = 0$, $w_2 = 0.5$ and $w_3 = 0.2$. This means that there are negative externalities associated with not offering products 2 and 3. Show that the e'-efficient sets are $E_0 = \emptyset$ and $E_i = S_i = \{1, \ldots, i\}$ for $i = 1, 2, 3$.
7. Consider the following mixture of BAM's with four products and three consumer classes. The revenues associated with the products are $(p_1, p_2, p_3) = (11.50, 11.00, 10.80, 10.75)$. Assume that an arriving consumer has 1/6 probability of being type 1, 1/3 probability of being type 2, and 1/2 probability of being type 3. The attraction values of the three types are, respectively, $(v_1^1, v_2^1, v_3^1, v_4^1) = (5, 2, 300, 1)$, $(v_1^2, v_2^2, v_3^2, v_4^2) = (6, 4, 300, 1)$, and

$(v_1^3, v_2^3, v_3^3, v_4^3) = (0, 1, 300, 7)$. For all consumer types assume that the attraction value of the no-purchase alternative is one. Show that the e'-efficient sets are $E_0 = \emptyset$, $E_1 = \{1\}$, $E_2 = \{1, 2\}$, $E_3 = \{1, 4\}$, $E_4 = \{1, 2, 4\}$, and $E_5 = \{1, 2, 3\}$.

8. A set E is said to be strongly A-efficient if $t(E) = 1$ is a solution to problem (5.25) for $\rho = A(E)$ and there is no other efficient set E' such that $t(E') = 1$ is a solution for $\rho = A(E)$. As an example of why this refinement may be needed, consider the BAM for the single resource problem where $\{1, \dots, n\}$ is e'-efficient for $A(\{1, \dots, n\})$, but $R(\{1, \dots, n-1\}) = \{1, \dots, n\}$ is e'-efficient for $A(\{1, \dots, n\})$. Then $t(\{1, \dots, n-1\}) = 1$ is also a solution for problem (5.25) for $\rho = A(\{1, \dots, n\}$ with $A(\{1, \dots, n-1\}) < \rho$. Show that if only strongly efficient sets are considered then the interiors of $O(E)$ and $O(E')$ for two different efficient sets must be disjoint, and so up to intersections at the boundary the sets $O(E)$, $E \in \mathcal{E}$ are mutually exclusive and collectively exhaustive partition of $\Re^m_+$.
9. Prove Theorem 5.13.
10. Consider a model with $m = n = 2$ with $A_1 = e_1$ and $A_2 = e_1 + e_2$, where e_1 and e_2 are the unit vectors in $\Re^2$. Find $M(\{1\})$, $M(\{2\})$ and $M(\{1, 2\})$ for an arbitrary choice model and find the characterization of each set through the extreme rays. Is $F(S) = M(S)$ in this case?
11. Show that for the BAM, the cone $F(S)$ is given by

$$F(S) = \{p \in \Re^n_+ : p_k \le R(S) \;\; \forall k \notin S,\, p_k \ge R(S-k) \;\;\; \forall k \in S\}.$$

Now fix p and assume without loss of generality that $p_1 > p_2 > \ldots > p_n$. Suppose that an optimal assortment for this p is $E_i = \{1, \dots, i\}$. Is $p \in F(E_i)$?

5.14 Bibliographic Remarks

Research on assortment optimization goes back to the work of van Ryzin and Mahajan (1999). Kok et al. (2008) provide a review of the related literature. Gallego et al. (2004) and Talluri and van Ryzin (2004a) independently show that nested-by-revenue assortments are optimal when customers choose according to the BAM. Rusmevichientong et al. (2010) studies assortment problems under the BAM, when there is a constraint on the number of offered products. Wang (2012) studies joint assortment and pricing problems under the MNL model with cardinality constraints, whereas Wang (2013) study assortment problems under the GAM with cardinality constraints. Truong (2014) studies the assortment problem under the BAM, when one maximizes a linear combination of the expected revenue and the expected utility of the customer from the purchase. Bront et al. (2009) and Rusmevichientong et al. (2014) establish that the assortment problem under a mixture of BAM's is NP-hard. Example 5.3 is taken from Rusmevichientong et al. (2014).

Davis et al. (2014) study assortment problems under the NL model. Theorem 5.4 is taken from Gallego and Topaloglu (2014), where the authors study the constrained assortment problem under the NL model when there is a limit on the number of products offered in each nest. Feldman and Topaloglu (2015b) generalize this work to the case where is a constraint on the total number of products offered over all nests. Li et al. (2015) and Wang and Shen (2017) discuss assortment problems under the NL model with a multi-level nest structure. Mendez-Diaz et al. (2014) study assortment problems under a mixture of BAM's and provide valid cuts for an integer programming formulation of the problem. Feldman and Topaloglu (2015a) and Kunnumkal (2015) provide approaches for computing upper bounds on the optimal expected revenue under a mixture of BAM's.

Gallego and Li (2016) provide an $O(n)$ algorithm to solve the assortment optimization problem for the RCS model. Blanchet et al. (2016) provide a linear programming formulation for the assortment optimization problem under the MC choice model. Feldman and Topaloglu (2017) provide some structural properties of the optimal assortment under the MC choice model. Desir et al. (2015) show that the assortment problem under the MC choice model is NP-hard when there are cardinality constraints and they provide approximation algorithms. Chung et al. (2019) focus on assortment problems under a choice model where a customer substitutes only a limited number of times and the probability of substitution in each trial depends on the past products visited. Theorem 5.9 is from Davis et al. (2013). A supporting result first appeared in Gallego et al. (2015). Rusmevichientong et al. (2009) and Desir and Goyal (2013) give approximation schemes for capacitated assortment optimization problems under the MNL and NL models.

Efficient sets were first developed in Talluri and van Ryzin (2004a) for single resource problems without using linear programming ideas. The extension to multiple resources as presented here is new and coincides with the definition in Talluri and van Ryzin (2004a) for the single resource case.

A general approach for representing choice models based on the random utility maximization principle is to use ordered preference lists. Each customer arrives into the system with a preference list of products and she purchases the highest-ranked product in her preference list that is available. Farias et al. (2013) focus on estimating such a choice model from the data. Honhon et al. (2012); Aouad et al. (2016, 2018a); Paul et al. (2018) and Bertsimas and Misic (2019) study assortment problems under the preference list-based choice model and give exact and approximation algorithms, along with integer programming formulations. See also Berbeglia and Joret (2017) for guarantees for nested-by-revenue assortments that apply to a broader set of discrete choice models that satisfy the so-called regularity assumption.

Jagabathula and Rusmevichientong (2017) study joint assortment and pricing problems under the preference list-based choice model. Desir et al. (2018) focus on assortment problems when there are a number of modal preference lists. There can be other preference lists as well, but the probability of observing such a preference list decreases exponentially fast when the list is far from a modal list. Feldman et al. (2019) give approximation algorithms when the preference lists are short. Farias et al. (2016) discuss the practical implementation of an assortment model that uses the preference list-based choice model. Pan and Honhon (2012) solve assortment under a choice model where the utility of a product is given by its quality minus a random price sensitivity times its price.

Caro and Gallien (2007) and Ulu et al. (2012) study dynamic assortment models, where the demand volume or the parameter of the choice model is learned through a Bayesian updating scheme. Rusmevichientong and Topaloglu (2012) and Bertsimas and Misic (2017) study robust assortment problems, where the attraction values are chosen by an adversary. Alptekinoglu and Grasas (2014) focus on an assortment optimization problem with product returns under the NL model. Caro et al. (2014) and Cinar and Martinez-de-Albeniz (2014) work with assortment models also under the BAM, where the attraction value of a product fades over time so the assortment needs to be refreshed periodically by adding new products. Jagabathula (2016) studies the effectiveness of local search heuristics for assortment problems. Gallego et al. (2016a) and Aouad and Segev (2018) study assortment problems where the assortment is displayed in multiple webpages and a customer views only a random number of webpages. Collado and Martinez-de-Albeniz (2017) give a variant of the BAM where the choices of the customers are influenced by the inventory levels and give an accompanying integer programming formulation to make inventory decisions. Kunnumkal and Martinez-de-Albeniz (2019) focus on the case where there is a cost associated with offering each product. Wang and Wang (2017) focus on assortment problems under the BAM, when the attraction value of a product depends on its relative market size with respect to the other products in the assortment. Feldman and Topaloglu (2018) work with a mixture of BAM's, where the different customer types have nested consideration sets but they associate the same preference weight with a product that is in their consideration set. Zhang et al. (2017) and Feldman (2018) study assortment problems under the paired combinatorial logit model.

Cachon et al. (2005) and Wang and Sahin (2018) study assortment optimization models when the customer has a search cost for the products. Dzyabura and Jagabathula (2018) work on an assortment problem when products can be offered in both online and offline stores and online shoppers can get a feel for the products by visiting the offline store. Feldman and Paul (2019) show that approximation schemes for assortment problems with a fixed cost of offering a product can be used to design approximation schemes under capacity constraints. Aouad et al. (2018b) give an approximation scheme for assortment optimization under the exponomial choice model.

Bernstein and Martinez-de-Albeniz (2017) study strategic assortment rotation problems, when customers time their purchase to maximize their utilities. Similarly, Ferreira and Goh (2018) study an assortment problem where the firm must choose between showing an assortment in one shot or incrementally with the understanding that a customer can make multiple purchases. Chong et al. (2001) discuss a practical implementation of an assortment optimization model to choose product varieties to offer in different food categories. Caro and Martinez-de-Albeniz (2009), Caro and Gallien (2010), Caro et al. (2010), Caro and Martinez-de-Albeniz (2014) and Caro and Martinez-de-Albeniz (2015) discuss a host of issues in fast fashion retail, including assortment optimization. Feldman et al. (2018) compare various assortment optimization approaches through a field experiment.

Appendix

Proof of Theorem 5.1 To see that a nested-by-revenue assortment solves problem (5.2), notice that (5.2) implies $\mathcal{R}^* \geq \sum_{j \in S} p_j\, v_j/(v_0+V(S))$ for all sets $S \subseteq N$ with equality holding at optimal assortments. The last inequality is equivalent to $v_0\, \mathcal{R}^* \geq \sum_{j \in S}(p_j - \mathcal{R}^*)\, v_j$. Therefore, $v_0\, \mathcal{R}^* \geq \sum_{j \in S}(p_j - \mathcal{R}^*)\, v_j$ for all sets $S \subseteq N$ with equality holding at optimal assortments. The last statement implies that an optimal assortment can be recovered by solving the problem

$$\max_{S \subseteq N} \left\{ \sum_{j \in S} (p_j - \mathcal{R}^*)\, v_j \right\}.$$

There is a simple solution to the problem above. It is optimal to offer each product whose revenue exceeds $\mathcal{R}^*$. Therefore, the optimal solution is $S^* = \{j \in N : p_j \geq \mathcal{R}^*\}$. Notice that the products are indexed such that $p_1 \geq p_2 \geq \ldots \geq p_n$, so S^* given in the last expression has be to one of the sets $E_0, E_1, \ldots, E_n$. □

Proof of Theorem 5.7 The fact that $H(E_i) \geq H(E_{i-1})$ follows directly from (5.12). Clearly $H(E_1) = \lambda_1 p_1 > 0$, so $\tilde{E}_1 = E_1 = \{1\}$ and $R(\tilde{E}_1) = H(E_1)$. Assume that the result holds for $i - 1$, so $H(E_{i-1}) = R(\tilde{E}_{i-1}) \geq R(E)$ for all $E \subseteq E_{i-1}$. We will show that $H(E_i) = R(\tilde{E}_i) \geq R(E)$ for all $E \subseteq E_i$. From the algorithm,

$$H(E_i) = H(E_{i-1}) + \lambda_i (p_i - H(E_{i-1}))^+ = R(\tilde{E}_{i-1}) + \lambda_i (p_i - R(\tilde{E}_{i-1}))^+.$$

If $p_i \leq H(E_{i-1})$, then we have $H(E_i) = H(E_{i-1}) = R(\tilde{E}_{i-1}) = R(\tilde{E}_i)$ on account of $\tilde{E}_i = \tilde{S}_{i-1}$. On the other hand, if $p_i > H(E_{i-1})$, then we have $\tilde{E}_i = \tilde{E}_{i-1} \cup \{i\}$, and

$$H(E_i) = (1 - \lambda_i) R(\tilde{E}_{i-1}) + \lambda_i p_i = R(\tilde{E}_i),$$

where the last equality follows from Eq. (5.11). To complete the proof, we need to show that $R(\tilde{E}_i) \geq R(E)$ for all $E \subseteq \tilde{E}_i$. If $i \notin E$, then we have $R(E) \leq R(\tilde{E}_{i-1}) \leq R(\tilde{E}_i)$. On the other hand, if $i \in E$, then we can write $E = T \cup \{i\}$ for some $T \subseteq E_{i-1}$. In this case, we get

$$\begin{aligned} R(E) &= R(T \cup \{i\}) = (1 - \lambda_i)\, R(T) + \lambda_i\, p_i \\ &\leq (1 - \lambda_i)\, R(\tilde{E}_{i-1}) + \lambda_i\, p_i = (1 - \lambda_i)\, H(E_{i-1}) + \lambda_i\, p_i \\ &\leq H(E_{i-1}) + \lambda_i\, (p_i - H(E_{i-1})^+ \\ &= H(E_i) = R(\tilde{E}_i), \end{aligned}$$

as desired. □

Proof of Theorem 5.4 For notational brevity, we let $V_i^* = V_i(x_i^*)$, $\Pi_i^* = R_i(x_i^*)$, $\hat{V}_i = V_i(\hat{x}_i)$ and $\hat{\Pi}_i = R(\hat{x}_i)$, where x_i^* is an optimal solution to problem (5.7) and $\hat{x}_i$ is an optimal solution to problem (5.9). It is enough to show that

$$\hat{V}_i^{\gamma_i}\,(\hat{\Pi}_i - z) \geq (V_i^*)^{\gamma_i}\,(\Pi_i^* - z).$$

First, assume that $z \geq \Pi_i^*$. By the definition of u_i^*, we have $u_i^* = z$. Also, the definition of $\hat{x}_i$ implies that $\hat{V}_i\,(\hat{\Pi}_i - u_i^*) \geq 0$ because not offering any of the products provides an objective value of zero for problem (5.9) so that the optimal objective value of this problem should at least be zero. Using the definition of u_i^* and the fact that $z \geq \Pi_i^*$, the last inequality yields $\hat{V}_i^{\gamma_i}\,(\hat{\Pi}_i - z) = \hat{V}_i^{\gamma_i}\,(\hat{\Pi}_i - u_i^*) \geq 0$. On the other hand, since $z \geq \Pi_i^*$, we have $0 \geq (V_i^*)^{\gamma_i}\,(\Pi_i^* - z)$. Thus, the last two inequalities yield $\hat{V}_i^{\gamma_i}\,(\hat{\Pi}_i - z) \geq (V_i^*)^{\gamma_i}\,(\Pi_i^* - z)$, as desired.

Second, assume that $\Pi_i^* > z$. By the definition of $\hat{x}_i$, we have $\hat{V}_i\,(\hat{\Pi}_i - u_i^*) \geq V_i^*\,(\Pi_i^* - u_i^*)$. Using the definition of u_i^* and the fact that $\Pi_i^* > z$, we have $u_i^* = \gamma_i\,z + (1-\gamma_i)\,\Pi_i^*$. Using the last equality, the last inequality can equivalently be written as $\hat{V}_i\,(\hat{\Pi}_i - \gamma_i\,z - (1-\gamma_i)\,\Pi_i^*) \geq \gamma_i\,V_i^*\,(\Pi_i^* - z)$. Multiplying both sides of this inequality by $\hat{V}_i^{\gamma_i - 1}$ yields $\hat{V}_i^{\gamma_i}\,(\hat{\Pi}_i - \gamma_i\,z - (1-\gamma_i)\,\Pi_i^*) \geq \gamma_i\,V_i^*\,\hat{V}_i^{\gamma_i - 1}\,(\Pi_i^* - z)$. Arranging the terms, we write this inequality as

$$\hat{V}_i^{\gamma_i}(\hat{\Pi}_i - z) \geq [\gamma_i\,V_i^*\,\hat{V}_i^{\gamma_i - 1} + (1-\gamma_i)\,\hat{V}_i^{\gamma_i}]\,(\Pi_i^* - z).$$

Since $\gamma_i \in [0, 1]$, u^{γ_i} is a concave function of u. Using the subgradient inequality for this concave function, we have $\hat{V}_i^{\gamma_i} + \gamma_i\,\hat{V}_i^{\gamma_i - 1}\,(V_i^* - \hat{V}_i) \geq (V_i^*)^{\gamma_i}$, which can also be written as $(1-\gamma_i)\,\hat{V}_i^{\gamma_i} + \gamma_i\,V_i^*\,\hat{V}_i^{\gamma_i - 1} \geq (V_i^*)^{\gamma_i}$. Using this inequality in the inequality displayed above, it follows that

$$\hat{V}_i^{\gamma_i}(\hat{\Pi}_i - z) \geq [\gamma_i\,V_i^*\,\hat{V}_i^{\gamma_i - 1} + (1-\gamma_i)\,\hat{V}_i^{\gamma_i}]\,(\Pi_i^* - z) \geq (V_i^*)^{\gamma_i}\,(\Pi_i^* - z),$$

which is the desired result. □

Proof of Theorem 5.9 The proof has two steps. First, we give an equivalent formulation of problem (5.16). In particular, using $\mathcal{R}^*$ to denote the optimal objective value of problem (5.16), we claim that problem (5.16) is equivalent to the linear program

$$\begin{aligned} \max \quad & \sum_{j \in N} (p_j - \mathcal{R}^*)\,\frac{v_j}{v_0}\,x_j \\ \text{s.t.} \quad & \sum_{j \in N} a_{ij}\,x_j \leq b_i \qquad \forall\, i \in L \\ & 0 \leq x_j \leq 1 \qquad \forall\, j \in N \end{aligned} \tag{5.27}$$

in the sense that an optimal solution to problem (5.16) can be recovered by using an optimal solution to problem (5.27) and the two problems share the same optimal objective value. To see the claim, we define the feasible set $\mathcal{F} = \{x \in \{0,1\}^n : \sum_{j \in N} a_{ij} x_j \le b_i \ \forall i \in L\}$ to capture the feasible set of problem (5.16). Since $\mathcal{R}^*$ is the optimal objective value of problem (5.16), we have $\mathcal{R}^* \ge \sum_{j \in N} p_j v_j x_j / (v_0 + \sum_{j \in N} v_j x_j)$ for all $x \in \mathcal{F}$ and the inequality holds as equality at the optimal solution to problem (5.16). Writing the last inequality equivalently as $\mathcal{R}^* \ge \sum_{j \in N} (p_j - \mathcal{R}^*) v_j x_j / v_0$, it follows that we have $\mathcal{R}^* \ge \sum_{j \in N} (p_j - \mathcal{R}^*) v_j x_j / v_0$ for all $x \in \mathcal{F}$ and the last equality holds as equality at the optimal solution. Therefore, we can obtain an optimal solution to problem (5.16) by solving

$$\max_{x \in \{0,1\}^n} \left\{ \sum_{j \in N} (p_j - \mathcal{R}^*) v_j x_j / v_0 : x \in \mathcal{F} \right\}$$

$$= \max_{x \in \{0,1\}^n} \left\{ \sum_{j \in N} (p_j - \mathcal{R}^*) v_j x_j / v_0 : \sum_{j \in N} a_{ij} x_j \le b_i \ \forall i \in L \right\}$$

and the optimal objective value of these problems would be $\mathcal{R}^*$. The objective function and constraints in the last problem are linear. Since the constraint matrix is TU, we can relax the binary constraints without loss of optimality. If we relax the binary constraints in the last problem, then we obtain problem (5.27), which establishes the claim. Thus, it is enough to show that problems (5.17) and (5.27) are equivalent to each other.

Let $y^* = \{y_j^* : j \in N\}$ and y_0^* be an optimal solution to problem (5.17) with the corresponding optimal objective value ζ^*. Let x^* be an optimal solution to problem (5.27). By the discussion above, the optimal objective value of problem (5.27) is $\mathcal{R}^*$. In the second part of the proof, we show that $\mathcal{R}^* = \zeta^*$. We construct a solution $(\hat{y}, \hat{y}_0)$ to problem (5.17) as follows. For all $j \in N$, $\hat{y}_j = v_j x_j^* / (v_0 + \sum_{i \in N} v_i x_i^*)$ and $\hat{y}_0 = 1 - \sum_{j \in N} \hat{y}_j = v_0 / (v_0 + \sum_{j \in N} v_j x_j^*)$. By definition, $(\hat{y}, \hat{y}_0)$ satisfies the first constraint in problem (5.17). Furthermore, we have

$$\sum_{j \in N} a_{ij} \frac{\hat{y}_j}{v_j} = \frac{\sum_{j \in N} a_{ij} x_j^*}{v_0 + \sum_{j \in N} v_j x_j^*} \le \frac{b_i}{v_0 + \sum_{j \in N} v_j x_j^*} = \frac{b_0}{v_0} \hat{y}_0$$

for all $i \in L$, where the equalities use the definition of $\hat{y}_j$ and $\hat{y}_0$, whereas the inequality uses the fact that x^* is a feasible solution to problem (5.27). Thus, $(\hat{y}, \hat{y}_0)$ satisfies the second set of constraints in problem (5.17). Also, we have $\hat{y}_j / v_j = x_j^* / (v_0 + \sum_{j \in N} v_j x_j^*) \le 1 / (v_0 + \sum_{j \in N} v_j x_j^*) = \hat{y}_0 / v_0$ for all $j \in N$, indicating that $(\hat{y}, \hat{y}_0)$ satisfies the third set of constraints in problem (5.17). Therefore, $(\hat{y}, \hat{y}_0)$ is a feasible solution to problem (5.17). In this case, the objective value provided by this feasible solution for problem (5.17) can at most be ζ^*, so that we obtain

$$\zeta^* \geq \sum_{j \in N} p_j \hat{y}_j = \frac{\sum_{j \in N} p_j v_j x_j^*}{v_0 + \sum_{j \in N} v_j x_j^*} = \mathcal{R}^*,$$

where the first equality is by the definition of $\hat{y}_j$ and the second equality uses the fact that x^* is an optimal solution to problem (5.27), whose optimal objective value is $\mathcal{R}^*$. So, we have $\zeta^* \geq \mathcal{R}^*$. To get a contradiction, assume that $\zeta^* > \mathcal{R}^*$. We construct the solution $\hat{x}$ to problem (5.27) as $\hat{x}_j = (y_j^*/v_j)/(y_0^*/v_0)$ for all $j \in N$. Noting the third set of constraints in problem (5.17), we have $0 \leq \hat{x}_j \leq 1$. Furthermore, we have

$$\sum_{j \in N} a_{ij} \hat{x}_j = \sum_{j \in N} a_{ij} \frac{y_j^*/v_j}{y_0^*/v_0} \leq b_i$$

for all $i \in L$, where the inequality uses the fact that y^* is a feasible solution to problem (5.17). Thus, $\hat{x}$ is a feasible solution to problem (5.27). In this case, the objective value provided by $\hat{x}$ for problem (5.27) can at most be $\mathcal{R}^*$, yielding

$$\begin{aligned}
\mathcal{R}^* &\geq \sum_{j \in N} (p_j - \mathcal{R}^*) \frac{v_j}{v_0} \hat{x}_j = \sum_{j \in N} (p_j - \mathcal{R}^*) \frac{y_j^*}{y_0^*} \\
&= \frac{1}{y_0^*} \sum_{j \in N} p_j y_j^* - \frac{1}{y_0^*} \mathcal{R}^* (1 - y_0^*) > \frac{\zeta^*}{y_0^*} - \frac{\zeta^*}{y_0^*}(1 - y_0^*) = \zeta^*,
\end{aligned}$$

where the second inequality uses the fact that $\sum_{j \in N} p_j y_j^* = \zeta^*$ and the assumption that $\zeta^* > \mathcal{R}^*$. The chain of inequalities contradict the assumption that $\zeta^* > \mathcal{R}^*$. Thus, we must have $\zeta^* = \mathcal{R}^*$ and the solutions $(\hat{y}, \hat{y}_0)$ and $\hat{x}$ must be optimal for problems (5.17) and (5.27), respectively. □

Proof of Theorem 5.11 Let z^1 and z^2 be two distinct vectors and let $\alpha \in [0, 1]$. Then

$$\begin{aligned}
\mathcal{R}(\alpha A' z^1 + (1 - \alpha) A' z^2) &= \max_{S \subseteq N} R(S, A'(\alpha z^1 + (1 - \alpha) z^2)) \\
&= \max_{S \subseteq N} [\alpha R(S, A' z^1) + (1 - \alpha) R(S, A' z^2)] \\
&\leq \alpha \max_{S \subseteq N} R(S, A' z^1) + (1 - \alpha) \max_{S \subseteq N} R(S, A' z^2) \\
&= \alpha \mathcal{R}(A' z^1) + (1 - \alpha) \mathcal{R}(A' z^2).
\end{aligned}$$

Now suppose that assortment S is optimal for each of $\{z^k : k = 1, \ldots, K\}$. Then $\mathcal{R}(A' z^k) = R(S, A' z^k)$ for $k = 1, \ldots, K$. Let α be a vector in the K-dimensional simplex, so $\alpha \geq 0$ and $e'\alpha = 1$. Let $z = \sum_{k=1}^{K} \alpha_k z^k$. So,

$$\begin{aligned}
R(S, A'z) &\leq \mathcal{R}(A'z) \\
&\leq \sum_{k=1}^{K} \alpha_k \mathcal{R}(A'z^k) \\
&= \sum_{k=1}^{K} \alpha_k R(S, A'z^k) \\
&= R(S, A'z).
\end{aligned}$$

The first inequality follows from the definition of $\mathcal{R}$. The second from the convexity of $\mathcal{R}$. The first equality is from the definition of S, and the last equality is from the definition of z. Thus, all inequalities above are equalities, which implies that S is optimal for $z = \sum_{k=1}^{K} \alpha^k z^k$ for any α in the K-dimensional simplex. □

Proof of Theorem 5.12 We will first show that for any $\rho \in \Re_+^m$, the problem that computes $Q(\rho)$ has a solution where $t(S) > 0$ only for sets $S \in \mathcal{E}$. To see this, suppose that $\rho \in \Re_+^m$ and there is a solution to $Q(\rho) = \sum_{S \subseteq N} R(S)t(S)$ with $t(S) > 0$ for some $S \notin \mathcal{E}$. Consider the problem that computes $Q(A(S))$. Then, we have $R(S) < Q(A(S)) = \sum_{U \subseteq N} R(U)t(U)$ with $\sum_{U \subseteq N} A(U)t(U) \leq A(S)$. We can then substitute S by this convex combination and strictly improve the solution without violating the capacity constraint contradicting the assumed optimality.

Suppose the problem of maximizing $R(S, A'z)$ has a solution, say S^* such that $R(S^*, A'z) > R(E, A'z)$ for all $E \in \mathcal{E}$. Then for all $E \in \mathcal{E}$, we have

$$R(E, A'z) < R(S^*, A'z) = R(S^*) - z'A(S^*) < Q(A(S^*)) - z'A(S^*),$$

where the first inequality follows from the assumption that all efficient sets are suboptimal, and the second inequality follows from the fact that S^* is not efficient. But then there exist a convex combination of efficient sets such that $R(S^*) < Q(A(S^*)) = \sum_{E \in \mathcal{E}} R(E)t(E)$ with $A(S^*) \geq \sum_{E \in \mathcal{E}} A(E)t(E)$. This implies that

$$R(S^*) - z'A(S^*) < \sum_{E \in \mathcal{E}} t(E)(R(E) - z'A(E)).$$

For this last inequality to be true, there must be an $E \in \mathcal{E}$ such that

$$\mathcal{R}(A'z) = R(S^*, A'z) = R(S^*) - z'A(S^*) < R(E) - z'A(E) = R(E, A'z)$$

contradicting the optimality of S^*. □

Chapter 6
Single Resource Revenue Management with Dependent Demands

6.1 Introduction

Revenue managers struggled for decades with the problem of finding optimal control mechanisms for fare class structures with dependent demands. In this context, a resource, such as seats on a plane, can be offered at different fares with potentially different restrictions and ancillary services, and the demand for those fares is interdependent. The question is what subset of the fares (or assortment of products) to offer for sale at any given time. Practitioners often use the term open, or open for sale, for a fare that is part of the offered assortment, and the term closed for fares that are not part of the offered assortment. For many years, practitioners preferred to model time implicitly by seeking extensions of Littlewood's rule and EMSR type heuristics to the case of dependent demands. Finding the right way to extend Littlewood's rule proved to be more difficult than anticipated. An alternative approach, favored by academics and gaining traction in industry, is to model time explicitly. In this chapter, we will explore both formulations but most of our attention is devoted to the more tractable model where time is treated explicitly.

In Sect. 6.2, we give a dynamic programming formulation for the revenue management problem with a single resource with dependent demands. In Sect. 6.2.2, we use a linear program to give an upper bound on the optimal total expected revenue and extract a bid-price heuristic from the linear program. In Sect. 6.2.3, we discuss a model where fares cannot be made available once they are closed. In Sect. 6.3, we focus on models that handle time implicitly. As we will see, these models are complicated by the fact that changing the protection level also changes the number of potential customers for higher fare classes. Nevertheless, we develop a heuristic that performs reasonably well.

G. Gallego, H. Topaloglu, *Revenue Management and Pricing Analytics*, International Series in Operations Research & Management Science 279, https://doi.org/10.1007/978-1-4939-9606-3_6

6.2 Explicit Time Models

In this section, we consider models where time is considered explicitly. Modeling time explicitly allows for time-varying arrival rates and time-varying discrete choice models. Customers arrive according to a time heterogeneous Poisson process with intensity $\{\lambda_t : 0 \le t \le T\}$, where T is the length of the sales horizon, and t represents the time-to-go. Thus, time T is the beginning of the sales horizon and time 0 is the end. The total expected number of customers that arrive during the last t units of time is $\Lambda_t := \int_0^t \lambda_s ds$.

The set of products is $N := \{1, \ldots, n\}$. We obtain a revenue of p_j from the sale of product j. There is a single resource with limited capacity. The sale of each product consumes one unit of the resource. A consumer arriving at time-to-go t selects from the offered assortment based on a discrete choice model, say $\{\pi_{tj}(\cdot) : j \in N\}$. More precisely, if we offer the subset S of products, then the customer arriving at time t selects product $j \in S$ with probability $\pi_{tj}(S)$. Let $V(t, x)$ denote the maximum total expected revenue that can be attained over the last t units of the sale horizon with x units of capacity. Assume that at time-to-go t, we offer set $S \subseteq N$ and keep this set of fares open for δt units of time. If $\lambda_t \delta t \ll 1$, then the probability that a customer arrives and requests product $j \in S$ is $\lambda_t \pi_{tj}(S)\delta t + o(\delta t)$, so

$$
\begin{aligned}
V(t,x) =& \max_{S\subseteq N} \left\{ \sum_{j\in S} \lambda_t\, \delta t\, \pi_{tj}(S)\, [p_j + V(t-\delta t, x-1)] \right. \\
&\left. + \left(1 - \lambda_t\, \delta t \sum_{k\in S} \pi_{tj}(S)\right) V(t-\delta t, x) \right\} + o(\delta t) \\
=& V(t-\delta t, x) + \lambda_t\, \delta t \max_{S\subseteq N} \sum_{j\in S} (p_j - \Delta V(t-\delta t, x))\, \pi_{tj}(S) + o(\delta t) \\
=& V(t-\delta t, x) + \lambda_t\, \delta t \max_{S\subseteq N} R_t(S, \Delta V(t-\delta t, x)) + o(\delta t)
\end{aligned} \tag{6.1}
$$

for $x \ge 1$, where $R_t(S, z) := \sum_{j\in S}(p_j - z)\, \pi_{tj}(S)$ and $\Delta V(t - \delta t, x) := V(t - \delta t, x) - V(t - \delta t, x - 1)$ for $x \ge 1$.

We can subtract $V(t - \delta t, x)$ from both sides of Eq. (6.1), divide by δt and take limits as $\delta t \downarrow 0$ to obtain the Hamilton–Jacobi–Bellman (HJB) equation

$$
\frac{\partial V(t,x)}{\partial t} = \lambda_t\, \mathcal{R}_t(\Delta V(t,x)) \tag{6.2}
$$

with boundary conditions that $V(t, 0) = 0$ and $V(0, x) = 0$, where $\mathcal{R}_t(z) := \max_{S\subseteq N} R_t(S, z)$.

The value function $V(t, x)$ is often computed approximately by solving a discrete-time dynamic program based on (6.1). This involves rescaling time and the arrival rates, using $\delta t = 1$, and dropping the $o(\delta t)$ term. Time can be rescaled by a positive real number, say a, such that $T \leftarrow aT$ is an integer by setting $\lambda_t \leftarrow \frac{1}{a}\lambda_{t/a}$, $\pi_{tj}(S) \leftarrow \pi_{t/a,j}(S)$. The resulting dynamic program is given by

$$V(t, x) = V(t-1, x) + \lambda_t \mathcal{R}_t(\Delta V(t-1, x)), \tag{6.3}$$

with the same boundary conditions.

The generic optimization problem in formulations (6.2) and (6.3) is of the form $\mathcal{R}_t(z) := \max_{S \subseteq N} R_t(S, z)$, where $z \in \Re_+$ is a non-negative scalar representing the marginal value of capacity. Since there are 2^n subsets $S \subseteq N$, solving the assortment optimization problem could require the evaluation of the objective function for an exponential number of subsets. Moreover, the problem has to be solved for different values of $z = \Delta V(t, x)$ as the marginal value of capacity changes with the state of the system (t, x).

As discussed at the end of chapter on assortment optimization, for any choice model there is a collection $\mathcal{E} = \{E_j : j \in K\}$, $K = \{0, 1, \ldots, k\}$ of efficient sets that can be ordered so that $\Pi_j := \sum_{i \in E_j} \pi_i(E_j)$ is increasing in $j \in K$, and $E_0 = \emptyset$. Letting $R_j := R(E_j) := \sum_{k \in E_j} p_k \pi_k(E_j)$ we can define the slopes $u_j := (R_j - R_{j-1})/(\Pi_j - \Pi_{j-1})$ for $j = 1, \ldots, k$. Then $u_0 > u_1 > \ldots > u_k > u_{k+1} = 0$, where for convenience we set $u_0 = \infty$ and $u_{k+1} = 0$. Then, the efficient set E_j is optimal to offer for all $z \in [u_{j+1}, u_j]$ for $j = 0, 1, \ldots, k$. At boundary points u_j, both E_j and E_{j-1} are optimal. This implies that the index that maximizes $R_j - z\Pi_j$ over $j \in K$ is given by $a(z) := \max\{j : u_j > z\}$.

If we apply the idea of efficient sets in the context of (6.2) where different choice models may apply at different values of t, we let $z = \Delta V(t, x)$, and $\{u_{tj} : j \in K_t\}$ be the slopes corresponding to the efficient sets in $\mathcal{E}_t = \{E_{tj} : j \in K_t\}$. In this case,

$$a(t, x) := \max\{j \in K_t : u_{tj} > \Delta V(t, x)\}$$

is the index of the efficient set in $\mathcal{E}_t$ that maximizes $R_{tj} - \Delta V(t, x)\Pi_{tj}$.[1] Consequently, it is optimal to offer assortment $E_{t,a(t,x)}$ at state (t, x), corresponding to efficient set E_{tj} with index $j = a(t, x)$. For formulation (6.3), the definition of $a(t, x)$ is the same except that we use $\Delta V(t-1, x)$ instead of $\Delta V(t, x)$ on the right side. The following result is valid for both formulations (6.2) and (6.3).

Theorem 6.1 *The index $a(t, x)$ is decreasing in t and increasing in x over every time interval where the choice model is time invariant.*

Sometimes it is convenient to refer to action j as shorthand for offering efficient set E_{tj}. Thus, it is optimal to offer action $a(t, x)$ at state (t, x). As t increases, $\Delta V(t, x)$ increases and the optimal solution shifts to efficient sets with a smaller

[1] Strictly speaking we should say an index, but the index is unique except at boundary points.

probability of a sale. In contrast, as x increases, $\Delta V(t,x)$ decreases, and the optimal solution shifts to efficient sets with larger sale probability. In general, we cannot say that we close lower fares when t is large (or open lower fares when x is large) because the efficient sets need not be nested-by-fare. For choice models for which the efficient sets enjoy the nested-by-fare property, we can talk of opening and closing fares as the state dynamics change with the understanding that if a fare is open, then all higher fares will be open at the same time.

6.2.1 Formulation as an Independent Demand Model

Consider formulation (6.2) for the dependent demand model. Is it possible to transform this into an independent demand model? The answer is yes, provided that the efficient sets have been properly identified. The transformation into an independent demand model requires creating artificial products that have artificial, but independent demands. Given λ_t, and (Π_{tj}, R_{tj}), $j \in K_t$ for the dependent demand model, the transformation is obtained by setting $\tilde{\lambda}_{tj} := \lambda_t[\Pi_{tj} - \Pi_{t,j-1}]$ and $\tilde{p}_{tj} := u_{tj}$ for $j \in K_t$. The set of $\tilde{p}_{tj}$ are known as transformed fares, and are equal to the slopes between efficient fares.

Then, the independent demand formulation

$$\frac{\partial V(t,x)}{\partial t} = \sum_{j \in K_t} \tilde{\lambda}_{tj}[\tilde{p}_{tj} - \Delta V(t,x)]^+ \tag{6.4}$$

generates the correct value function. The proof of equivalence for the formulation (6.4) is left as an exercise.

The transformation is part of folklore and has appeared in many papers. The fact that the transformation works for the dynamic program has led some practitioners to conclude that the transformed data can be used as an input to Littlewood's rule or to heuristics such as the EMSR, as the transformed demands are independent. There are two problems with this approach. First, the transformation is often used without first identifying the efficient sets. More serious, perhaps, is the fact that Littlewood's rule and its variants require the low-before-high demand arrival pattern. This is tantamount to assuming that Poisson demands with parameters $\tilde{\lambda}_{tj}$ will arrive low-before-high, but these are artificial demands from customers willing to buy under action j but not under action $j-1$. When capacity is allocated to this marginal customer, we cannot prevent some degree of demand cannibalization from customers willing to buy under action $j-1$ into some of the fares in action j. We will return to this issue in Sect. 6.3.

6.2.2 Upper Bound and Bid-Price Heuristic

We will now present an upper bound on the value functions (6.2) for the case where the choice models are time invariant and later explain how to deal with the time variant case. The upper bound is based on approximate dynamic programming with affine value function approximations.

It is well known that a dynamic program can be solved as a mathematical program by making the value function at each state a decision variable. This leads to the formulation $V(T, c) = \min F(T, c)$ subject to the constraints $\partial F(t, x)/\partial t \geq \lambda_t[R_j - \Delta F(t, x)\Pi_j] \;\; \forall(t, x)$ for all $j \in K$, where the decision variables are the class of non-negative functions $F(t, x)$ that are differential in x with boundary conditions $F(t, 0) = F(0, x) = 0$ for all $t \in [0, T]$ and all $x \in \{0, 1, \ldots, c\}$. At optimality $F(t, x) = V(t, x)$ for all $t \in [0, T]$ and all $x \in \{0, 1, \ldots, c\}$.

While this formulation is daunting, it becomes easier once we restrict the functions to be of the affine form

$$\tilde{F}(t, x) = \int_0^t \beta_s(x)ds + xz_t \;\text{ and }\; z_t \geq 0.$$

We will further restrict ourselves to the invariant case: $\beta_s(x) = \beta_s$ for $x > 0$, $\beta_s(0) = 0$, $z_t = z$ for $t > 0$ and $z_0 = 0$. With this restriction, the partial derivative and marginal value of capacity have simple forms and the boundary conditions are satisfied. More precisely,

$$\partial\tilde{F}(t, x)/\partial t = \beta_t \;\text{ and }\; \Delta\tilde{F}(t, x) = z \;\; \forall t > 0, x > 0,$$

with $\tilde{F}(t, 0) = \tilde{F}(0, t) = 0$.

This reduces the program to $\tilde{V}(T, c) = \min \tilde{F}(T, c) = \min \int_0^T \beta_t dt + cz$, subject to $\beta_t \geq \lambda_t[R_j - z\Pi_j] \, \forall j \in K$. Since we have restricted the set of functions $F(t, x)$ to be affine we obtain an upper bound $\bar{V}(T, c) \geq V(T, c)$.

Since this is a minimization problem, the optimal choice for β_t is $\beta_t = \lambda_t \max_{j\in K}[R_j - z\Pi_j] = \lambda_t\mathcal{R}(z)$, where $\mathcal{R}(z) := \max_{j\in K}[R_j - z\Pi_j]$ is a decreasing, convex, non-negative and piecewise linear function of z. Consequently, the overall problem reduces to

$$\bar{V}(T, c) = \min_{z\geq 0}\left[\int_0^T \lambda_t\mathcal{R}(z)dt + cz\right] = \min_{z\geq 0}[\Lambda\mathcal{R}(z) + cz], \tag{6.5}$$

where $\Lambda := \int_0^T \lambda_t dt$ is the aggregate arrival rate over the sales horizon $[0, T]$. Notice that $\Lambda\mathcal{R}(z) + cz$ is convex in z.

We next link the upper bound to the function $Q(\rho)$ that was used in the previous chapter to define efficient sets. We reproduce the definition of $Q(\rho)$ here for convenience. Let $\Pi(S) := \sum_{i\in S} \pi_i(S)$ be the probability of a sale when assortment $S \subset N$ is offered, and let

$$
\begin{aligned}
Q(\rho) := \max & \sum_{S \subseteq N} R(S)t(S) \\
\text{subject to} \quad & \sum_{S \subseteq N} \Pi(S)t(S) \leq \rho \\
& \sum_{S \subseteq N} t(S) = 1 \\
& t(S) \geq 0 \quad \forall S \subseteq N,
\end{aligned}
$$

denote the maximum expected revenue from selecting a convex combination of assortments whose sale rate is bounded by the scalar $\rho \geq 0$. We are now ready to link $\bar{V}(T, c)$ and $Q(\rho)$.

Proposition 6.2

$$
\bar{V}(T, c) = \Lambda Q(c/\Lambda).
$$

Having established the upper bound, we now turn to finding an optimal solution to problem (6.5), which we will denote by $z(T, c)$. We will show that $z(T, c)$ is one of the slopes $u_j := (R_j - R_{j-1})/(\Pi_j - \Pi_{j-1})$ between consecutive efficient sets $\mathcal{E} = \{E_j, j \in K\}$, $K = \{0, 1, \ldots, k\}$. Let $\rho := c/\Lambda$ and define

$$
a(T, c) := \min\{j \leq k : \Pi_j > \rho\},
$$

and set $a(T, c) := k + 1$ if $\rho \geq \Pi_k$.

If $a(T, c) = k + 1$, then the marginal value of capacity is $z(T, c) = u_{k+1} := 0$, and it is optimal to offer the efficient set E_k. If $a(T, c) = j \leq k$, then $\Pi_{j-1} \leq \rho < \Pi_j$, and the marginal value of capacity is $z(T, c) = u_{a(T,c)} = u_j$, with the primal solution offering a convex combination of E_{j-1} and E_j, where the weight on set E_j positive unless $\rho = \Pi_{j-1}$ in which case it is optimal to offer set E_{j-1} all the time. In summary, $z(T, c) = u_{a(T,c)}$. If $z(T, c) = u_{k+1} = 0$, it is optimal to offer set $A(T, c) = E_k$. Otherwise, it is optimal to offer a convex combination of sets $E_{a(T,c)-1}$ and $E_{a(T,c)}$.

We now define a bid-price heuristic that offers set $E_j = k$ if $a(T, c) = k+1$ and offers set $E_{a(T,c)}$ otherwise. This heuristic offers the efficient set with the highest sales probability that is part of the optimal solution to the primal problem. We can express this heuristic more succinctly by offering at state (T, c) the set

$$
A(T, c) = E_{\min(a(T,c),k)}
$$

while capacity is positive, and switching to $E_0 = \emptyset$ when capacity is exhausted. When $z(T, c) > 0$, we have $\Lambda \Pi_{a(T,c)} \geq c$, so the bid-price heuristic is likely to exhaust capacity before the end of the horizon. An obvious refinement is to compute $a(t_j, x_j)$ at reading dates $T = t_1 > t_2 > \ldots > t_J > t_{J+1} = 0$ and to offer the set

$$
E_{\min(a(t_j,x_j),k)} \quad \forall x_j > 0,
$$

Table 6.1 Efficient sets in Example 6.3

Index	Efficient set	Π_i	R_i	u_i
0	∅	0	0	
1	{1}	0.50	500.00	1000
2	{1, 2}	$0.6\bar{6}$	$533.3\bar{3}$	200

Table 6.2 Upper bound and optimal actions in Example 6.3

c	ρ	$\bar{V}(T,c)$	$Z(T,c)$	t_1	t_2	Sales E_1	Sales E_2	Fare 1 sales	Fare 2 sales
12	0.30	12,000	1000	0.6	0.0	12	0	12	0
16	0.40	16,000	1000	0.8	0.0	16	0	16	0
20	0.50	20,000	1000	1.0	0.0	20	0	20	0
22	0.55	20,400	200	0.7	0.3	14	8	18	4
24	0.60	20,800	200	0.4	0.6	8	16	16	8
26	0.65	21,200	200	0.1	0.9	2	24	14	12
28	0.70	21,333	0	0.0	1.0	0	26.6	13.3	13.3
32	0.80	21,333	0	0.0	1.0	0	26.6	13.3	13.3

over time interval $(t_{j+1}, t_j]$, where $\rho_j := x_j/\Lambda_{t_j}$, and

$$a(t_j, x_j) := \min\{i \in K : \Pi_i > \rho_j\},$$

and $a(t_j, x_j) := k + 1$ if $\rho_j > \Pi_k$. This refinement helps curb sales at marginal fares.

Example 6.3 Suppose that $p_1 = 1000$, $p_2 = 600$, and a BAM with $v_0 = v_1 = v_2 = e^1$. Table 6.1 shows the efficient sets, together with the sale and revenue rates, and the slopes between efficient sets. We will assume that the aggregate arrival rate over the sales horizon $[0, T] = [0, 1]$ is $\lambda = 40$, so the expected number of customers to arrive over $[0, T]$ is $\Lambda = 40$. Table 6.2 provides the upper bound $\bar{V}(T, c)$ for different values of c. The table also provides $z(T, c)$ and the solution to the problem $Q(\rho)$ in terms of the proportion of time sets t_1 and t_2 that the efficient sets $E_1 = \{1\}$ and $E_2 = \{1, 2\}$ are offered. Notice that sales under action E_1 first increase and then decrease as c increases. While this may not be intuitive, the logical explanation is that when we have sufficient capacity we exclusively use E_2 because this is the efficient set that maximizes the revenue rate (since $R_2 > R_1$). When $\rho = c/\Lambda < \Pi_2$, we have insufficient capacity to sustain sales at E_2 and that is why we have $t_1 > 0$ for $c \leq 26 < \Lambda\Pi_2$.

If the discrete choice model is time varying, then we have $\mathcal{R}_t(z) = \max_{j \in K_t}[R_{tj} - z\Pi_{tj}]$, resulting in

$$\bar{V}(T, c) = \min_{z \geq 0}\left[\int_0^T \lambda_t \mathcal{R}_t(z)dt + cz\right],$$

where the objective function is also convex in z. For this model, it is also possible to find a bid-price heuristic but it is important to update the dual variable at least as frequently as the changes in the underlying choice model.

6.2.3 Monotone Fares

Formulations (6.2) and (6.3) implicitly assume that the capacity provider can offer any subset of fares at any state (t, x). This flexibility works well if customers are not strategic. Otherwise, customers may anticipate the possibility of lower fares and postpone their purchases in the hope of being offered lower fares at a later time. If customers act strategically, the capacity provider may counter by imposing restrictions that do not allow lower fares to reopen once they are closed. Actions to limit strategic customer behavior are commonly employed by revenue management practitioners, although competitive pressures sometimes force them to deviate from this goal.

Let $V_S(t, x)$ be the optimal expected revenue when the state is (t, x), and we are restricted to use only assortments that are subsets of S. The system starts at state (T, c) and $S = N$. If a strict subset U of S is used then all fares in $U' := \{j \in N : j \notin U\}$ are permanently closed and cannot be offered at a later state regardless of the evolution of sales. To obtain a discrete-time counterpart to (6.3), let

$$W_U(t, x) := V_U(t-1, x) + \lambda_t[R_t(U) - \pi_t(U)\Delta V_U(t-1, x)].$$

Then the dynamic program is given by

$$V_S(t, x) := \max_{U \subseteq S} W_U(t, x) \tag{6.6}$$

with boundary conditions $V_S(t, 0) = V_S(0, x) = 0$ for all $t \geq 0$, $x \geq 1$ and $S \subseteq N$. The goal of this formulation is to find $V_N(T, c)$ and the corresponding optimal control policy.

Notice that formally the state of the system has been expanded to (S, t, x) where S is the last offered set and (t, x) are, as usual, the time-to-go and the remaining inventory. Formulation (6.3) requires optimizing over all subsets $S \subseteq N$, while formulation (6.6) requires an optimization over all subsets $U \subseteq S$ for any given $S \subseteq N$. Obviously the complexity of these formulations is large if the number of fares is more than a handful. Airlines typically have over twenty different fares so the number of possible subsets gets large very quickly. Fortunately, in many cases, we do not need to do the optimization over all possible subsets. As we have seen, the optimization can be reduced to the set of efficient fares. For the p-GAM, we know that the collection of efficient sets is contained by the nested-by-fare sets $\{E_0, E_1, \ldots, E_n\}$ where $E_0 = \emptyset$ and $E_j := \{1, \ldots, j\}$ for $j = 1, \ldots, n$. For the p-GAM, and any other model for which the nested-by-fare property holds, the state

Table 6.3 Value functions for dynamic allocation policies in Example 6.5

c	$\rho = c/\Lambda$	$V_3(T,c)$	$V(T,c)$	$\bar{V}(T,c)$
4	0.16	3769	3871	4000
6	0.24	5356	5534	6000
8	0.32	6897	7013	7477
10	0.40	8259	8335	8950
12	0.48	9304	9382	10,423
14	0.56	9976	10,111	10,846
16	0.64	10,418	10,583	11,146
18	0.72	10,803	10,908	11,447
20	0.80	11,099	11,154	11,504
22	0.88	11,296	11,322	11,504
24	0.96	11,409	11,420	11,504
26	1.04	11,466	11,470	11,504
28	1.12	11,490	11,492	11,504
30	1.20	11,498	11,500	11,504
32	1.27	11,502	11,503	11,504

of the system reduces to (j,t,x) where E_j is the last offered set at (t,x). For such models, the formulation (6.6) reduces to

$$V_j(t,x) = \max_{k \le j} W_k(t,x) \tag{6.7}$$

where $V_j(t,x) := V_{E_j}(t,x)$ and

$$W_k(t,x) = V_k(t-1,x) + \lambda_t[R_{kt} - \Pi_{kt}\Delta V_k(t-1,x)],$$

$R_{kt} := \sum_{l \in S_k} p_l \pi_{lt}(E_k)$ and $\Pi_{kt} := \sum_{l \in S_k} \pi_{lt}(E_k)$. Let

$$a_j(t,x) := \max\{k \le j : W_k(t,x) = V_j(t,x)\}. \tag{6.8}$$

Theorem 6.4 *At state (j,t,x), it is optimal to offer efficient set*

$$E_{a_j(t,x)} := \{1,\ldots,a_j(t,x)\},$$

where $a_j(t,x)$ given by (6.8) is decreasing in t and increasing in x over time intervals where the choice model is time invariant.

The proof of this result follows the sample path arguments of the corresponding proof in the independent demand chapter. Clearly $V_1(t,x) \le V_2(t,x) \le V_n(t,x) \le V(t,x) \le \bar{V}(t,x)$.

Example 6.5 Table 6.3 presents the value functions $V(T,c)$ that results from solving the dynamic program (6.3), the upper bound $\bar{V}(T,c) = \Lambda Q(c/\Lambda)$, as well as the performance $V_3(T,c)$ corresponding to the dynamic program (6.7). All of these quantities are computed for the MNL model with fares $p_1 = \$1000$, $p_2 = \$800$, $p_3 = \$500$ with price sensitivity $\beta_p = -0.0035$, schedule quality $s_i = 200$ for $i = 1, 2, 3$ with quality sensitivity $\beta_s = 0.005$, and an outside alternative with $p_0 = \$1100$ and schedule quality $s_0 = 500$, Gumbel parameter $\phi = 1$, arrival rate $\lambda = 25$ and $T = 1$. Recall that for the MNL model, the attractiveness $v_i = e^{\phi\mu_i}$ where μ_i is the mean utility. In this case $\mu_i = \beta_p p_i + \beta_s s_i$. The computations were done with time rescaled by a factor $a = 10{,}000$. Not surprisingly $V_3(T, c) \leq V(T, c)$ as $V_3(T, c)$ constrains fares to remain closed once they are closed for the first time. However, the difference in revenues is relatively small except for small values of c.

6.3 Implicit Time Models

We now turn to models where the notion of time is implicit. The effort is mostly directed to finding extensions of Littlewood's rule to the case of dependent demands. We will assume that we are working with a choice model with efficient sets that are nested: $E_0 \subseteq E_1 \ldots \subseteq E_k$, even if they are not nested-by-fare. We continue using the notation $\Pi_j := \sum_{k \in E_j} \pi_k(E_j)$ and $R_j := \sum_{k \in E_j} p_k \pi_k(E_j)$, so the slopes $u_j := (R_j - R_{j-1})/(\Pi_j - \Pi_{j-1})$, $j = 1, \ldots, k$ are positive and decreasing. We will denote by $q_j := R_j/\Pi_j$ the average fare, conditioned on a sale, under efficient set E_j (action j) for $j \geq 1$ and define $q_0 = 0$.

Suppose that the total number of potential customers over the selling horizon is a random variable D. For example, D can be Poisson with parameter Λ. Let D_j be the total demand if only set E_j is offered. Then D_j is conditionally binomial with parameters D and Π_j, so if D is Poisson with parameter Λ, then D_j is Poisson with parameter $\Lambda\Pi_j$.

We will present an exact solution for the two-fare class problem and a heuristic for the multi-fare case. The solution to the two-fare class problem is, in effect, an extension of Littlewood's rule for discrete choice models. The heuristic for the multi-fare problem applies the two-fare result to each pair of consecutive actions, say j and $j + 1$, and selects the best j.

6.3.1 *Two Fare Classes*

For the two-fare case, while capacity is available, provider will offer either action 2 (associated with efficient set $E_2 = \{1, 2\}$) or action 1 (associated with efficient set $E_1 = \{1\}$). If the provider runs out of inventory, he offers action 0, corresponding to $E_0 = \emptyset$. Action 2 is optimal for ample capacity, while action 1 is optimal when capacity is scarce. Our task is to find an optimal number, say $y(c) \in \{0, \ldots, c\}$ of units to protect for sale under action 1.

To find $y(c)$, we start with an arbitrary protection level $y \in \{0, \ldots, c\}$. The expected revenue under action 2 is $q_2 \mathbb{E}[\min(D_2, c - y)]$ where q_2 is the average fare per unit sold under action 2. Of the $(D_2 - c + y)^+$ customers denied bookings, a fraction $\beta := \Pi_1/\Pi_2$ will be willing to purchase under action 1. Thus, the demand under action 1 will be a conditionally binomial random variable, say $U(y)$, with a random number $(D_2 - c + y)^+$ of trials and success probability β. The expected revenue that results from allowing up to $c - y$ bookings under action 2 is given by

$$W_2(y, c) := q_1 \mathbb{E}[\min(U(y), \max(y, c - D_2))] + q_2 \mathbb{E}[\min(D_2, c - y)],$$

where the first term corresponds to the revenue under action 1. Conditioning the first term on the event $D_2 > c - y$, allows us to write

$$W_2(y, c) = q_1 \mathbb{E}[\min(U(y), y)|D_2 > c-y)] \mathbb{P}(D_2 > c-y)+q_2 \mathbb{E}[\min(D_2, c-y)].$$

The reader may be tempted to follow the marginal analysis idea presented in Chap. 1 for the independent demand case. In the independent demand case, the marginal value of protecting one more unit of capacity is realized only if the marginal unit is sold. The counterpart here would be $\mathbb{P}(U(y) \geq y|D_2 > c - y)$, and a naive application of marginal analysis would protect the y-th unit whenever $q_1 \mathbb{P}(U(y) \geq y|D_2 > c - y) > q_2$.

However, with dependent demands, protecting one more unit of capacity *also* increases the potential demand under action 1 by one unit. This is because an additional customer is denied capacity under action 2 (when $D_2 > c - y$) and this customer may end up buying a unit of capacity under action 1 even when not all the y units are sold. Ignoring this can lead to very different results in terms of protection levels. The correct analysis is to acknowledge that an extra unit of capacity is sold to the marginal customer with probability $\beta \mathbb{P}(U(y-1) < y-1|D_2 > c-y)$. This suggests protecting the y-th unit whenever

$$q_1 [\mathbb{P}(U(y) \geq y|D_2 > c - y) + \beta \mathbb{P}(U(y - 1) < y - 1|D_2 > c - y)] > q_2.$$

To simplify the left-hand side, notice that conditioning on the decision of the marginal customer results in

$$\mathbb{P}(U(y) \geq y|D_2 > c - y) = \beta \mathbb{P}(U(y - 1) \geq y - 1|D_2 > c - y) \\ + (1 - \beta) \mathbb{P}(U(y - 1) \geq y|D_2 > c - y).$$

Combining terms leads to protecting the y-th unit whenever

$$q_1 [\beta + (1 - \beta) \mathbb{P}(U(y - 1) \geq y|D_2 > c - y)] > q_2.$$

Let

$$r := u_2/q_1 = \frac{q_2 - \beta q_1}{(1 - \beta)q_1}, \tag{6.9}$$

denote the critical fare ratio. In industry, the ratio r given by (6.9) is known as fare adjusted ratio, in contrast to the unadjusted ratio q_2/q_1 that results when $\beta = 0$.

The arguments above suggest that the optimal protection level can be obtained by selecting the largest $y \in \{1, \ldots, c\}$ such that $\mathbb{P}(U(y-1) \geq y | D_2 > c - y) > r$ provided that $\mathbb{P}(U(0) \geq 1 | D_2 \geq c) > r$ and to set $y = 0$ otherwise.

To summarize, an optimal protection level can be obtained by setting $y(c) = 0$ if $\mathbb{P}(U(0) \geq 1 | D_2 > c) \leq r$; otherwise setting

$$y(c) = \max\{y \in \{1, \ldots, c\} : \mathbb{P}(U(y-1) \geq y | D_2 > c - y) > r\}. \tag{6.10}$$

One important observation is that for dependent demands the optimal protection level $y(c)$ is first increasing and then decreasing in c. The reason is that for low capacity it is optimal to protect all the inventory for sale under action 1. However, for high capacity, it is optimal to allocate all the capacity to action 2. The intuition is that action 2 has a higher revenue rate, so with high capacity we give up trying to sell under action 1. This is clearly seen in Table 6.2 of Example 6.3. Heuristic solutions that propose protection levels of the form $\min(y^h, c)$, which are based on independent demand logic, are bound to do poorly when c is close to $\Lambda \Pi_2$.

One can derive Littlewood's rule for discrete choice models (6.10) formally by analyzing $\Delta W_2(y, c) := W_2(y, c) - W_2(y - 1, c)$, the marginal value of protecting the y-th unit of capacity for sale under action 1.

Proposition 6.6

$$\Delta W_2(y, c) = [q_1(\beta + (1 - \beta)\,\mathbb{P}(U(y-1) \geq y | D_2 > c - y) - q_2]\,\mathbb{P}(D_2 > c - y). \tag{6.11}$$

Moreover, the expression in brackets is decreasing in $y \in \{1, \ldots, c\}$.

Consequently, $\Delta W_2(y, c)$ has at most one sign change. If it does, then it must be from positive to negative. $W_2(y, c)$ is then maximized by the largest integer $y \in \{1, \ldots, c\}$, say $y(c)$, such that $\Delta W_2(y, c)$ is positive, and by $y(c) = 0$ if $\Delta W_2(1, c) < 0$. This confirms Littlewood's rule for discrete choice models (6.10).

6.3.2 Heuristic Protection Levels

While the computation of $y(c)$ and $V_2(c) = W_2(y(c), c)$ is not numerically difficult, the conditional probabilities involved may be difficult to understand conceptually. Moreover, the formulas do not provide intuition and do not generalize easily to multiple fares. In this section, we develop a simple heuristic to find near-optimal protection levels that provides some of the intuition that is lacking in the computation of optimal protection levels $y(c)$. In addition, the heuristic can easily be extended to multiple fares.

The heuristic consists of approximating the conditional binomial random variable $U(y-1)$ with parameters $(D_2 - c + y - 1)^+$ and β by its conditional expectation, namely by $(\text{Bin}(D_2, \beta) - \beta(c + 1 - y))^+$. Since $\text{Bin}(D_2, \beta)$ is just D_1, the approximation yields $(D_1 - \beta(c + 1 - y))^+$. We expect this approximation to be reasonable if $\mathbb{E}[D_1] \geq \beta(c + 1 - y)$. This is equivalent to the condition

$$c < y^p + \mathbb{E}[D_2 - D_1] = y^p + \Lambda\pi_2(1 - \beta),$$

where we have $y^p = \max\{y \in \mathcal{N} : \mathbb{P}(D_1 \geq y) > r)\}$. In this case, $\mathbb{P}(U(y-1) \geq y | D_2 > c - y)$ can be approximated by the expression $\mathbb{P}(D_1 \geq (1-\beta)y + \beta(c+1))$. We think of $y^p = (1 - \beta)y + \beta(c + 1)$ as a pseudo-protection level that will be modified to obtain a heuristic protection level when the approximation is reasonable, e.g., when $c < y^p + \mathbb{E}[D_2 - D_1]$, by setting

$$y^h(c) = \max\left\{y \in \mathcal{N} : y \leq \frac{y^p - \beta(c+1)}{(1-\beta)}\right\} \wedge c.$$

If $c > y^p + \mathbb{E}[D_2 - D_1]$, we set $y^h(c) = 0$. Thus, the heuristic will stop protecting capacity for action 1 when c is sufficiently large! This makes sense since action 2 maximizes the expected revenue per customer and this is optimal when capacity is sufficiently abundant.

Notice that the heuristic involves three modifications to Littlewood's rule for independent demands. First, instead of using the first choice demand for fare 1, when both fares are open, we use the stochastically larger demand D_1 for fare 1, when it is the only open fare. Second, instead of using the ratio of the fares p_2/p_1 we use the modified fare ratio $r = u_2/q_1$ based on sell-up adjusted fare values. From this we obtain a pseudo-protection level y^p that is then modified to obtain $y^h(c)$. Finally, we keep $y^h(c)$ if capacity is scarce, e.g., if $c < y^p + E[D_2 - D_1]$ and set $y^h(c) = 0$ otherwise. In summary, the heuristic involves a different distribution, a fare adjustment, and a modification to the pseudo-protection level. The following example illustrates the performance of the heuristic.

Example 6.7 Suppose that $p_1 = 1000$, $p_2 = 600$, and a BAM with $v_0 = v_1 = v_2 = e^1$ and that $\Lambda = 40$ as in Example 6.3. We report the optimal protection level $y(c)$, the heuristic protection level $y^h(c)$, the upper bound $\bar{V}(c)$, the optimal expected revenue $V(c)$ of the uni-directional formulation (6.6), the performance $V_2(c)$ of $y(c)$ and the performance $V_2^h(c) = W_2(y^h(c), c)$ of $y^h(c)$, and the percentage gap between $(V_2(c) - V_2^h(c))/V_2(c)$ in Table 6.4. Notice that the performance of the static heuristic, $V_2^h(c)$, is almost as good as the performance $V_2(c)$ of the optimal policy under formulation (6.6).

Table 6.4 Performance of the heuristic for two-fare problem in Example 6.7

c	$y(c)$	$y^h(c)$	$\bar{V}(c)$	$V(c)$	$V_2(c)$	$V_2^h(c)$	Gap (%)
12	12	12	12,000	11,961	11,960	11,960	0.00
16	16	16	16,000	15,610	15,593	15,593	0.00
20	20	20	20,000	18,324	18,223	18,223	0.00
24	21	24	20,800	19,848	19,526	19,512	0.07
28	9	12	21,333	20,668	20,414	20,391	0.11
32	4	0	21,333	21,116	21,036	20,982	0.26
36	3	0	21,333	21,283	21,267	21,258	0.05
40	2	0	21,333	21,325	21,333	21,322	0.01

6.3.3 *Theft Versus Standard Nesting and Arrival Patterns*

The types of inventory controls used in the airline's reservation system along with the demand order of arrival are additional factors that must be considered in revenue management optimization. If $y(c) < c$, we allow up to $c - y(c)$ bookings under action 2 with *all* sales counting against the booking limit $c - y(c)$. In essence, the booking limit is imposed on action 2 (rather than on fare 2). This is known as theft nesting. Implementing theft nesting controls may be tricky if a capacity provider needs to exert controls through the use of standard nesting, i.e., when booking limits are only imposed on the lowest open fare. This modification may be required either because the system is built on the philosophy of standard nesting or because users are accustomed to thinking of imposing booking limits on the lowest open fare. Here we explore how one can adapt protection levels and booking limits for the dependent demand model to situations where controls must be exerted through standard nesting.

A fraction of sales under action 2 corresponds to sales under fare p_2. This fraction is given by $\omega := \pi_2(E_2)/\Pi_2$. So if booking controls need to be exerted directly on the sales at fare p_2, we can set booking limit $\omega(c - y(c))$ on sales at fare p_2. This is equivalent to using the larger protection level

$$\hat{y}(c) := (1 - \omega)c + \omega y(c) \tag{6.12}$$

for sales at fare 1. This modification makes implementation easier for systems designed for standard nesting controls, and it performs very well under a variety of demand arrival patterns.

It is possible to combine demand choice models with fare arrival patterns by sorting customers through their first choice demand and then assuming a low-before-high demand arrival pattern. For the two-fare case, the first choice demands for fare 1 and fare 2 are Poisson random variables with rates $\Lambda\pi_1(E_2)$ and $\Lambda\pi_2(E_2)$. Assume now that customers whose first choice demand is for fare 2 arrive first, perhaps because of purchasing restrictions associated with this fare. customers whose first choice is fare 2 will purchase this fare if available. They will consider upgrading to fare 1 if fare 2 is not available. One may wonder what kind of control is effective to deal with this arrival pattern. It turns out that setting protection level $\hat{y}(c)$ given by (6.12) for fare 1, with standard nesting, is optimal for this arrival pattern and is very robust to other (mixed) arrival patterns.

6.3.4 Multiple Fare Classes

For multiple fare classes, finding optimal protection levels can be very complex. However, if we limit our search to the best two consecutive efficient sets we can easily adapt the results from the two-fare class to deal with the multiple-fare class problem. For any $j \in \{1, \dots, n-1\}$, consider the problem of allocating capacity between action j (corresponding to efficient set E_j) and action $j+1$ (corresponding to efficient set E_{j+1}) where action $j+1$ is offered first. In particular, suppose we want to protect $y \leq c$ units of capacity for action j against action $j+1$. We will then sell $\min(D_{j+1}, c-y)$ units under action $j+1$ at an average fare q_{j+1}. We will then move to action j with $\max(y, c-D_{j+1})$ units of capacity and residual demand $U_j(y)$, where $U_j(y)$ is conditionally binomial with parameters $(D_{j+1}-c+y)^+$ and $\beta_j := \Pi_j/\Pi_{j+1}$. Assuming we do not restrict sales under action j, the expected revenue under actions $j+1$ and j will be given by

$$W_{j+1}(y,c) := q_j \,\mathbb{E}\min(U_j(y), \max(y, c-D_{j+1})) \\ + q_{j+1}\,\mathbb{E}\min(D_{j+1}, c-y). \quad (6.13)$$

Notice that under action j we will either run out of capacity or will run out of customers. Indeed, if $U_j(y) \geq y$ then we run out of capacity, and if $U_j(y) < y$ then we run out of customers. Let $W_{j+1}(c) := \max_{y \leq c} W_{j+1}(y,c)$ and set $W_1(c) := q_1\,\mathbb{E}\min(D_1, c)$. Clearly,

$$V_n(c) \geq \max_{1 \leq j \leq n} W_j(c), \quad (6.14)$$

so a simple heuristic is to compute $W_j(c)$ for each $j \in \{1, \dots, n\}$ and select j to maximize $W_j(c)$. To find an optimal protection level for E_j against E_{j+1}, we need to compute $\Delta W_{j+1}(y,c) = W_{j+1}(y,c) - W_{j+1}(y-1,c)$. For this we can repeat the analysis of the two-fare case to show that an optimal protection level for action E_j against action E_{j+1} is given by $y_j(c) = 0$ if $\Delta W_{j+1}(1,c) < 0$ and by

$$y_j(c) = \max\{y \in \{1, \dots, c\} : \mathbb{P}(U_j(y-1) \geq y | D_{j+1} > c-y) > r_j\}, \quad (6.15)$$

where

$$r_j := \frac{u_{j+1}}{q_j} = \frac{q_{j+1} - \beta_j q_j}{(1-\beta_j)q_j}.$$

Alternatively, we can use the heuristic described in the two-fare section to approximate $U_j(y-1)$ by $D_j - \beta_j(c+1-y)$ and use this in turn to approximate the conditional probability in (6.15) by $\mathbb{P}(D_j \geq y + \beta(c-y+1))$. This involves finding the pseudo-protection level

$$y_j^p = \max\{y \in \mathcal{N} : \mathbb{P}(D_j \geq y) > r_j\}.$$

If $c < y_j^p + d_{j+1}$, then

$$y_j^h(c) = \max\left\{y \in \mathcal{N}_+ : y \leq \frac{y_j^p - \beta_j(c+1)}{1-\beta_j}\right\} \wedge c, \tag{6.16}$$

and set $y^h(c) = 0$ if $c \geq y_j^p + d_{j+1}$.

We will let $V_n^h(c)$ be the expected revenues resulting from applying the protection levels.

Example 6.8 Consider now a three fare example with fares $p_1 = 1000$, $p_2 = 800$, $p_3 = 500$, schedule quality $s_i = 200$, $i = 1, 2, 3$, $\beta_p = -0.0035$, $\beta_s = 0.005$, $\phi = 1$. Then $v_1 = 0.082$, $v_2 = 0.165$, $v_3 = 0.472$. Assume that the outside alternative is a product with price $p_0 = 1100$ and schedule quality $s_0 = 500$ and that the expected number of potential customers is Poisson with parameter $\Lambda = 25$. Table 6.5 reports the protection levels $y_j(c)$ and $y_j^h(c)$ as well as $V_3(c)$ and $V_3^h(c)$ for $c \in \{4, 6, \ldots, 26, 28\}$. As shown in table, the heuristic performs very well with a maximum gap of 0.14% relative to $V_3(c)$ which was computed through exhaustive search. It is also instructive to see that $V_3^h(c)$ is not far from $V_3(c, T)$, as reported in Table 6.3, for the dynamic model. In fact, the average gap is less than 0.5% while the largest gap is 1.0% for $c = 18$.

Example 6.8 suggests that the heuristic for the static model works almost as well as the optimal dynamic program $V_n(T, c)$ for the case where efficient sets are nested-by-fare and fares cannot be opened once they are closed for the first time. Thus, the multi-fare heuristic described in this section works well to prevent strategic customers from gaming the system provided that the efficient fares are nested-by-fare as they are in a number of important applications. While the heuristic for the static model gives up a bit in terms of performance relative to the dynamic model, it has several advantages. First, the static model does not need the overall demand to be Poisson. Second, the static model does not need as much detail in terms of the arrival rates. These advantages are part of the reason why people in industry have a preference for static models, even though dynamic models are easier to understand, easier to solve to optimality, and just as easy to implement.

Table 6.5 Performance of Heuristic for three-fare problem in Example 6.8

c	ρ	$y_1(c)$	$y_2(c)$	$y_1^h(c)$	$y_2^h(c)$	$V_3(c)$	$V_3^h(c)$	Gap (%)
4	0.16	4	4	4	4	3769	3769	0.00
6	0.24	3	6	3	6	5310	5310	0.00
8	0.32	1	8	1	8	6845	6845	0.00
10	0.40	0	10	0	10	8217	8217	0.00
12	0.48	0	12	0	12	9288	9288	0.00
14	0.56	0	14	0	14	9971	9971	0.00
16	0.64	0	13	0	14	10,357	10,354	0.02
18	0.72	0	9	0	10	10,700	10,694	0.05
20	0.80	0	5	0	6	11,019	11,019	0.00
22	0.88	0	4	0	2	11,254	11,238	0.14
24	0.96	0	3	0	0	11,391	11,388	0.03
26	1.04	0	2	0	0	11,458	11,450	0.08
28	1.12	0	2	0	0	11,488	11,485	0.03

6.4 End of Chapter Problems

1. For the MNL model, let $\Pi_j = \sum_{k \in S_j} \pi_k(S_j)$ and $R_j = \sum_{k \in S_j} p_k \pi_k(S_j)$ for $j = 0, \ldots, n$. Consider the dynamic program

$$V(t, x) = V(t-1, x) + \lambda_t \max_{j \in K} \left[R_j - \Pi_j \Delta V(t-1, x) \right],$$

with boundary condition $V(t, 0) = V(0, x) = 0$ for $t \geq 0$ and $x \in \mathcal{N}$, where $M = \{0, 1, \ldots, n\}$. Let

$$u_j = \frac{R_j - R_{j-1}}{\Pi_j - \Pi_{j-1}}.$$

(a) Show that $u_j = (p_j - r_{j-1})/(1 - \pi_{j-1})$ and in particular that $u_1 = p_1$.
(b) Show that it is optimal to offer set $S_{a(t,x)}$ at state (t, x) where

$$a(t, x) = \max\{j : u_j \geq \Delta V(t, x)\}.$$

Hint: You may want to use the following two facts: 1) $\Delta V(t, x) \leq p_1$ and 2) u_j is decreasing in j for the MNL model.

2. Code the following dynamic program:

$$V(t, x) = V(t-1, x) + \lambda_t \max_j \left[R_j - \Pi_j \Delta V(t-1, x) \right] \tag{6.17}$$

with boundary condition $V(t, 0) = V(0, x) = 0$ for $t \geq 0$ and $x \in \mathcal{N}$.

Run the code for a flight with 3 fares $p_1 = 1150$, $p_2 = 950$, $p_3 = 650$, quality attributes $q_1 = 1000$, $q_2 = 850$, $q_3 = 750$, price sensitivity $\beta_p = -1$ and quality sensitivity $\beta_q = 1.25$. Suppose that the utility of fare i is $U_i = \mu_i + \epsilon_i$ where $\mu_i = \beta_p p_i + \beta_q q_i, i = 1, 2, 3$ and the ϵ_is are independent Gumbel random variables with parameter $\phi = 0.01$. Assume $\lambda_t = \lambda = 0.01$, $T = 10{,}000$. Find $V(T, c)$ for $c \in \{35, 40, 55, 60, 65, 70, 75, 80, 85, 90\}$.

3. Prove that the transformation that leads to the independent demand formulation (6.4) provides the correct value function.
4. Show that an alternative formulation is given by

$$\frac{\partial V(t,x)}{\partial t} = \max_{j \in M_t} \lambda_t \Pi_{tj}[q_{tj} - \Delta V(t,x)]$$

where $q_{tj} = R_{tj}/\Pi_{tj}$, and for convenience we define $q_{0t} = 0$. We can think of $\lambda_t \Pi_{tj}$ as the demand rate associated with average fare q_{tj}, which reduces the dynamic revenue management model with dependent demands to a dynamic pricing model with a finite price menu.

5. Consider a two-fare problem with dependent demands governed by a BAM with parameters $v_0 = 1$, $v_1 = 1.1$, $v_2 = 1.2$. Suppose that the fares are $p_1 = 1000$ and $p_2 = 720$ and that the total number of potential customers is Poisson with parameter $\Lambda = 55$.

   (a) Determine the sale rate Π_i and the revenue rate R_i per arriving customer under action $i = 1, 2$, where $E_1 = \{1\}$ and $E_2 = \{1, 2\}$.
   (b) For capacity values $c \in \{16, 17, \ldots, 35\}$ solve the linear problem

$$\begin{aligned} \Lambda R(c/\Lambda) = & \max \quad \Lambda[R_1 t_1 + R_2 t_2] \\ & \text{s.t.} \quad \Lambda[\Pi_1 t_1 + \Pi_2 t_2] \le c \\ & \qquad t_1 + t_2 + t_0 = 1 \\ & \qquad t_i \ge 0, \quad \forall i = 0, 1, 2, \end{aligned}$$

   and determine the expected number of units $\Lambda \Pi_i t_i$ sold under action $i = 1, 2$.
   (c) From your answer to part b, determine the optimal number of units sold for each fare $i = 1, 2$ for each value of $c \in \{16, \ldots, 35\}$. What happens to optimal number of sales for each fare $1 = 1, 2$ as c increases?
   (d) Find the largest integer, say y^p, such that $P(D_1 \ge y) > r$ where D_1 is Poisson with parameter $\Lambda_1 = \Lambda \Pi_1$, $r = u_2/q_1$, $u_2 = (R_2 - R_1)/(\Pi_2 - \Pi_1)$ and $q_1 = R_1/\Pi_1 = p_1$.
   (e) Let $\Lambda_2 = \Lambda \Pi_2$ and $\beta = \Lambda_1/\Lambda_2$. For each $c \in \{16, 11, \ldots, 35\}$, check if $c < y^p + \Lambda(\Pi_2 - \Pi_1)$ and if so, let

$$y^h(c) = \max\left\{ y \in \mathcal{N} : y \le \frac{y^p - \beta(c+1)}{1-\beta} \right\} \wedge c,$$

   and set $y^h(c) = 0$ otherwise.

(f) For each $c \in \{16, 11, \ldots, 36\}$, use simulation to compute the expected revenue using protection level $y^h(c)$ for action 1 against action 2. Compare the expected revenues to the upper bound $\Lambda R(c/\Lambda)$. For what value of c do you find the largest gap?

6.5 Bibliographical Remarks

Formulation in (6.3) and Theorem 6.1 are due to Talluri and van Ryzin (2004a). The formulation in that paper reduces to the one in Lee and Hersh (1993) when demands are independent. The fare and demand transformations that map λ_t and $(\Pi_{tj}, R_{tj}), j \in K_t$ into $(\hat{p}_{tj}, \hat{\lambda}_{tj}), j \in K_t$ as discussed in Sect. 6.2.1 appeared first in Kincaid and Darling (1963), as documented by Walczak et al. (2010). Fiig et al. (2010) and Walczak et al. (2010) proposed feeding the transformed data into a static model and using the EMSR-b heuristic to compute protection levels. Sierag et al. (2015) and Ge and Pan (2010) extend the work of Talluri and van Ryzin (2004a) to incorporate cancellations and overbooking into a single-resource revenue management problem.

The protection level formula in (6.10) is due to Gallego et al. (2009a). This formula is a reinterpretation of the main result in Brumelle et al. (1990). Efforts to transform the problem into an independent demand model include Belobaba and Weatherford (1996), and more recently by Fiig et al. (2010) and Walczak et al. (2010). Gallego et al. (2009b) show that setting protection level $\hat{y}(c)$ given by (6.12) with standard nesting is optimal and quite robust to other arrival patterns.

Cooper et al. (2006) and Cooper and Li (2012) develop models to study the consequences of specifying a simple customer behavior for choosing among the fare classes, when, in fact, the customer behavior is more complicated.

Appendix

Proof of Proposition 6.2 We can linearize (6.5) by introducing a new variable, say y, such that $y \geq R_j - z\Pi_j$ for all $j \in K$ and $z \geq 0$, which results in the linear program:

$$\begin{aligned} \bar{V}(T, c) = &\min_{z \geq 0}[\Lambda y + cz], \\ \text{subject to} \quad &\Lambda y + \Lambda \Pi_j z \geq \Lambda R_j \quad j \in K \\ &z \geq 0, \end{aligned}$$

where for convenience we have multiplied the constraints $y + \Pi_j z \geq R_j$, $j \in K$ by $\Lambda > 0$. The dual of this problem is given by

$$\bar{V}(T,c) = \Lambda \max \sum_{j \in K} R_j t_j$$

$$\text{subject to} \quad \Lambda \sum_{j \in K} \Pi_j t_j \leq c$$

$$\sum_{j \in K} t_j = 1$$

$$t_j \geq 0 \quad \forall j \in K.$$

This linear program decides the proportion of time, $t_j \in [0, 1]$, that each efficient set E_j is offered to maximize the revenue subject to the capacity constraint. Dividing the constraint by Λ and defining $\rho = c/\Lambda$ we see that $\bar{V}(T,c)/\Lambda = Q(\rho)$, or equivalently $\bar{V}(T,c) = \Lambda Q(c/\Lambda)$. □

Chapter 7
Network Revenue Management with Dependent Demands

7.1 Introduction

Network revenue management models have traditionally been developed under the independent demand assumption. In the independent demand setting, customers arrive into the system with the intention to purchase a particular product. If this product is available, they purchase it. Otherwise, they leave the system. This model is reasonable when products are well differentiated so that customers do not substitute between products. The independent demand model is harder to justify when there are few differences, other than price, between fares. Indeed, a more general setting is needed when the demand for each product depends heavily on whether or not other products are available for sale. This setting gives the firms the opportunity to shape the demand for each product by adjusting the offer set made available to the customer.

In this chapter, we start by giving a dynamic programming formulation of the network revenue management problem under dependent demands. Similar to the other dynamic programming formulations that we give for network revenue management problems, the state variable in this dynamic program is high dimensional. Thus, solving the dynamic programming formulation is intractable for realistic networks. We show that a large scale deterministic linear program provides an upper bound on the value function. This linear program has one decision variable for each subset of origin-destination-fares (ODF's) that can be offered to customers. Therefore, the number of decision variables can get large and it is common to solve the linear program by using column generation. We show how to extend some of the heuristics from the independent demand to the dependent demand setting. In particular, we show how to extract heuristic control policies from the solution to the linear program that are similar in spirit to those derived for the independent demand model. As an example, we can use the dual variables to approximate the displacement costs to drive a bid-price heuristic. However, unlike the independent demand model, an

G. Gallego, H. Topaloglu, *Revenue Management and Pricing Analytics*,
International Series in Operations Research & Management Science 279,
https://doi.org/10.1007/978-1-4939-9606-3_7

assortment problem needs to be solved to determine which ODF's to offer at any state of the system. As in the independent demand model, we can obtain more refined heuristics by solving single-resource dynamic programs either by netting the fares using the dual variables of the capacity constraint or by allocating the fares among the resources. These refined heuristics often result in tighter upper bounds and better performance. For the basic attraction model (BAM) and the Markov chain (MC) choice model, we show that it is actually possible to reduce the size of the deterministic linear program to roughly the same size as the independent demand model.

In Sect. 7.2, we give a dynamic programming formulation of the network revenue management problem with dependent demands. In Sect. 7.3, we give a deterministic linear program that provides an upper bound on the optimal total expected revenue. In Sect. 7.4, we discuss how to obtain approximate solutions to the dynamic programming formulation by decomposing the formulation by the resources. In Sect. 7.3.2, we show that if the customer choices are governed by the BAM or the MC choice model, then the size of the deterministic linear program can be reduced to the size of the linear program in the independent demand case.

7.2 Formulations

There are m resources in the network, and the capacities of the resources are given by $c = (c_1, \ldots, c_m)$. We let $M := \{1, \ldots, m\}$ denote the set of resources. As in our work for network revenue management with independent demands, we will use the single index model to capture the ODF's and the set of ODF's is $N := \{1, \ldots, n\}$. For each ODF $j \in N$, we associate a fare p_j and a resource consumption vector $A_j := (a_{1j}, \ldots, a_{mj})$. We let A be the $m \times n$ matrix whose j-th column is A_j. The length of the sales horizon is T. Time is measured backwards, so the time-to-go is $t = T$ at the beginning of the sales horizon and $t = 0$ at the end. Customers arrive over the sales horizon according to a Poisson process with rate $\{\lambda_t : 0 \leq t \leq T\}$. For economy of notation, we will assume that the choice model is time invariant. However, most of the results are easily extended to the case of time-variant choice models.

To capture our decisions at any time point, we use $\{u_j : j \in N\} \in \{0, 1\}^n$, where $u_j = 1$ if ODF j is offered and $u_j = 0$ otherwise. Given $u = (u_1, \ldots, u_n)$, an arriving customer purchases ODF j with probability $\pi_j(u)$. The choice probability $\pi_j(u)$ may be governed by any choice model, including the ones that are discussed in the chapter on introduction to choice modeling.

For any time-to-go t, we let (t, x) to represent the state of the system, where $x = (x_1, \ldots, x_m)$ denotes the vector of remaining capacities. The set of feasible decisions are given by $\mathcal{U}(x) := \{u \in \{0, 1\}^n : A_j u_j \leq x\}$, indicating that we cannot offer ODF's for which we do not have sufficient capacity.

Let $V(t, x)$ be the maximum total expected revenue that can be collected starting from state (t, x). To compute $V(t, x)$, consider a time increment δt that is small enough so that we can approximate the probability of a customer arrival over this time increment by $\lambda_t \, \delta t$. If the ODF offer decisions are given by u, then a customer arrives with probability $\lambda_t\,\delta t$ and is interested in ODF $j \in N$ with probability $\pi_j(u)$. For a customer who selects product j, with $u_j = 1$ we collect p_j in revenues and provide them A_j in resources. With probability $1 - \sum_{j\in N} \pi_j(u)$, the arriving customer leaves without purchasing any of the ODF's. Finally, with probability $1 - \lambda_t\,\delta t$ there are no arrivals. Following this argument, we have the dynamic program

$$V(t,x) = \max_{u\in\mathcal{U}(x)} \left\{ \sum_{j\in N} \lambda_t\,\delta t\,\pi_j(u)\Big[p_j + V(t-\delta t, x - A_j)\Big]\right.$$

$$\left. + \left(1 - \lambda_t\,\delta t + \lambda_t\,\delta t\left(1 - \sum_{j\in N}\pi_j(u)\right)\right)V(t-\delta t, x)\right\} + o(\delta t)$$

$$= V(t-\delta t, x)$$

$$+ \max_{u\in\mathcal{U}(x)} \left\{ \sum_{j\in N} \lambda_t\,\delta t\,\pi_j(u)\Big[p_j + V(t-\delta t, x - A_j) - V(t-\delta t, x)\Big]\right\} + o(\delta t),$$

where the equality follows by arranging the terms. Subtracting $V(t-\delta t, x)$, dividing by δt, and taking the limit as $\delta t \to 0$, we obtain the Hamilton–Jacobi–Bellman (HJB) equation

$$\frac{\partial V(t,x)}{\partial t} = \max_{u\in\mathcal{U}(x)} \sum_{j\in N} \lambda_t\,\pi_j(u)\,(p_j - \Delta_j V(t,x))$$

with the boundary conditions $V(t,0) = V(0,x) = 0$ or all $t \geq 0$ and $x \geq 0$. Similar to our notation in the chapter on network revenue management with independent demands, $\Delta_j V(t,x)$ denotes the displacement cost $V(t,x) - V(t, x - A_j)$ when $x \geq A_j$, and we set $\Delta_j V(t,x) = \infty$ otherwise.

The right side of the HJB equation is an assortment problem of the type studied in the chapter on assortment optimization. Notice that we can write any assortment either as a subset, say $S \subseteq N$, or as the incidence vector $u \in \{0,1\}^n$. Thus, when necessary, it is reasonable to abuse the notation and write $\pi_j(S)$ as the probability of selecting j when the set S is offered.

For any vector $\theta \in \Re^n_+$, let

$$R_t(S,\theta) := \lambda_t \sum_{j\in S}(p_j - \theta_j)\,\pi_j(S),$$

and

$$\mathcal{R}_t(\theta) := \max_{S\subseteq N} R_t(S,\theta).$$

We can write the HJB equation as

$$\frac{\partial V(t,x)}{\partial t} = \mathcal{R}_t(\Delta V(t,x)) \tag{7.1}$$

with boundary conditions $V(t, 0) = V(0, x) = 0$, where we let $\Delta V(t, x) = (\Delta_1 V(t, x), \ldots, \Delta_n V(t, x))$. Notice that if S^* is an optimal assortment to offer at state (t, x) and $j \in S^*$, then $x \geq A_j$, because otherwise $j \notin \mathcal{U}(x)$. We can also handle this by defining $\Delta_j V(t, x) = \infty$ whenever $j \notin \mathcal{U}(x)$ as this avoids the inclusion of j in the assortment.

We close this section by providing a discrete-time dynamic programming formulation that will be useful in numerous places in this chapter. The approach that we use to obtain this discrete-time dynamic programming formulation closely mirrors the one in the chapter with independent demands. In particular, the discrete-time formulation is given by

$$V(t, x) = V(t - 1, x) + \mathcal{R}_t(\Delta V(t - 1, x)), \tag{7.2}$$

with the boundary conditions $V(t, 0) = V(0, x) = 0$ for all $t \geq 0$ and all $x \geq 0$. As before, the discrete-time formulation requires rescaling the arrival rates and the sales horizon so that the probability that two or more customers arrive in a single period is negligible.

Both in the continuous and discrete-time formulations, the state variable (t, x) is a high-dimensional vector and it is difficult to compute the value functions, so we focus on building tractable approximations. For any approximation, say $\tilde{V}(t, x)$ of $V(t, x)$, there is a natural heuristic, which consists of solving the assortment problem

$$\mathcal{R}_t(\Delta \tilde{V}(t, x)) = \max_{S \subseteq N} R_t(S, \Delta \tilde{V}(t, x)), \tag{7.3}$$

for the continuous time formulation (7.1), or solving the assortment problem

$$\mathcal{R}_{t-1}(\Delta \tilde{V}(t - 1, x)) = \max_{S \subseteq N} R_t(S, \Delta \tilde{V}(t - 1, x)), \tag{7.4}$$

for the discrete-time formulation (7.2). Thus, for any approximation $\tilde{V}(t, x)$, solving problem (7.3) or (7.4) yields heuristic policies for the continuous or discrete-time formulation. As a rule of thumb, tighter approximations lead to better heuristics. Thus, we will seek for progressively tighter upper bounds on $V(t, x)$ in our quest for heuristics.

7.3 Linear Programming-Based Upper Bound on $V(T, c)$

In this section, we give a linear program to obtain an upper bound $\bar{V}(T, c)$ on $V(T, c)$. The upper bound is similar in spirit to that obtained for the independent demand model, but requires an exponential number of variables. We show how column generation can be used to solve the linear program for general choice models. The column generation step requires solving an assortment problem. Later in this section, we will present some formulations for discrete choice models that do not require an exponential number of variables.

To formulate the linear program, we let $\Lambda := \int_0^T \lambda_t \, dt$ denote the expected number of customer arrivals over the sales horizon. We will also use the notation

$$R(S) := \sum_{j \in S} p_j \pi_j(S) \text{ and } A(S) := \sum_{j \in S} A_j \pi_j(S) \ \forall S \subseteq N.$$

These quantities represent the gross revenue and the expected consumption rate associated with offering set S.

Let $\tau(S)$ be the proportion of customers offered assortment S over the sales horizon. We consider the linear program

$$\begin{aligned} \bar{V}(T, c) := \max \quad & \Lambda \sum_{S \subseteq N} R(S)\, \tau(S) && (7.5) \\ \text{s.t.} \quad & \Lambda \sum_{S \subseteq N} A(S)\, \tau(S) \leq c \\ & \sum_{S \subseteq N} \tau(S) = 1 \\ & \tau(S) \geq 0 \qquad \forall S \subseteq N. \end{aligned}$$

The linear program above is commonly known as the choice-based deterministic linear program. The objective function accumulates the total expected revenue over all customers and over all assortments. The first set of constraints ensures that the expected consumptions of the resources do not exceed the capacities. The second constraint ensures that a subset is offered to each customer. In the next theorem, we show that the optimal objective value of the linear program above is an upper bound on the optimal total expected revenue $V(T, c)$.

Theorem 7.1 $V(T, c) \leq \bar{V}(T, c)$.

The proof of the theorem above is based on constructing a feasible solution to the linear program in (7.5) by using the decisions of the optimal policy.

7.3.1 Column Generation Procedure

There are 2^n decision variables in problem (7.5). Solving this problem directly by using linear programming software can be difficult or impossible if n is large. In practice, problem (7.5) is commonly solved by using column generation. The idea behind column generation is to iteratively solve a master problem that has the same objective function and constraints as problem (7.5), but the master problem includes only a small fraction of the decision variables $\{\tau(S) : S \subseteq N\}$. Let $\{\hat{z}_i : i \in M\}$ and

$\hat{\beta}$ denote, respectively, the optimal dual variables associated with the first and second set of constraints in the master problem. The reduced cost of the decision variable $\tau(S)$ is given by $R_t(S, A'\hat{z}) - \hat{\beta}$. Suppose we can efficiently solve the assortment optimization problem

$$\mathcal{R}_t(A'\hat{z}) = \max_{S \subseteq N} R_t(S, A'\hat{z}) = \max_{S \subseteq N} \sum_{j \in N} \pi_j(S) \left(p_j - \sum_{i \in N} a_{ij} \hat{z}_i \right)$$

If $\mathcal{R}_t(A'\hat{z}) \le \hat{\beta}$, then all the reduced costs are non-positive, so there are no decision variables $\{\tau(S) : S \subseteq N\}$ that can improve the current solution to the master problem. Consequently, the optimal solution to the master problem is optimal for problem (7.5) and we can stop. On the other hand, if there is a subset $\hat{S}$ such that $R_t(\hat{S}, A'\hat{z}) - \hat{\beta} > 0$, then adding the column associated with $\hat{S}$ will improve the objective value of the master problem. We add this decision variable to the master problem and resolve the master problem. Consequently, a critical step in the column generation idea is to check whether there is some subset S such that the corresponding reduced cost above is strictly positive.

7.3.2 *Sales-Based Linear Program*

The linear program (7.5) involves one decision variable for each subset of ODF's, corresponding to the frequency of offering each subset of ODF's. As a result, the number of decision variables can be large and we propose solving the linear program by using column generation. We now show that if the customers choose under the BAM or the MC choice model, then the linear program (7.5) can be reduced to an equivalent linear program with a manageable number of decision variables. This result eliminates the need for using column generation when the customers choose according to the BAM or the MC choice model.

Basic Attraction Model

Under the BAM, a customer associates the attraction value v_j with ODF j and the attraction value v_0 with the no-purchase alternative. A customer then selects ODF $j \in S$ with probability

$$\pi_j(S) = \frac{v_j}{v_0 + V(S)},$$

where $V(S) = \sum_{j \in S} v_j$. Naturally, we have $\pi_j(S) = 0$ when $j \notin S$. Here, we show that if the customers choose according to the BAM model, then problem (7.5) is equivalent to the linear program

$$\begin{aligned} \tilde{V}(T, c) = \max \quad & \Lambda \sum_{j \in N} p_j \, x_j && && (7.6) \\ \text{s.t.} \quad & \Lambda \sum_{j \in N} a_{ij} \, x_j \leq c_i && \forall i \in M \\ & \sum_{j \in N} x_j + x_0 = 1 \\ & \frac{x_j}{v_j} \leq \frac{x_0}{v_0} && \forall j \in N \\ & x_j, x_0 \geq 0 && \forall j \in N. \end{aligned}$$

In the problem above, we interpret the decision variable x_j as the fraction of customers that purchase ODF j, whereas we interpret the decision variable x_0 as the fraction of customers that leave without making a purchase. The objective function accumulates the total expected revenue over the sales horizon. The first set of constraints ensures that the expected capacity consumption of each resource does not violate the capacity available of the resource. The second constraint ensures that each customer either purchases an ODF or leaves without a purchase. The third set of constraints scales the purchase probability of each ODF and the no-purchase probability to ensure that they are consistent with the BAM. Note that the parameters of the BAM only appear in these constraints.

We emphasize that the decision variables in problem (7.6) are the fractions of customers that purchase different ODF's, rather than the probabilities of offering different subsets of ODF's. In other words, the sales for different ODF's are decision variables in problem (7.6). For this reason, problem (7.6) is sometimes referred to as the sales-based linear program. The appealing aspect of problem (7.6) is that this problem has $n + 1$ decision variables, whereas problem (7.5) has 2^n decision variables. Thus, problem (7.6) can be solved directly through linear programming software without using column generation, even for relatively large-scale airline networks.

In the next theorem, we establish that problem (7.6) is equivalent to problem (7.5). Note that $\tilde{V}(T, c)$ corresponds to the optimal objective value of problem (7.6).

Theorem 7.2 $\tilde{V}(T, c) = \bar{V}(T, c)$.

One of the useful aspects of the proof of Theorem 7.2 is that it shows how to recover an optimal solution to problem (7.5) by using an optimal solution to problem (7.6). This is helpful if primal-based heuristics are to be used. As will be discussed later in this chapter, the dual variables $z^* = \{z_i^* : i \in M\}$ of the capacity constraints can be used to develop bid-price heuristics. The dual variables of the capacity constraints in the sales-based linear program and in the original linear program match.

Markov Chain Choice Model

We now provide a compact formulation of problem (7.5) for the MC choice model. Under the MC choice model, a customer arriving into the system is interested in ODF j with probability γ_j. If this ODF is available, then the customer purchases it. Otherwise, the customer transitions from ODF j to ODF k with probability ρ_{jk}. In this way, the customer transitions between the ODF's, until she reaches an ODF that is offered or she reaches the no-purchase option. In this section, we show that if the customers choose according to the MC choice model, then problem (7.5) is equivalent to the linear program

$$\begin{aligned}
\tilde{V}(T,c) = \max \quad & \Lambda \sum_{j\in N} p_j\, x_j && && (7.7)\\
\text{s.t.} \quad & \Lambda \sum_{j\in N} a_{ij}\, x_j \le c_i && \forall\, i \in M\\
& x_j + y_j = \gamma_j + \sum_{k\in N} \rho_{kj}\, y_k && \forall\, j \in N\\
& x_j, y_j \ge 0 && \forall\, j \in N.
\end{aligned}$$

In the problem above, we interpret the decision variable x_j as the fraction of customers that consider purchasing product j during the course of their choice process and purchase this product. On the other hand, we interpret the decision variable y_j as the fraction of customers that consider purchase product j during the course of their choice process but do not purchase this product because of the unavailability of this product. The objective function accounts for the total expected revenue over the sales horizon. The first set of constraints ensures that the expected capacity consumption of each resource does not violate the capacity of the resource. The second set of constraints can be interpreted as a balance constraint. Noting the definitions of the decision variables x_j and y_j, $x_j + y_j$ corresponds to the fraction of customers that consider product j during the course of their choice process. For a customer to consider product j, either she should arrive with the intention of purchasing product j or she should consider purchasing some product i, not purchase this product and transition from product i to product j.

In the next theorem, we establish that problem (7.7) is equivalent to problem (7.5) when customers choose according to the MC choice model.

Theorem 7.3 $\tilde{V}(T,c) = \bar{V}(T,c)$.

By Theorem 7.3, problems (7.5) and (7.7) have the same optimal objective value. It is straightforward to show that the optimal values of the dual variables associated with the first set of constraints in the two problems are the same. Thus, letting $z^* = \{z_i^* : i \in M\}$ be the optimal values of the dual variables associated with the first set of constraints in problem (7.7), we can use these dual variables as input to the heuristics we will discuss next.

7.3.3 Heuristics Based on the Linear Program

We can use the solution to the linear program (7.5) to obtain heuristic control policies. One approach for doing this is to use the primal solution to problem (7.5). In this heuristic, letting $\{\tau^*(S) : S \subseteq N\}$ be an optimal solution to problem (7.5), we offer assortment S to a random fraction $\tau^*(S)$ of customers, subject to capacity availability. Another approach is to use the dual solution to problem (7.5). In particular, letting $z^* = (z_1^*, \ldots, z_m^*)$ be the optimal values of the dual variables associated with the first set of constraints, the idea is to use z_i^* as an opportunity cost of resource i. This is equivalent to using the approximate value function $\tilde{V}(t, x) = x'z^*$ and results in $\tilde{V}(t, x) - \tilde{V}(t, x - A_j) = A'_j z^*$. Consequently, the continuous time heuristic (7.3) solves the assortment problem $\mathcal{R}_t(A'z^*) = \max_{S \subseteq N} R_t(S, A'z^*)$, and the discrete-time heuristic (7.4) solves the assortment problem $\mathcal{R}_{t-1}(A'z^*) = \max_{S \subseteq N} R_{t-1}(S, A'z^*)$. Notice that unlike the independent demand model, it is not necessarily optimal to offer all ODF's j with non-negative net contributions $p_j - A'_j z^*$.

Example 7.4 Consider a network with two connecting resources and assume that two fares are available for each origin-destination pair. Thus, there are six ODF's with capacity consumption vectors

$$A_1 = A_2 = \begin{pmatrix} 1 \\ 0 \end{pmatrix} \quad A_3 = A_4 = \begin{pmatrix} 0 \\ 1 \end{pmatrix} \quad A_5 = A_6 = \begin{pmatrix} 1 \\ 1 \end{pmatrix}.$$

The fares for the different ODF's are given by

$$(p_1, \ldots, p_6) = (30, 150, 60, 120, 100, 200).$$

The length of the sales horizon is $T = 100$. The customer arrival rate is constant at $\lambda_t = 1$ for $0 \le t \le T$. The capacities on the resources are $c = (30, 30)$. To specify the choice process of the customers, we assume that there are three classes of customers. Customers of class 1 are interested in ODF's 1 and 2, customers of class 2 are interested in ODF's 3 and 4, and customers of class 3 are interested in ODF's 5 and 6. An arriving customer is of class 1, 2, and 3, respectively, with probabilities 0.3, 0.3, and 0.4. The probability that a customer of each class chooses a particular ODF as a function of the subset of ODF's offered to this customer is given in Table 7.1. For example, if both ODF's 1 and 2 are available, then a customer of class 1 chooses ODF 1 with probability 0.64 and ODF 2 with probability 0.03. With probability 0.33, this customer leaves without a purchase. For this problem instance, the optimal objective value of problem (7.5) is \$5740. Thus, the total expected revenue obtained by the optimal policy is upper bounded by \$5740. The dual solution is given by $z^* = (18, 120)$.

Given this vector of dual variables, we can solve the assortment problem for each of the three customer classes. For example, the net fares for customer class one are

Table 7.1 Choice probabilities for the three customer classes

Class 1			Class 2			Class 3		
Offer Set	Choice Prob.		Offer Set	Choice Prob.		Offer Set	Choice Prob.	
	ODF 1	ODF 2		ODF 3	ODF 4		ODF 5	ODF 6
{1}	0.66	0	{3}	0.75	0	{5}	0.75	0
{2}	0	0.09	{4}	0	0.5	{6}	0	0.5
{1, 2}	0.64	0.03	{3, 4}	0.6	0.2	{5, 6}	0.6	0.2

$p_1 - z_1^* = 12$ and $p_2 - z_1^* = 132$. Given this net fares, it is optimal to offer only ODF 2 to this customer class. For customer class two, all ODF's have non-positive contribution, so it is not optimal to offer ODF 3 or 4. Finally, for customer class three, the net fares are $p_5 - z_1^* - z_2^* = -38$ and $p_6 - z_1^* - z_2^* = 62$, so it is optimal to offer only ODF 6. In summary, the bid-price heuristic offers ODF's 2 and 6 only, even if ODF 1 has a positive net contribution.

In practice, the performance of the policy obtained from problem (7.5) tends to improve if this problem is periodically resolved over the sales horizon. In particular, we can resolve the linear program (7.5) at state (t, x) after replacing Λ with $\int_0^t \lambda_s ds$ and c_i with x_i for all $i \in M$. If we want to use the policy from the primal solution, then letting $\{\tau^*(S) : S \subseteq N\}$ be the optimal solution, we offer the subset S of ODF's with probability $\tau^*(S)$, subject to capacity availability. If, on the other hand, we want to use the policy from the dual solution, then letting $z^* = (z_1^*, \ldots, z_m^*)$ be the optimal values of the dual variables associated with the capacity constraints, we offer the assortment that maximizes $R_t(S, A'z^*)$.

Similar to our analysis for network revenue management problems with independent demands, it is possible to show that several policies extracted from the linear program (7.5) are asymptotically optimal.

7.4 Dynamic Programming Decomposition

In this section, we describe two approaches for obtaining tractable approximations to the value functions in the dynamic programming formulation of the network revenue management problem. These approaches are based on decomposing the network revenue management problem by resources. To make our exposition simpler, we work with the discrete-time dynamic program given in (7.2).

7.4.1 Exploiting the Deterministic Linear Program

In the chapter on network revenue management with independent demands, we showed how we can leverage the linear programming approximation to decompose the dynamic programming formulation of the network revenue management problem by the resources. In this section, we extend the idea to the dependent demand setting. Assume that we solve problem (7.5) and let $(z_1^*, \ldots, z_m^*)$ be the optimal values of the dual variables associated with the first set of constraints. We choose an arbitrary resource i and relax the first set of constraints for all other resources by associating the dual multipliers z_k^*, $k \neq i$ with them. Thus, the objective function of problem (7.5) reads

$$\Lambda \sum_{S \subseteq N} \sum_{j \in N} p_j \, \pi_j(S) \, \tau(S) + \sum_{k \neq i} z_k^* \Big[c_k - \Lambda \sum_{S \subseteq N} \sum_{j \in N} a_{kj} \, \pi_j(S) \, \tau(S) \Big]$$
$$= \Lambda \sum_{S \subseteq N} \sum_{j \in N} \Big[p_j - \sum_{k \neq i} a_{kj} \, z_k^* \Big] \pi_j(S) \, \tau(S) + \sum_{k \neq i} c_k \, z_k^*.$$

Since we relax the first set of constraints by using the optimal values of the dual variables associated with these constraints as multipliers, problem (7.5) has the same optimal objective value as the problem

$$\begin{aligned} \bar{V}(T, c) \quad = \quad \max \quad & \Lambda \sum_{S \subseteq N} \sum_{j \in N} \Big[p_j - \sum_{k \neq i} a_{kj} \, z_k^* \Big] \pi_j(S) \, \tau(S) + \sum_{k \neq i} c_k \, z_k^* \\ \text{s.t.} \quad & \Lambda \sum_{S \subseteq N} \sum_{j \in N} a_{ij} \, \pi_j(S) \, \tau(S) \leq c_i \\ & \sum_{S \subseteq N} \tau(S) = 1 \\ & \tau(S) \geq 0 \qquad \forall \, S \subseteq N. \end{aligned}$$

If we ignore for the moment the constant term $\sum_{k \neq i} c_k \, z_k^*$ in the objective function, then the problem above is the linear programming approximation for a single-resource revenue management problem that takes place over resource i only. In this single-resource problem, product j generates net revenue $p_j(z^*) = p_j - \sum_{k \neq i} a_{kj} \, z_k^*$. By Theorem 7.1, the optimal objective value of the linear programming approximation provides an upper bound on the optimal total expected revenue. Therefore, $\bar{V}(T, c) - \sum_{k \neq i} c_k \, z_k^*$ is an upper bound on the optimal total expected revenue in the single-resource revenue management problem that takes place over resource i.

We can solve a dynamic program to compute the optimal total expected revenue in the single-resource revenue management problem that takes place over resource i with revenues $p_j(z^*)$ for each ODF $j \in N$. Let $v_i(t, x_i)$ denote the optimal expected revenue for this dynamic program for resource i when the time-to-go is t and the remaining inventory of resource i is x_i. Then, the discrete time version of the dynamic program is given by

$$v_i(t,x_i) = v_i(t-1,x_i) + \lambda_t \max_{S \in \mathcal{U}_i(x_i)} \sum_{j \in N} \pi_j(S)\,(p_j(z^*) - \Delta_j v_i(t-1,x_i)) \tag{7.8}$$

with boundary condition $v_i(0,x_i) = v_i(t,0) = 0$, where $\Delta_j v_i(t,x_i) = v_i(t,x_i) - v_i(t,x_i - a_{ij})$ if $x_i \geq a_{ij}$ and $\Delta_j v_i(t,x_i) = \infty$ otherwise. In the dynamic program (7.8), $\mathcal{U}_i(x_i) = \{u \in \{0,1\}^n : a_{ij}u_j \leq x_i \;\; \forall j \in N\}$ corresponds to the set of feasible subsets of ODF's to offer given that the remaining capacity on resource i is x_i. By the discussion at the end of the previous paragraph, $v_i(T,c_i) \leq \bar{V}(T,c) - \sum_{k \neq i} c_k\, z_k^*$.

On the other hand, by using induction over time, it is possible to show that $v_i(t,x_i) + \sum_{k \neq i} z_k^*\, x_k$ provides an upper bound on the exact value function $V(t,x)$. The proof of this result is identical to the corresponding result under independent demands. Computing this upper bound at the beginning of the sales horizon with the initial capacities, we obtain

$$V(T,c) \leq \min_{i \in M} \Big\{ v_i(T,c_i) + \sum_{k \neq i} z_k^*\, c_k \Big\} \leq \bar{V}(T,c).$$

By solving the single-resource dynamic program for all choices of $i \in M$, we can approximate $V(t,x)$ by $v(t,x) = \sum_{i \in M} v_i(t,x_i)$. This allow us to approximate $\Delta_j V(t,x)$ by $\Delta_j v(t,x) = \sum_{i \in M}[v_i(t,x_i) - v_i(t,x_i - a_{ij})]$ on the right side of (7.8). This is equivalent to solving the assortment problem $\max_{S \subseteq N} R_t(S, \Delta v(t,x))$ corresponding to the continuous time heuristic (7.3) with a similar counterpart for the discrete-time heuristic (7.4).

Example 7.5 For the problem instance in Example 7.4, we compute the value functions $\{v_i(t,\cdot) : t = 1,\ldots,T\}$ by using the dynamic program in (7.8). Thus, we can obtain an upper bound on the optimal total expected revenue by using $\min_{i \in M}\{v_i(T,c_i) + \sum_{k \neq i} z_k^*\, c_k\} = \$5622$, which tightens the upper bound of \$5740 provided by the linear program in (7.5).

7.4.2 Decomposition by Fare Allocation

In this section, we provide an alternative approach for obtaining approximate solutions to the dynamic programming formulation of the network revenue management problem. The main idea behind this approach is to allocate the fare of an ODF over the different resources. Once we allocate the fare of each ODF over the different resources, we solve a single-resource revenue management problem for each resource in the network with the allocated fares. We show that the sum of the value functions over all of the resources is an upper bound on the optimal total expected revenue. We convert the bound into a heuristic by using the bound as an approximation to the value function. The approach that we develop in this section can be viewed as an analogue of Lagrangian relaxation under independent demands.

We use α_{ij} to denote the fare allocation of ODF j over resource i. We do not yet specify how the fare allocations are chosen. For the moment, we only assume that $\sum_{i \in M} \alpha_{ij} = p_j$, so that the total fare allocation of ODF j over all resources is equal to the original fare for ODF j. Using the fare allocations $\alpha = \{\alpha_{ij} : i \in M, j \in N\}$, we solve a single-resource revenue management problem for each resource $i \in M$. The dynamic programming formulation of the single-resource revenue management problem that takes place over resource i is given by

$$v_i^\alpha(t, x_i) = v_i^\alpha(t-1, x_i) + \lambda_t \max_{u \in \mathcal{U}_i(x_i)} \sum_{j \in N} \pi_j(u) \left(\alpha_{ij} - \Delta_j v_i^\alpha(t-1, x_i)\right), \tag{7.9}$$

where $\Delta_j v^\alpha(t, x_i) = v_i^\alpha(t, x_i) - v_i^\alpha(t, x_i - a_{ij})$ if $x_i \geq a_{ij}$ and $\Delta_j v_i^\alpha(t, x_i) = \infty$ otherwise. The superscript α in the value functions emphasizes that the value functions obtained from this dynamic program depends on our choice of the fare allocations. In the single-resource revenue management problem above, which takes place over resource i, if we sell ODF j, then the revenue that we obtain is α_{ij}. In the next theorem, we show that we can obtain upper bounds on the exact value functions by solving the dynamic program above for all resources.

Theorem 7.6 *If the fare allocations $\alpha = \{\alpha_{ij} : i \in M\ , j \in N\}$ satisfy $\sum_{i \in M} \alpha_{ij} = p_j$ for all $j \in N$, then $V(t, x) \leq \sum_{i \in M} v_i^\alpha(t, x_i)$.*

By Theorem 7.6, for any fare allocations $\alpha = \{\alpha_{ij} : i \in M,\ j \in N\}$ that satisfy $\sum_{i \in M} \alpha_{ij} = p_j$ for all $j \in N$, $\sum_{i \in M} v_i^\alpha(t, x_i)$ provides an upper bound on the exact value function $V(t, x)$. Computing this upper bound with the initial capacity at the beginning of the sales horizon, it follows that $\sum_{i \in M} v_i^\alpha(T, c_i)$ provides an upper bound on the optimal total expected revenue. To obtain the tightest possible upper bound on the optimal total expected revenue, we can solve the problem

$$\begin{aligned} \min \quad & \sum_{i \in M} v_i^\alpha(T, c_i) && (7.10) \\ \text{s.t.} \quad & \sum_{i \in M} \alpha_{ij} = p_j \qquad \forall\, j \in N, \end{aligned}$$

where the decision variables are the fare allocations $\alpha = \{\alpha_{ij} : i \in M,\ j \in N\}$. It is possible to show that $v_i^\alpha(T, c_i)$ is a convex function of α. Thus, the problem above has a convex objective function and linear constraints, so we can solve this problem using standard convex optimization methods. We can use problem (7.10) not only to obtain the tightest possible upper bound on the optimal expected revenue, but also to choose the fare allocations. Let α^* be the optimal solution to problem (7.10), and assume we solve the dynamic program (7.9) by using these fare allocations. Then, we can replace $\Delta_j V(t, x)$ with $\Delta_j v^{\alpha^*}(t, x) = \sum_{i \in M} [v_i^{\alpha^*}(t, x_i) - v_i^{\alpha^*}(t, x_i - a_{ij})]$ to solve the assortment problem (7.3) or (7.4) to pick the assortment of ODF's to offer.

Example 7.7 Consider the problem instance in Example 7.4. We solve problem (7.10) to obtain the optimal solution α^*. Then, $\sum_{i \in M} v_i^{\alpha^*}(T, c_i)$ provides an upper bound on the optimal total expected revenue. This upper bound comes out to be \$5606. This is the tightest upper bound among the upper bounds provided by the different solution methods in this chapter.

We described two approaches to obtain approximate solutions to the dynamic programming formulation. Both approaches are based on solving a sequence of single-resource revenue management problems. The first approach exploits the linear program, whereas the second approach is based on allocating the fare of an ODF over different resources. Both approaches can benefit from periodic resolving. For the first approach, if we are at time t with remaining capacities x, then we can solve the linear program in (7.5) after replacing Λ with $\int_0^t \lambda_s \, ds$ and c_i with x_i. Let $z^* = (z_1^*, \ldots, z_m^*)$ be the optimal values of the dual variables associated with the first set of constraints. Using these dual variables in (7.8), we can obtain a heuristic solution to the dynamic programming formulation by approximating $V(t, x)$ by $\sum_{i \in M} v_i(t, x_i)$ until the time point where we carry out the next periodic resolve. For the second approach, if we are at time t with remaining capacities x, then we solve a variant of problem (7.10), where we minimize $\sum_{i \in M} v_i^{\alpha}(t, x)$ subject to the constraint that $\sum_{i \in M} \alpha_{ij} = p_j$ for all $j \in N$. Letting α^* be the optimal solution to this problem, we solve the dynamic program in (7.9) with these fare allocations. We then use $\sum_{i \in M} v_i^{\alpha^*}(t, x_i)$ as an approximation to $V(t, x)$ until the time point where we carry out the next periodic resolve.

Example 7.8 Table 7.2 provides the total expected revenues obtained by the different solution strategies discussed in this chapter, namely the bid-price heuristic based on linear programming and the two decomposition methods: (a) by netting fares via the dual of the linear program and (b) by selecting a fare allocation. We use 1, 4, and 10 resolves over the sales horizon. The first line corresponds to the bid-price policy obtained from the dual solution to the linear program in (7.5). The second line corresponds to the value function approximation approach that exploits the linear program. The third line corresponds to decomposition by fare allocation. The performance of the bid-price policy improves significantly when we resolve the linear program. On the other hand, the value function approximation methods can provide good revenues without resolving too many times. Furthermore, Example 7.7 shows that the optimal total expected revenue is no larger than \$5606. The largest total expected revenue in Table 7.2 is quite close to \$5606, indicating that the performance of the best policy is near-optimal.

As mentioned at the beginning of this chapter, almost every revenue management model built under independent demands has an analogue under dependent demands. So far, we focused our attention on the linear programming formulation and two approaches that approximate the value functions by decomposing the dynamic programming formulation by the resources. There are other solution approaches

Table 7.2 Performance of the solution methods considered in this chapter

Control	1	4	10
Linear program	\$4416	\$5226	\$5329
Decomposition (a)	\$5584	\$5583	\$5584
Decomposition (b)	\$5545	\$5548	\$5532

under dependent demands. For example, we can generalize the extended linear programming formulation under independent demands to cover dependent demands. To see how this works, we discretize time so that there is at most one customer arrival at each time period. Assuming that there are T time periods in the selling horizon, we use the decision variable $\alpha_t(S)$ to capture the probability of offering the subset S of ODF's at time period t. Assuming there is a customer arrival at time period t with probability λ_t, it is possible to show that the optimal objective value of the linear program

$$\max \quad \sum_{t=1}^{T} \sum_{S \subseteq N} \sum_{j \in N} p_j \lambda_t \pi_j(S) \alpha_t(S) \tag{7.11}$$

$$\begin{aligned} \text{s.t.} \quad & x_{Ti} = c_i && \forall i \in M \\ & x_{t-1,i} = x_{ti} - \sum_{S \subseteq N} \sum_{j \in N} a_{ij} \lambda_t \pi_j(S) \alpha_t(S) && \forall i \in M,\ t = 2, \ldots, T \\ & \sum_{S \subseteq N} \Theta_i(S) \alpha_t(S) \leq x_{ti} && \forall i \in M,\ t = 1, \ldots, T \\ & \sum_{S \subseteq N} \alpha_t(S) = 1 && \forall t = 1, \ldots, T \\ & \alpha_t(S),\ x_{ti} \geq 0 && \forall i \in M,\ S \subseteq N,\ t = 1, \ldots, T \end{aligned}$$

provides an upper bound on the optimal total expected revenue. In the linear program above, we have $\Theta_i(S) = \max\{a_{ij} : i \in M,\ j \in S\}$, corresponding to the maximum capacity consumption on resource i by one of the ODF's in the subset S. In addition to obtaining upper bounds on the optimal total expected revenue, if we let $z_t^* = \{z_{it}^* : i \in M,\ t = 1, \ldots, T\}$ be the optimal values of the dual variables associated with the first two sets of constraints, then we can use $\sum_{i \in M} z_{it}^* x_i$ as an approximation to the value function $V(t, x)$. In this case, we can replace $\Delta V(t, x)$ with $A' z_t^*$ to decide which subset of ODF's to make available at state (t, x).

Last, it is possible to show that a heuristics solution based on the primal solution where assortment S is offered with probability $\tau(S)$ is asymptotically optimal. However, all of the heuristics presented here have better performance particularly when the number of times the heuristic is updated during the sales horizon is large.

7.4.3 Overbooking

Here, we consider a sales-based network revenue management model that allows for overbooking under the MC choice model. We assume a discrete-time model over T time periods, where λ_t is the probability of a customer arrival at time period t. A customer that arrives at time-to-go t makes a choice among the offered ODF's according to the MC choice model with first choice probabilities $\{\gamma_{tj} : j \in N\}$ and transition probabilities $\{\rho^t_{ij} : i, j \in N\}$. If we make a sale for ODF j at time-to-go t, then this sale is not canceled by the departure time, i.e. retained, with probability Q_{tj}. The net expected revenue obtained for each booking of ODF j at time-to-go t is p_{tj}. As an example, if a refund of c with $0 \leq c \leq p_j$ is given to bookings that do not survive until the departure time, then we have $p_{jt} = p_j - c(1 - Q_{jt})$. Finally, we let θ_j to be the overbooking cost for each unit of ODF j that we cannot accommodate at the departure time. We use the decision variable x_{tj} to capture the probability that a customer arriving at time period t visits ODF j during the course of her choice process and purchases this ODF, whereas we use the decision variable y_{tj} to capture the probability that a customer arriving at time period t visits ODF j during the course of her choice process and does not purchases this ODF due to unavailability. Also, we use the decision variable z_j to capture the number of ODF j bookings that are denied during boarding. In this case, we can obtain an upper bound on the optimal total expected profit by using the optimal objective value of the linear program

$$
\begin{aligned}
\tilde{V}(T, c) \;=\; \max \;& \sum_{t=1}^{T} \lambda_t \sum_{j \in N} p_{tj}\, x_{tj} - \sum_{j \in N} \theta_j z_j \\
\text{s.t.} \;& \sum_{j \in N} a_{ij} \sum_{t=1}^{T} Q_{tj}\, \lambda_t\, x_{tj} - \sum_{j \in N} a_{ij} z_j \leq c_i \quad \forall\, i \in M \\
& x_{tj} + y_{tj} = \gamma_{tj} + \sum_{k \in N} \rho^t_{kj}\, y_{tk} \qquad \forall\, j \in N,\ t = 1, \ldots, T \\
& z_j \leq \sum_{t=1}^{T} Q_{tj}\, \lambda_t\, x_{tj} \qquad \forall\, j \in N \\
& x_{tj},\ y_{tj},\ z_j \geq 0 \qquad \forall\, j \in N,\ t = 1, \ldots, T.
\end{aligned}
$$

7.5 End of Chapter Problems

1. An airline operates a flight network among three locations A, B, and C. There are two flights. The first flight is from A to B, and the second flight is from B to C. There are six ODF's. The fares associated with the ODF's are given in Table 7.3. A customer interested in flying between a certain origin-destination pair chooses among the two ODF's that serve that destination pair according to the basic attraction model. The attraction values of the six ODF's are, respectively, 5, 3, 1, 0.5, 6, and 2. The attraction value of the no-purchase option is always 1. For example, a customer interested in going from location A to B chooses among the first two ODF's according to the basic attraction model with the attraction values

Table 7.3 ODF's and fares

ODF	Fare
A–B, direct, cheap	400
A–B, direct, expensive	700
B–C, direct, cheap	300
B–C, direct, expensive	600
A–C, through B, cheap	900
A–C, through B, expensive	1600

Table 7.4 Value function at the next time period

(Cap. 1, Cap. 2)	$V(t+1, (\text{Cap. } 1, \text{Cap. } 2))$
(1,1)	1500
(1,0)	1400
(0,1)	1250
(0,0)	0

5 and 3, with the attraction value of the no-purchase option being 1. Assume that we are at time period t and the remaining capacities on the flights are $(1, 1)$. The value function at the next time period is given in Table 7.4. For example, if the remaining capacities at the next time period are given by $(1, 0)$, then the value function takes value 25.

(a) For each ODF, decide whether it is optimal to open or close it.
(b) At time period t, it is equally likely to observe a customer wanting to go from each origin to each destination, with each probability being $1/3$. If the capacities at time period t are given by $x = (1, 1)$, then compute the value function at time period t at the current capacities. That is, compute $V(t, x)$ at $x = (1, 1)$.

2. In this problem, you will show that a policy obtained from problem (7.5) is asymptotically optimal. Consider the revenue management problem discussed in Sect. 7.2 but under the assumption that there is a single resource so c is a scalar and the arrival rates of the customers are stationary so $\lambda_t = \lambda$ for all $t \in [0, T]$ for some fixed λ. Let $\{\tau^*(S) : S \subseteq N\}$ be an optimal solution to problem (7.5). Consider a heuristic policy where we offer subset S with probability $\tau^*(S)$ at each time point, as long as we have capacity on the resource. As soon as we run out of capacity of the resource, we stop offering any products.
Let H^k be the total expected revenue obtained by the heuristic policy above when the capacity of the resource is kc and the length of the selling horizon is kT. Similarly, let V^k be the total expected revenue obtained by the optimal policy. Show that

$$\lim_{k \to \infty} \frac{H^k}{V^k} = 1.$$

(Hint: Follow an approach similar to the one that we used in the asymptotic optimality result under the independent demand model.)

3. Show that the optimal objective value of problem (7.11) is an upper bound on the optimal total expected revenue.
4. Consider the random consideration set model discussed in Chap. 2.
   (a) Show that this choice model is a special case of the Markov chain choice model. In other words, assume that we are given a random consideration set model with n products, where the products are labeled such that $1 \prec 2 \prec \ldots \prec n$ and the attention probabilities are $\{\beta_i : i = 1, \ldots, n\}$. Show that we can construct a single Markov chain choice model such that the choice probability for each product out of each assortment is the same under the two choice models.
   (b) Building on Part a gives a sales-based linear programming formulation under the random consideration set model.
5. We want to solve an assortment optimization problem when customers choose according to the basic attraction model, each product occupies a certain amount of space and we want to ensure that the total amount of space occupied by the offered assortment does not violate the space availability in the store. There are n products indexed by $1, \ldots, n$. The revenue and the attraction value associated with product j are, respectively, r_j and v_j. The attraction value of the no-purchase option is normalized to 1. Product j occupies c_j units of space. The total space consumed by the offered assortment cannot exceed C. Show that we can obtain an upper bound on the optimal expected revenue from a customer by using the optimal objective value of the linear program

$$
\begin{aligned}
\max \quad & \sum_{j=1}^{n} r_j x_j \\
\text{s.t.} \quad & \sum_{j=1}^{n} x_j + x_0 = 1 \\
& \frac{x_j}{v_j} \leq x_0 \qquad \forall\, j = 1, \ldots, n \\
& \sum_{j=1}^{n} c_j \frac{x_j}{v_j} \leq C\, x_0 \\
& x_j \geq 0,\ x_0 \geq 0 \quad \forall\, j = 1, \ldots, n.
\end{aligned}
$$

7.6 Bibliographical Remarks

The linear program in (7.5) is from Gallego et al. (2004). The authors show that the optimal objective value of this linear program provides an upper bound on the optimal total expected revenue. Liu and van Ryzin (2008a) show that a heuristic resulting from the linear program is asymptotically optimal. Balseiro et al. (2010) use the linear program for pricing tickets in tournaments, where the tickets are sold before knowing which teams will advance to the final. Jasin and Kumar (2012) show that resolving the linear program yields bounded revenue loss. Bront et al. (2009) study the same linear program when there are multiple customer segments choosing according to the BAM with different parameters and show that the column generation subproblem is NP-hard. Meissner et al. (2013), Talluri (2014) and Strauss and Talluri (2017) give tractable relaxations of the same linear program that avoids an NP-hard column generation subproblem.

Zhang and Adelman (2009) extend the approximate dynamic programming approach described for independent demands to the dependent demand case. The authors of that paper use linear value function approximations. Kunnumkal and Talluri (2015b) also construct linear approximations to the value functions. Zhang (2011) gives an approach to obtain nonseparable approximations to the value functions. Kunnumkal and Topaloglu (2011a) and Kunnumkal (2014) give variants of the linear programming formulation that are directed towards capturing the randomness in customer choice process more accurately. Kunnumkal and Topaloglu (2008) and Zhang and Vossen (2015) generalize the extended linear programming formulation that we discussed for the independent demands to cover the case with dependent demands.

The idea behind the dynamic programming decomposition method described in Sect. 7.4.1 rests on Liu and van Ryzin (2008a). Zhang and Adelman (2009) establish that this idea provides upper bounds on the exact value functions. The dynamic programming decomposition approach in Sect. 7.4.2 is based on Kunnumkal and Topaloglu (2010a). Kunnumkal and Talluri (2015a) give a dynamic programming decomposition approach that provides tighter upper bounds on the optimal total expected revenue. The sales-based linear program under the BAM appears in Gallego et al. (2015), whereas the sales-based linear program under the MC choice model appears in Feldman and Topaloglu (2017). The papers by van Ryzin and Vulcano (2008b) and Chaneton and Vulcano (2011) use stochastic approximation methods to tune bid-price and protection-level policies under customer choice behavior. Vulcano et al. (2010) and Dai et al. (2015) discuss applications of choice-based network revenue management in practice.

There are several other models for network revenue management problems with dependent demands. Zhang and Cooper (2005) consider the case where customers choose among single-resource ODF's that operate between the same origin-destination pair. They show how to obtain upper bounds on the value functions in the dynamic programming formulation of the problem. Golrezaei et al. (2014), Bernstein et al. (2015), Gallego et al. (2016b) and Chen et al. (2016d) consider related problems arising in the retail and resource allocation settings, where the firm dynamically adjusts the assortment of products or resources offered to its customers.

Another related class of problems occur when the firm chooses the assortment of products to offer to the customers, along with their initial inventory levels and the customers arriving over time make a choice among the offered products. For this stream of literature, we point the reader to van Ryzin and Mahajan (1999), Mahajan and van Ryzin (2001), Gaur and Honhon (2006), Honhon et al. (2010), Topaloglu (2013), Honhon and Seshadri (2013), Goyal et al. (2016), Aouad et al. (2018c), Segev (2019) and Aouad et al. (2019).

Appendix

Proof of Theorem 7.1 Under the optimal policy, let $\mathcal{T}(S)$ be the set of time points over the sales horizon during which set S is offered, and for each $j \in S$, let $X_j(S)$ denote the number of sales for ODF j during the time that the set S is offered. Note that $X_j(S)$ is a thinned Poisson random variable with mean $\pi_j(S) \int_{\mathcal{T}(S)} \lambda_t \, dt$, so $\mathbb{E}[X_j(S)]$ is equal to $\Lambda \pi_j(S) \, \mathbb{E}[\int_{\mathcal{T}(S)} \lambda_t \, dt / \Lambda] = \Lambda \pi_j(S) \, \tau^*(S)$, where $\tau^*(S) := \mathbb{E}[\int_{\mathcal{T}(S)} \lambda_t \, dt / \Lambda]$ is the expected proportion of customers offered set S during the sales horizon by the optimal policy. By construction $\tau^*(S) \geq 0$ and $\sum_{S \subseteq N} \tau^*(S) = 1$. The total expected revenue obtained by the optimal policy is given by

$$V(T, c) = \sum_{S \subseteq N} \sum_{j \in N} p_j \, \mathbb{E}[X_j(S)] = \Lambda \sum_{S \subseteq N} \sum_{j \in N} p_j \pi_j(S) \tau^*(S) = \Lambda \sum_{S \subseteq N} R(S) \tau^*(S).$$

Since the optimal policy has to obey the capacity constraints, we have $\sum_{S \subseteq N} \sum_{j \in N} A_j \, X_j(S) \leq c$. Taking expectations, it follows that

$$\sum_{S \subseteq N} \sum_{j \in N} A_j \, \mathbb{E}[X_j(S)] = \Lambda \sum_{S \subseteq N} \sum_{j \in N} A_j \pi_j(S) \tau^*(S) = \Lambda \sum_{S \subseteq N} A(S) \tau^*(S) \leq c.$$

It follows that $\{\tau^*(S) : S \subseteq N\}$ is a feasible solution to problem (7.5) providing the objective value $V(T, c)$ for this problem. This implies that the optimal objective value of problem (7.5) is at least $V(T, c)$. □

Proof of Theorem 7.6 We show the result by using induction. One can check that the result holds at time period 1, which is the time period in the discrete-time dynamic programming formulation right before the departure time. Assuming that the result holds at time period $t - 1$, we show that the result holds at time period t as well. We let S^* be the optimal subset of ODF's to offer in state (t, x) in the exact dynamic programming formulation of the network revenue management problem. In this case, noting the dynamic program in (7.2), it follows that

$$
\begin{aligned}
V(t,x) &= \sum_{j\in N} \lambda_t\, \pi_j(S^*)\, [p_j - \Delta_j V(t-1,x)] + V(t-1,x) \\
&= \sum_{j\in N} \lambda_t\, \pi_j(S^*) \Big[\sum_{i\in M} \alpha_{ij} + V(t-1, x - A_j) \Big] \\
&\qquad + \Big[1 - \sum_{j\in N} \lambda_t\, \pi_j(S^*) \Big] V(t-1,x) \\
&\le \sum_{j\in N} \lambda_t\, \pi_j(S^*) \Big[\sum_{i\in M} \alpha_{ij} + \sum_{i\in M} v_i^\alpha(t-1, x_i - a_{ij}) \Big] \\
&\qquad + \Big[1 - \sum_{j\in N} \lambda_t\, \pi_j(S^*) \Big] \sum_{i\in M} v_i^\alpha(t-1, x_i),
\end{aligned}
$$

where the second equality follows from the fact that $\sum_{i\in M} \alpha_{ij} = p_j$ for all $j \in N$ and arranging the terms, whereas the inequality is by the induction assumption. Rearranging the terms on the right side of the chain of inequalities above, the right side of the chain of inequalities is given by

$$
\begin{aligned}
&\sum_{i\in M} \Big\{ \sum_{j\in N} \lambda_t\, \pi_j(S^*) \Big[\alpha_{ij} + v^\alpha(t-1, x_i - a_{ij}) - v_i^\alpha(t-1, x_i) \Big] + v_i^\alpha(t-1, x_i) \Big\} \\
&\le \sum_{i\in M} \max_{S\in \mathcal{U}_i(x_i)} \Big\{ \sum_{j\in N} \lambda_t\, \pi_j(S) \Big[\alpha_{ij} + v^\alpha(t-1, x_i - a_{ij}) - v_i^\alpha(t-1, x_i) \Big] \\
&\qquad\qquad + v_i^\alpha(t-1, x_i) \Big\} \\
&= \sum_{i\in M} v_i^\alpha(t, x_i).
\end{aligned}
$$

Thus, we have $V(t,x) \le \sum_{i\in M} v_i^\alpha(t, x_i)$, completing the induction argument. □

Proof of Theorem 7.2 Let $\tau^* = \{\tau^*(S) : S \subseteq N\}$ be an optimal solution to problem (7.5). We consider the solution $\hat{x} = \{\hat{x}_j : j = 0, 1, \ldots, n\}$ for problem (7.6), where $\hat{x}_j$ is defined as $\hat{x}_j = \sum_{S\subseteq N} \pi_j(S)\, \tau^*(S)$ for all $j = 0, 1, \ldots, n$. First, we will now show that $\hat{x}$ is a feasible solution to problem (7.6) and provides an objective value of $\bar{V}(T,c)$ for this problem, which implies that $\tilde{V}(T,c) \ge \bar{V}(T,c)$.

By the definition of $\hat{x}$, $\Lambda \sum_{j\in N} a_{ij}\, \hat{x}_j = \Lambda \sum_{j\in N} \sum_{S\subseteq N} a_{ij}\, \pi_j(S)\, \tau^*(S) \le c_i$ for all $i \in M$, where the inequality uses the fact that τ^* is a feasible solution to problem (7.5). Thus, the solution $\hat{x}$ satisfies the first set of constraints in problem (7.6). Similarly, using the fact that $\sum_{j\in N} \pi_j(S) + \pi_0(S) = 1$ for all $S \subseteq N$ and $\sum_{S\subseteq N} \tau^*(S) = 1$, we obtain $\sum_{j\in N} \hat{x}_j + x_0 = \sum_{j\in N} \sum_{S\subseteq N} \pi_j(S)\, \tau^*(S) + \sum_{S\subseteq N} \pi_0(S)\, \tau^*(S) = \sum_{S\subseteq N} (\sum_{j\in N} \pi_j(S) + \pi_0(S))\, \tau^*(S) = 1$, which implies that the solution $\hat{x}$ satisfies the second constraint in problem (7.6). Finally, if $j \in S$, then $\pi_j(S)/v_j = 1/(v_0 + V(S)) = \pi_0(S)/v_0$, whereas if $j \notin S$, then $\pi_j(S)/v_j = 0$. Thus, we obtain $\hat{x}_j/v_j = \sum_{S\subseteq N} \pi_j(S)\,\tau^*(S)/v_j = \sum_{S\subseteq N} \mathbf{1}(j \in S)\,\pi_0(S)\,\tau^*(S)/v_0 \le \sum_{S\subseteq N} \pi_0(S)\,\tau^*(S)/v_0 = \hat{x}_0/v_0$ for all $j = 1, \ldots, n$, which shows that the solution $\hat{x}$ satisfies the third set of constraints in problem (7.6). The discussion so far implies that $\hat{x}$ is a feasible solution to problem (7.6). Furthermore, we have $\Lambda \sum_{j\in N} p_j\,\hat{x}_j = \Lambda \sum_{j\in N} \sum_{S\subseteq N} p_j\,\pi_j(S)\,\tau^*(S) = \bar{V}(T, c)$, where the last equality uses the fact that τ^* is an optimal solution to problem (7.5).

Second, let $x^* = \{x_j^* : j = 0, 1, \ldots, n\}$ be an optimal solution to problem (7.6). We reindex the products such that $x_1^*/v_1 \ge x_2^*/v_2 \ge \ldots \ge x_n^*/v_n$. Noting the third set of constraints in problem (7.6), we also have $x_0^*/v_0 \ge x_1^*/v_1 \ge x_2^*/v_2 \ge \ldots \ge x_n^*/v_n$. Label the sets $S_0 = \emptyset$ and $S_j = \{1, 2, \ldots, j\}$ for all $j = 1, \ldots, n$. Construct a solution $\hat{\tau} = \{\hat{\tau}(S) : S \subseteq N\}$ to problem (7.5) by setting

$$\hat{\tau}(S_j) = \left[\frac{x_j^*}{v_j} - \frac{x_{j+1}^*}{v_{j+1}}\right] V(S_j)$$

for all $j = 0, 1, \ldots, n$ with the convention that $\hat{\tau}(S_n) = [x_n^*/v_n]\,V(S_n)$. Set $\hat{\tau}(S) = 0$ for all $S \notin \{S_0, S_1, \ldots, S_n\}$. We will show that $\hat{\tau}$ is a feasible solution to problem (7.5) and provides an objective value of $\tilde{V}(T, c)$ for this problem, in which case, we obtain $\bar{V}(T, c) \ge \tilde{V}(T, c)$.

Using the definition of $\hat{\tau}(S)$ and noting that ODF j is in the sets $S_j, S_{j+1}, \ldots, S_n$ but not in $S_0, S_1, \ldots, S_{j-1}$, we have

$$\sum_{S\subseteq N} \pi_j(S)\,\hat{\tau}(S) = \pi_j(S_j)\,\hat{\tau}(S_j) + \pi_j(S_{j+1})\,\hat{\tau}(S_{j+1}) + \ldots + \pi_j(S_n)\,\hat{\tau}(S_n)$$

$$= v_j \left[\frac{x_j^*}{v_j} - \frac{x_{j+1}^*}{v_{j+1}}\right] + v_j \left[\frac{x_{j+1}^*}{v_{j+1}} - \frac{x_{j+2}^*}{v_{j+2}}\right] + \ldots + v_j \left[\frac{x_n^*}{v_n}\right] = x_j^*, \tag{7.12}$$

where use the definition of $\pi_j(S)$ in the second equality above. In this case, we have $\Lambda \sum_{S\subseteq N} \sum_{j\in N} a_{ij}\,\pi_j(S)\,\hat{\tau}(S) = \Lambda \sum_{j\in N} a_{ij}\,(\sum_{S\subseteq N} \pi_j(S)\,\hat{\tau}(S)) = \Lambda \sum_{j\in N} a_{ij}\,x_j^* \le c_i$, where the last inequality uses the fact that x^* is a feasible solution to problem (7.6). So, the solution $\hat{\tau}$ satisfies the first set of constraints in problem (7.5). On the other hand, we have

$$\sum_{S\subseteq N} \hat{\tau}(S) = \hat{\tau}(S_0) + \hat{\tau}(S_1) + \ldots + \hat{\tau}(S_n)$$

$$= V(S_0)\left[\frac{x_0^*}{v_0} - \frac{x_1^*}{v_1}\right] + V(S_1)\left[\frac{x_1^*}{v_1} - \frac{x_2^*}{v_2}\right] + \ldots + V(S_n)\left[\frac{x_n^*}{v_n}\right]$$

$$= \frac{x_0^*}{v_0}\,V(S_0) + \frac{x_1^*}{v_1^*}\,(V(S_1) - V(S_0)) + \frac{x_2^*}{v_2^*}\,(V(S_2) - V(S_1)) + \ldots$$

$$+ \frac{x_n^*}{v_n} (V(S_n) - V(S_{n-1})) = x_0^* + x_1^* + x_2^* + \ldots + x_n^* = 1,$$

where the third equality follows by arranging the terms, the fourth equality is by the fact that $V(S_j) - V(S_{j-1}) = v_j$, and the fifth equality uses the fact that x^* satisfies the second constraint in problem (7.6). Thus, the solution $\hat{\tau}$ satisfies the second constraint in problem (7.5). So, $\hat{\tau}$ is a feasible solution to problem (7.5). Finally, we have $\Lambda \sum_{S \subseteq N} \sum_{j \in N} p_j \pi_j(S) \hat{\tau}(S) = \Lambda \sum_{j \in N} p_j x_j^* = \tilde{V}(T, c)$, where the first equality uses (7.12) and the second equality uses the fact that x^* is an optimal solution to problem (7.6). □

Proof of Theorem 7.3 Letting $z = \{z_i : i \in M\}$ and β, respectively, be the dual variables associated with the first and second sets of constraints in problem (7.5) and expanding the values of $R(S)$ and $A(S)$ by using their definitions, the dual of problem (7.5) is

$$\bar{V}(T, c) = \min \quad \sum_{i \in M} c_i z_i + \beta \tag{7.13}$$

$$\text{s.t.} \quad \Lambda \sum_{i \in M} \sum_{j \in N} a_{ij} \pi_j(S) z_i + \beta \geq \Lambda \sum_{j \in N} p_j \pi_j(S) \qquad \forall S \subseteq N$$

$$z_i \geq 0, \ \beta \text{ is free} \qquad \forall i \in M.$$

Arranging the terms, we can write the first set of constraints above succinctly as

$$\beta \geq \Lambda \max_{S \subseteq N} R(S, A'z)$$

whereas before, $R(S, A'z) = \sum_{j \in S} (p_j - \sum_{i \in M} a_{ij} z_i) \pi_j(S)$.

Because problem (7.13) is a minimization problem and the objective function coefficient of the decision variable β is positive, the decision variable β takes the value $\Lambda \mathcal{R}(A'z)$. Thus, problem (7.13) is equivalent to

$$\bar{V}(T, c) = \min_{z \in \Re_+^m} \left\{ c'z + \Lambda \max_{S \subseteq N} R(S, A'z) \right\}. \tag{7.14}$$

The maximization problem above is an assortment optimization problem, where the revenue associated with ODF j is $p_j - \sum_{i \in M} a_{ij} z_i$ and the customers make their choices according to the MC choice model. Using the discussion in assortment optimization problems under the MC choice model, we know that the optimal objective value of this assortment optimization problem can be obtained by solving the linear program

$$\Lambda \max_{S \subseteq N} R(S, A'z)$$

$$= \min_{v \in \Re^n} \left\{ \Lambda \sum_{j \in N} \gamma_j \, v_j : v_j \geq p_j - \sum_{i \in M} a_{ij} \, z_i \; \forall \, j \in N, \; v_j \geq \sum_{k=1}^{n} \rho_{jk} \, v_k \; \forall \, j \in N \right\},$$

where the second problem above is a linear program involving the decision variables $v = \{v_j : j \in N\}$. Thus, problem (7.14) can be written as a two-level problem, where minimize at both levels, the decision variables at the outer level are z and the decision variables at the inner level are v. Since we minimize at both levels, we can solve this problem as a single-level problem, simultaneously minimizing over the decision variables z and v to obtain the problem

$$\begin{aligned} \bar{V}(T, c) \;=\; \min \quad & \sum_{i \in M} c_i \, z_i + \Lambda \sum_{j \in N} \gamma_j \, v_j \\ \text{s.t.} \quad & v_j \geq p_j - \sum_{i \in M} a_{ij} \, z_i && \forall \, j \in N \\ & v_j \geq \sum_{k=1}^{n} \rho_{jk} \, v_k && \forall \, j \in N \\ & z_i \geq 0, \;\; v_j \text{ is free} && \forall \, i \in M, \;\; j \in N \end{aligned}$$

Letting $\{\Lambda \, x_j : j \in N\}$ and $\{\Lambda \, y_j : j \in M\}$, respectively, be the dual variables associated with the two sets of constraints above and writing down the dual of this problem, we immediately obtain problem (7.7). Therefore, the optimal objective value of problem (7.7) is equal to the optimal objective value of the problem above, as desired. □

Part III
Pricing Analytics

Chapter 8
Basic Pricing Theory

8.1 Introduction

This chapter provides an introduction to multi-product monopoly pricing when the variable costs are linear. Profit maximization problems with linear variable costs arise from capacity constraints, where the firm maximizes the expected profit net of the opportunity costs of the capacities used. We argue that under mild assumptions, both the optimal profit function and the expected consumer surplus are convex functions of the variable costs. Consequently, when variable costs are random, both the firm and the representative consumer benefit from prices that dynamically respond to changes in variable costs. Randomness in variable cost is often driven by randomness in demand in conjunction with capacity constraints, and this accounts for some of the benefits of dynamic pricing. We explore conditions for the existence and uniqueness of maximizers of the expected profit and analyze in detail problems with capacity constraints both when prices are set for the entire sales horizon a priori, and when prices are allowed to change during the sales horizon. The firm's problem is discussed in Sect. 8.2, while the representative consumer's problem is presented in Sect. 8.3. The case with finite capacity is discussed in Sect. 8.4. Details about existence and uniqueness for single product problems are discussed in Sect. 8.5. This section also includes applications to priority pricing, social planning, multiple market segments, and peak-load pricing. Multi-product pricing problems are discussed in Sect. 8.6.

G. Gallego, H. Topaloglu, *Revenue Management and Pricing Analytics*,
International Series in Operations Research & Management Science 279,
https://doi.org/10.1007/978-1-4939-9606-3_8

8.2 The Firm's Problem

Consider a firm with variable cost vector $z = (z_1, \ldots, z_n)$ for n products. The firm's profit function is given by

$$R(p, z) := (p - z)'d(p) = \sum_{i=1}^{n}(p_i - z_i)\, d_i(p_1, \ldots, p_n), \tag{8.1}$$

where p and $d(p)$ are the vector of prices and expected demands as a function of prices, all of dimension $n \geq 1$. The goal is to find a prices, say $p(z)$, that maximizes $R(p, z)$. The profit function (8.1) models situations with linear variable costs. Linear costs arise as dual variables of capacity constraints, and this is our primary motivation for the study of this model. The maximum profit, as a function of z, is given by

$$\mathcal{R}(z) := \max_{p \in X} R(p, z), \tag{8.2}$$

where X is the set of allowable prices.

The set of allowable prices $X = X_1 \times \ldots X_n$ defines different type of optimization problems. The *assortment* optimization problem arises when $X_i = \{r_i, \bar{r}_i\}$, where r_i is the regular price of product i and $\bar{r}_i$ is the choke-off price, also known as the null-price for product i, so demand for product i is zero whenever it is priced at or above $\bar{r}_i$. The choke price $\bar{r}_i$ may be finite or infinity. On occasions, we will find it convenient to write ∞ instead of $\bar{r}_i$, and this should be interpreted as not offering product i. The *joint assortment and pricing* problem can be modeled by setting $X_i = \{r_{i1}, \ldots, r_{in_i}, \bar{r}_i\}$, where there is a finite price menu that includes the option of not offering product i. As an example, a product may be offered at the regular price, at a discounted price, or not offered at all. When the set X is finite, a maximizer $p(z)$ is guaranteed to exist. These are combinatorial problems and even relatively simple versions can be NP-hard. Fortunately, as we have seen in the chapter on assortment optimization, there are instances of practical importance that can be solved efficiently, sometimes by linear programming.

In this chapter, we are mainly concerned with the *continuous pricing* problem where $X_i = \Re_+ = [0, \infty)$ for all $i = 1, \ldots, n$. For this problem, we should more formally write $\mathcal{R}(z) = \sup_{p \in X} R(p, z)$ as the maximum may not be attained. We will later investigate conditions that guarantee the existence and uniqueness of a finite maximizer $p(z)$, as well as comparative statics that inform us of how $p(z)$ changes with z. To facilitate the exposition, we would use max instead of sup, but all the arguments except where noted would continue to hold for sup. Regardless of whether an optimizer exists or is unique, we can show that the profit function $\mathcal{R}(z)$ is a decreasing convex function of the cost vector z.[1]

Theorem 8.1 *$\mathcal{R}(z)$ is decreasing convex in z.*

8.2.1 Random Costs

We now investigate the impact of randomness in the variable cost vector. The motivation for this is that in dynamic pricing, the variable cost vector depends on the remaining capacity and time-to-go, and since the remaining capacity depends on random realizations of arrivals and sales, it follows that the variable costs change randomly over time. Let Z denote a vector of random unit costs. Then, by Jensen's inequality $\mathbb{E}[\mathcal{R}(Z)] \geq \mathcal{R}(\mathbb{E}[Z])$. The difference between $\mathbb{E}[\mathcal{R}(Z)]$ and $\mathcal{R}(\mathbb{E}[Z])$ can be interpreted as the difference between a dynamic pricing policy $p(Z)$ that responds to changes in Z and a static pricing policy $p(\mathbb{E}[Z])$ that does not. The following proposition allows us to assess the difference between $\mathbb{E}[\mathcal{R}(Z)]$ and $\mathcal{R}(\mathbb{E}[Z])$.

Proposition 8.2 *If the function $\mathcal{R} : \Re^n \rightarrow \Re$ is twice continuously differentiable, and H is the Hessian of $\mathcal{R}(z)$ evaluated at $\mathbb{E}[Z]$, then*

$$\mathbb{E}[\mathcal{R}(Z)] - \mathcal{R}(\mathbb{E}[Z]) \simeq \frac{1}{2}\mathbb{E}[(Z - \mathbb{E}[Z])'H(Z - \mathbb{E}[Z])] \geq 0.$$

This suggests that the difference between static and dynamic pricing is large when the variance of Z is large and $\mathcal{R}$ has significant curvature at $\mathbb{E}[Z]$. On the other hand, in situations where there is little variance in Z or little curvature in $\mathcal{R}$ around $\mathbb{E}[Z]$, we expect dynamic pricing to be of little help in improving profits. Proposition 8.2 follows by taking a second-order Taylor expansion of $\mathcal{R}(z)$ and using convexity.

Example 8.3 Suppose $d(p) = 100(1 - p)$ for $p \in [0, 1]$ and $d(p) = 0$ for $p > 1$. Then $R(p, z) = 100(p - z)(1 - p)$ is maximized at $p(z) = 0.5(1 + z)$ for $z \in [0, 1)$, $z \leq p(z) \leq 1$, and $\mathcal{R}(z) = 25(1 - z)^2$. If Z is random, with $P(Z = 1/3) = P(Z = 2/3) = 0.5$, then by Jensen's inequality $\$6.94 = \mathbb{E}[\mathcal{R}(Z)] \geq \mathcal{R}(\mathbb{E}[Z]) = R(p(\mathbb{E}[Z]), \mathbb{E}[Z]) = \mathbb{E}[R(p(\mathbb{E}[Z]), Z)] = \$6.25$, so a firm that responds to changes in Z makes 11.1% more profits than one who prices based on $\mathbb{E}[Z]$. Furthermore, a firm who can respond to changes in variable costs benefits from randomness in costs.

[1] We use the terms increasing, decreasing, concave and convex in the weak sense unless stated otherwise.

The following corollary pushes the idea a bit further.

Corollary 8.4 *If* $g(z) : \Re^m \to \Re^n_+$ *is increasing in* z, *then* $\mathcal{R}(g(z))$ *is decreasing in* z. *If* $g(z)$ *is also concave, then* $\mathcal{R}(g(z))$ *is convex in* z. *Moreover, if* $Z \in \Re^m$ *is random, then* $\mathbb{E}[\mathcal{R}(g(Z))] \geq \mathcal{R}(g(\mathbb{E}[Z]))$.

We can interpret $g(z)$ as the vector of unit costs for the products and z as the vector of unit costs of the resources that are used to build the products. As an example, if $g(z) = A'z$ and $A \geq 0$ is an $m \times n$ matrix, with A_{ij} the number of units of resource i required for product j, then $\mathcal{R}(A'z)$ is convex in z. This shows that a risk-neutral firm is better off with random component costs Z than with deterministic component costs equal to $\mathbb{E}[Z]$, provided it can charge prices $p(A'Z)$. The case $A = I$ represents the case where product i only uses component i, and results in the price vector $p(Z)$.

8.3 The Representative Consumer's Problem

While the firm is better off using dynamic pricing $p(Z)$, the reader may wonder whether consumers are better off with $p(Z)$ or with $p(\mathbb{E}[Z])$. In other words, do consumers prefer dynamic or static prices?

To answer this question, we will use the framework of utility theory, where we frame the question in terms of the surplus of the representative consumer. Suppose that a representative consumer derives utility $U(q)$ from purchasing a non-negative vector $q = (q_1, \ldots, q_n)$ of products. It is typically assumed that $U(q)$ is an increasing concave function of q. The *net utility* or consumers' surplus at (q, p) is given by

$$S(q, p) := U(q) - q'p,$$

which is simply the utility $U(q)$ minus the cost of purchasing the bundle q at prices $p = (p_1, \ldots, p_n)$. The *optimal surplus*, also known as the net indirect utility, in the absence of a budget constraint is given by

$$\mathcal{S}(p) := \max_{q \geq 0} S(q, p).$$

The solution, say $q = d(p)$, if it exists, gives us the bundle demanded at p. The first-order condition for optimality (ignoring the non-negativity constraints) is $\nabla U(q) - p = 0$, which is also sufficient given the assumed concavity. If there exists an inverse function, say $\nabla^{-1}U(p)$, then $d(p) = \nabla^{-1}U(p)$ for all p in the set $\mathcal{P} = \{p : \nabla^{-1}U(p) \geq 0\}$. The demand function $d(p)$ can be extended to a set larger than $\mathcal{P}$, but then at least one of the non-negativity constraints on q will be binding, so the problem needs to be projected into a subspace of non-negative demands. The following theorem shows that $\mathcal{S}(p)$ is decreasing convex in p and how $\mathcal{S}(p)$ changes with $p \in \mathcal{P}$.

Theorem 8.5 *$\mathcal{S}(p)$ is decreasing convex in p. Moreover, if $d(p)$ is differentiable in $p \in \mathcal{P}$, then $\nabla \mathcal{S}(p) = -d(p) \leq 0$ for all p in the interior of the set $\mathcal{P}$.*

One implication from Theorem 8.5 is that if p is one dimensional, then $\mathcal{S}(p) = \int_p^\infty d(x)dx$. In particular, if Ω is a random variable with finite mean, and $d(p) = P(\Omega \geq p)$ then

$$\mathcal{S}(p) = \int_p^\infty P(\Omega \geq x)dx = \mathbb{E}[(\Omega - p)^+],$$

so the expected surplus for a consumer with willingness to pay Ω is the expectation of the gain, $(\Omega - p)^+$, from the transaction if it happens.

Returning to the n-dimensional case, Jensen's inequality and the convexity of $\mathcal{S}(p)$ imply that if the price vector P is random with finite expectation $\mathbb{E}[P]$, then $\mathbb{E}[\mathcal{S}(P)] \geq \mathcal{S}(\mathbb{E}[P])$, so consumers prefer random prices P over expected prices $\mathbb{E}[P]$. Notice that this result does not assume that the representative consumer is risk-neutral, as we have already taken into account risk preference through the utility function. This result gives us hope that customers may prefer dynamic prices $p(Z)$ over static prices $p(\mathbb{E}[Z])$. The next result gives sufficient conditions for this to be true.

Corollary 8.6 *If $p(z) : \Re^n \to \Re^n_+$ is increasing in z, then $\mathcal{S}(p(z))$ is decreasing in z. If $p(z)$ is increasing concave in z, then $\mathcal{S}(p(z))$ is decreasing convex in z. Moreover, if $Z \in \Re^m$ is random and $\mathcal{S}(p(z))$ is convex in z, then*

$$\mathbb{E}[\mathcal{S}(p(Z))] \geq \mathcal{S}(p(\mathbb{E}[Z])).$$

This result follows directly from Corollary 8.4. Intuitively, if $p(Z)$ is increasing concave, then $\mathbb{E}[p(Z)] \leq p(\mathbb{E}[Z])$, so prices are lower on average under dynamic pricing. If we combine this with the fact that consumers prefer random prices, we obtain a weaker sufficient condition, namely that $\mathbb{E}[p(Z)] \leq p(\mathbb{E}[Z])$ implies $\mathbb{E}[\mathcal{S}(p(Z)) \geq \mathcal{S}(p(\mathbb{E}[Z]))]$. To verify this, notice that

$$\mathbb{E}[\mathcal{S}(p(Z))] \geq \mathcal{S}(\mathbb{E}[p(Z)]) \geq \mathcal{S}(p(\mathbb{E}[Z])),$$

where the first inequality follows from the convexity of $\mathcal{S}(p)$, and the second from the assumption that $\mathbb{E}[p(Z)] \leq p(\mathbb{E}[Z])$, and the fact that $\mathcal{S}(p)$ is decreasing in p.

We will now argue that $\mathcal{S}(p(z))$ is convex up to a quadratic approximation of an increasing concave utility function.

Theorem 8.7 *The function $\mathcal{S}(p(z))$ is convex in z up to a quadratic approximation of any increasing concave utility function.*

For the single product case, it is possible to show that $p(z)$ is linear if and only if the demand function belongs to one of the following three classes:

- $d(p) = \lambda \exp(-p/\theta)$ for $\lambda, \theta > 0$: the exponential demand.

- $d(p) = ((a - bp)^+)^c$ for $a, b, c > 0$: root-linear demand (linear for $c = 1$).
- $d(p) = (a+bp)^{-c}$ for $a, b > 0, c > 1$: constant elasticity of substitution demand ($a = 0$).

The class of single product demand functions for which $p(z)$ is linear contains many of the demand functions that appear commonly in the literature. For any of these demand functions, $\mathcal{S}(p(z))$ is convex in z, and consequently $\mathbb{E}[\mathcal{S}(p(Z))] \geq \mathcal{S}(p(\mathbb{E}[Z]))$. There are cases, where $p(z)$ is increasing convex and yet $\mathcal{S}(p(z))$ is still convex in z provided that $p(z)$ is not "too" convex.

Example 8.8 Suppose that $n = 1$, and $U(q) = q - q^2/200$, then $d(p) = 100(1-p)$ over $p \in [0, 1]$, $p(z) = 0.5(1 + z)$ for $z \in [0, 1]$ and $\mathcal{S}(p(z)) = 12.5(1 - z)^2$. Assume again, as in Example 8.3, that $\mathbb{P}(Z = 1/3) = \mathbb{P}(Z = 2/3) = 1/2$. By Jensen's inequality, we have $\$3.47 = \mathbb{E}[\mathcal{S}(p(z))] \geq \mathcal{S}(p(\mathbb{E}[Z])) = \$3.125$, so consumers are better off by 11.1% when prices are dynamic and driven by Z compared to static prices $p(\mathbb{E}[Z])$. From Example 8.3, we see that the firm is also 11.1% better off using dynamic pricing, resulting in a win-win situation.

8.4 Finite Capacity

Finite capacity is a central theme for both revenue management and dynamic pricing. We will assume that $d(p) : \Re^n_+ \to \Re^n_+$ is continuous function in p and that variable costs are zero. Let c be an m-dimensional vector of resources available for the n products. Let A be an $m \times n$ matrix where the j-th column, say A_j, is the vector of resources consumed by each unit of product j. For $X = \Re^n_+$, the optimal revenue as a function of c is given by

$$\bar{V}(c) := \max_{p \in X} R(p, 0) \text{ subject to } A\,d(p) \leq c. \tag{8.3}$$

The interpretation is that there is a sunk investment in capacity c, the firm wants to maximize the revenue that can be obtained from this capacity, and no variable costs are incurred. The objective function of problem (8.3) may not be concave and the constraint set may not be convex, which makes solving problem (8.3) potentially difficult. There are two techniques that we can use to try to solve problem (8.3). First, we can work with the inverse demand function $p(q)$ assuming it exists. In this case, problem (8.3) can be written as maximizing $p(q)'q$ subject to $A\,q \leq c, q \geq 0$. Now the constraint set is convex and if the objective function is concave, or quasi-concave, then standard techniques such as the KKT conditions and its extensions can be used to solve the problem. The second technique is based on Lagrangian relaxation, as outlined next.

8.4.1 Lagrangian Relaxation

We will explore the Lagrangian relaxation approach on problem (8.3) under the assumption that $R(p, w) := (p - w)'d(p)$ has a finite maximizer $p(w)$ for all $w \in \Re^n_+$ with $d(p(w))$ continuous in w. Although these are strong conditions, they turn out to hold for many important applications. A sufficient condition for the existence of a finite maximizer is that $R(p, w)$ is upper semi-continuous (USC)[2] in p, and that for some $\alpha(w) \in \Re_+$, the upper contour set $\{p : R(p, w) \geq \alpha(w)\}$ is non-empty and compact. In this case, by the extreme value theorem (EVT),[3] the function $R(p, w)$ achieves its maximum over the compact upper contour set. Sharper conditions for the existence of $p(w)$ will be presented later.

Let

$$L(p, z) := R(p, 0) + z'(c - Ad(p)) = R(p, A'z) + z'c$$

be the Lagrangian corresponding to the dual vector $z \in \Re^m_+$. The Lagrangian program is $\min_{z\geq 0} \max_{p\geq 0} L(p, z)$. The inner optimization yields

$$L(z) := \max_{p\geq 0} L(p, z) = \mathcal{R}(A'z) + z'c,$$

where we have taken advantage of the assumption that there exists a price, say $p(A'z)$, that maximizes $R(p, A'z)$, and $\mathcal{R}(A'z) = R(p(A'z), A'z)$.

By weak duality $L(z) \geq \bar{V}(c)$ for all $z \geq 0$. The Lagrangian dual problem is

$$\Gamma(c) := \min_{z\geq 0} L(z) = \min_{z\geq 0} \left[\mathcal{R}(A'z) + z'c\right] \geq \bar{V}(c),$$

which is a convex minimization problem in z subject to non-negativity constraints $z \in \Re^m_+$.

If $c \geq A\,d(p(0))$, then $z(c) = 0 \in \Re^m$ is optimal. This follows because $p(0)$ is feasible and maximizes $R(p, 0)$. Otherwise $z(c)$ has at least one positive component. We next investigate conditions under which there is no duality gap and we can assert that $\bar{V}(c) = \Gamma(c)$. If $p(A'z(c))$ satisfies the capacity constraint, and the complementary slackness condition $z(c)'(c - A\,d(p(A'z(c)))) = 0$ holds, then there is no duality gap, since then $L(z(c)) = R(p(A'z(c)), 0) = \bar{V}(c)$, and consequently $p(A'z(c))$ is an optimal solution to problem (8.3). We summarize this result in the following proposition.

---

[2]A function $f : X \to [-\infty, \infty]$ is upper semi-continuous if and only if $\{x \in X : f(x) \geq a\}$ is closed for every $a \in \Re$.

[3]The EVT is also known as the Bolzano-Weierstrass theorem.

Proposition 8.9 *Assume that* $R(p, A'z)$ *has a solution* $p(A'z)$ *for any* $z \geq 0$*, and let* $z(c) \geq 0$ *solve the convex optimization problem* $\min_{z \geq 0}[\mathcal{R}(A'z) + z'c]$*. If* $A\,d(p(A'z(c))) \leq c$ *and* $z(c)'[c - A\,d(p(A'z(c)))] = 0$*, then* $p(A'z(c))$ *is a solution to problem* (8.3).

8.4.2 Finite Capacity and Finite Sales Horizon

Another central theme for dynamic pricing and revenue management is the existence of a finite sales horizon over which the products can be sold. Let t be the time-to-go, and assume a sales horizon of length T. At the end of the sales horizon, no further sales are possible. Let c be the initial inventory, and assume that inventory replenishments are not possible. This situation is typical in fashion retailing and in revenue management applications. We will assume that the demand *rate* $d_t(p)$ at price p at time-to-go t is continuous in p for all $t \in [0, T]$. The profit contribution over the sales horizon from using price path $p_t, t \in [0, T]$ is given by $\int_0^T R_t(p_t, 0)dt := \int_0^T p_t' d_t(p_t)dt$. The optimal revenue as a function of T and c is given by

$$\bar{V}(T, c) := \int_0^T \max_{p_t \in X} R_t(p_t, 0)dt \tag{8.4}$$

$$\text{s.t.} \quad \int_0^T A d_t(p_t)dt \leq c,$$

where as before $X = \Re_+^n$ and $A \geq 0$ is an $m \times n$ matrix representing the consumption of resources by products.

The Lagrangian penalizes component shortfalls at rate $z \in \Re_+^m$. We will assume that there exists a price $p_t(A'z)$ that maximizes $R_t(p, A'z)$ for every $t \in [0, T]$. The inner optimization of the Lagrangian function yields $\int_0^T \mathcal{R}_t(A'z) + z'c$, so the outer optimization is given by

$$\Gamma(T, c) := \min_{z \geq 0} \left[\int_0^T \mathcal{R}_t(A'z)dt + z'c \right] \geq \bar{V}(T, c) \tag{8.5}$$

whose objective function is convex in z. Let $z(T, c)$ be the optimal solution to this convex optimization problem. If the price path $p_t(z(T, c)), t \in [0, T]$ is feasible and the complementary slackness condition

$$z(T, c)' \left[c - A \int_0^T d_t(A' p_t(z(T, c))dt \right] = 0$$

holds, then there is no duality gap and the price path $p_t(z(T, c)), t \in [0, T]$ is an optimal solution to problem (8.4), turning the inequality in (8.5) into an equality.

It is instructive to compare formulation (8.4) to a formulation based on aggregate demand $D(p) = \int_0^T d_t(p)dt$ that yields

$$\bar{V}_f(T, c) := \max_{p \in X} p' D(p) \qquad (8.6)$$
$$\text{s.t.} \quad A\, D(p) \leq c,$$

Notice that in formulation (8.6) we seek a single vector of prices that maximizes revenue over the entire horizon subject to an aggregate capacity constraint. Clearly $\bar{V}(T, c) \geq \bar{V}_f(T, c)$, as the ability to respond to changes in demand $d_t(p)$ over the sales horizon $t \in [0, T]$ gives formulation (8.4) an important advantage over (8.6). Of course, this advantage can materialize only if consumers that arrive at time-to-go t either purchase at $p_t(z(T, c))$ or leave the system. This model assumes that the firm can do price discrimination over time. The model may break down when consumers are strategic and they face no disutility from waiting for a lower price, except when prices $p_t(z(T, c))$ are monotone so there is no incentive for waiting. In retailing, for example, some consumers are strategic and prefer to wait for lower prices, but they are exposed to rationing risks and the disutility of waiting.

8.5 Single Product Pricing Problems

In this section, we investigate issues of existence and uniqueness for single product pricing problems. We next study real options and bargaining as mechanisms to improve profits and reduce the dead weight loss. We end this section with a look at multiple market segments and direct price discrimination.

8.5.1 Existence and Uniqueness

For the single product case with $n = 1$, $d(p)$ is the demand for the single product at price $p \geq 0$. We seek sufficient conditions for the existence of a finite maximizer of $R(p, z) = (p-z)\, d(p)$ over $p \in \Re_+$. Let $\bar{d}(p) := \sup_{\tilde{p} \geq p} d(\tilde{p})$. Notice that $\bar{d}(p) \geq d(p)$ is a decreasing function even if $d(p)$ is not. Let $\bar{R}(p, z) := (p - z)\, \bar{d}(p)$. We next show that if $d(p)$ is USC and $p\bar{d}(p) \to 0$ as $p \to \infty$ (so $\bar{d}(p) = o(1/p)$), then $\bar{R}(p, z)$ has a finite maximizer $p(z)$ that also maximizes $R(p, z)$.

Theorem 8.10 *If $d(p)$ is USC in $p \geq 0$, and $\bar{d}(p) = o(1/p)$, then there exists a finite maximizer $p(z)$, increasing in $z \geq 0$, that simultaneously maximizes $R(p, z)$ and $\bar{R}(p, z)$, so $\mathcal{R}(z) = \bar{\mathcal{R}}(z)$.*

A formal proof of Theorem 8.10 is in the appendix. Notice that Theorem 8.10 does not require $d(p)$ to be decreasing or eventually decreasing in p. While the conditions of Theorem 8.10 may seem technical, they imply the existence of finite maximizers for pricing problems that are typically encountered in practice. For example, if we have a finite population of λ potential consumers with independent and identically distributed (IID) willingness to pay Ω, then the expected demand at price p is $d(p) = \lambda P(\Omega \geq p)$. Then $d(p)$ is USC, and if $E[\Omega^+] < \infty$, then $d(p) = o(1/p)$, so there exist a maximizers $p(z)$ of $R(p, z)$. As an example, assume that Ω is exponential with mean θ. Then $d(p) = \lambda e^{-p/\theta}$, and $p(z) = z + \theta$ maximizes $R(p, z)$, so $\mathcal{R}(z) = \theta\lambda e^{-1-z/\theta}$.

We now turn to conditions on the demand function $d(p)$ that guarantee that $R(p, z)$ does not have local, non-global, maximizers or more succinctly that $R(p, z)$ is unimodal in $p \geq z$. This is equivalent to $R(p, z)$ being quasi-concave in $p \geq z$ and to $R(p, z)$ having convex upper level sets: $\{p \geq z : R(p, z) \geq \alpha\}$ for all $\alpha \geq 0$. If $d(p)$ is continuous and differentiable, then we define the *hazard rate* at p to be $h(p) := -d'(p)/d(p)$ where $d'(p)$ is the derivative of $d(p)$ at p. The hazard rate function $h(p)$ is defined for all $p < \bar{r}$, where $\bar{r}$ is the choke-off price. The hazard rate is the event rate at price p, conditional on $\Omega \geq p$. Taking the derivative of $R(p, z)$ with respect to p leads to first-order condition for optimality:

$$f(p, z) = 1 - (p - z)h(p) = 0.$$

Let $p(z)$ be a root of $f(p, z) = 0$. Then, $p(z)$ is a maximizer of $R(p, z)$, and $R(p, z)$ is quasi-concave if $f(p, z)$ is non-negative for all $p < p(z)$ and non-positive for all $p > p(z)$. The following result provides conditions on the hazard rate that guarantee the existence and uniqueness of a finite maximizer $p(z)$, as well as some results about the optimal mark-up $\Delta(z) := p(z) - z$.

Theorem 8.11

(a) If $h(p)$ is continuous and increasing in p and $h(z) > 0$, then there is a unique optimal price $p(z)$, strictly increasing in z, satisfying $z < p(z) \leq z + 1/h(z)$, with $\Delta(z) = p(z) - z$ decreasing in z. The upper bound is attained by the exponential demand function.

(b) If $ph(p)$ is continuous and strictly increasing in p and there exists a finite $\tilde{z} \geq z > 0$ such that $1 < \tilde{z}h(\tilde{z})$, then there is a unique optimal price $p(z)$, strictly increasing in z, satisfying $z < p(z) \leq z/(1 - 1/\tilde{z}h(\tilde{z}))$. The upper bound is attained by the constant elasticity of demand function. Moreover, if $1/h(z)$ is concave in z, then $p(z)$ is concave in z.

(c) If $\tilde{d}(p)$ is a demand function with hazard rate $\tilde{h}(p)$ and $\tilde{h}(p) \geq h(p)$ for all p, then $\tilde{p}(z) \leq p(z)$.

The condition $ph(p)$ increasing in p is weaker than $h(p)$ increasing in p, and leads to weaker results as we cannot claim that $\Delta(z)$ is decreasing in z. As an example, for the constant elasticity of demand model, $d(p) = \lambda p^{-b}$, $b > 1$, we have $\Delta(z) = z/(b - 1)$, which is increasing in z.

Economists often write the solution to the first-order condition $f(p, z) = 0$ in terms of the (absolute) price *elasticity of demand* $e(p) := -pd'(p)/d(p) = ph(p)$ resulting in

$$p(z) = \frac{e(p(z))}{e(p(z)) - 1}z.$$

This formula suggests that the mark-up on marginal cost should be equal to $e(p(z))/(e(p(z)) - 1)$. Notice that both the left and the right hand sides depend on $p(z)$ except for the constant elasticity demand model, so the mark-up interpretation needs to be taken with a grain of salt. Nevertheless, this mark-up formula provides some guidelines that link elasticities to prices via the mark-up on marginal costs.

The solution to the first-order condition is sometimes written as

$$\frac{\Delta(z)}{p(z)} = \frac{1}{e(p(z))},$$

with the left hand side known as the Lerner index, so the Lerner index is equal to one over the elasticity of demand. If $z = 0$, then $\Delta(z) = p(z)$, so $e(p(0)) = 1$. The last equation is often written as

$$p(z) = z + 1/h(p(z)).$$

It can be shown that if $1/h(p)$ is concave (respectively, convex) in p then $p(z)$ is increasing concave (respectively, convex) in z.

The problem of maximizing $R(p, z)$ can sometimes be transformed so that demand rather than price is the decision variable. This can be done if there is an inverse demand function, say $\tilde{p}(q)$, that yields demand $q \leq d(0)$ at price $\tilde{p}(q)$. The problem is to maximize $(\tilde{p}(q) - z)q$ over $q \geq 0$. It can be shown that the concavity of $\tilde{p}(q)q$ in q is equivalent to the convexity of $1/d(p)$ in p, so from this we surmise that another sufficient condition for $R(p, z)$ to be quasi-concave in p is that $1/d(p)$ is convex. A weaker condition for the quasi-concavity of $(\tilde{p}(q) - z)q$ is that $\tilde{p}(q)$ is log-concave in q. It is interesting to note that there are demand functions for which $R(p, z)$ is concave in p without $(\tilde{p}(q) - z)q$ being concave in q.

8.5.2 *Priority Pricing*

Consider the finite capacity problem for the single product case. We will assume that there is a unique solution, say $p(z(c))$, to the problem of maximizing $R(p, 0)$ subject to $d(p) \leq c$, where $z(c)$ is the dual variable associated with the capacity constraint and that $d(p)$ is continuous in p. Let $\bar{c}$ be the smallest integer at which the dual variable is zero. Then $p(z(c)) = p(0)$ for all $c \geq \bar{c}$.

Suppose that capacity is a random variable, say C, and that the firm prices at $p(z(C))$. Since the price is the same for all $C \geq \bar{c}$, it is convenient to redefine C to be $\min(C, \bar{c})$, so its support is in $\{0, 1, \ldots, \bar{c}\}$. With this notation, the expected profit to the firm is equal to

$$\mathbb{E}[Cp(z(C))] = \sum_{c=1}^{\bar{c}} cp(z(c))\mathbb{P}(C = c).$$

Changing the order of summation, we see that the average price paid for the cth unit of capacity is equal to $\mathbb{E}[p(z(C))|C \geq c]$. This pricing policy applies to situations where yields are random, and the firm can pass the price signal $p(z(C))$ to consumers who select whether or not they want to buy at that price. The policy is somewhat controversial as it calls for the disposal of capacity when yields are high and can be perceived as price gouging when yields are low. In some instances, such as the consumption of power, consumers cannot react to the changes in capacity in real time. Therefore, the application of this scheme requires a priority matching to consumers who value the service the most, and this is why this is called a priority pricing schedule.

8.5.3 Social Planning and Dead Weight Loss

A social planner is interested in selecting p to maximize the sum of the consumers' surplus $\mathcal{S}(p)$ and the firm's profit $R(p, z)$. The sum of these two quantities is known as the *social welfare* function, given by

$$W(p, z) := \mathcal{S}(p) + R(p, z).$$

Optimizing over p, we obtain the *optimal welfare* function

$$\mathcal{W}(z) := \max_{p \geq z} W(p, z).$$

Proposition 8.12 *If $d(p)$ is differentiable and decreasing in p, then $\mathcal{W}(z) = \mathcal{S}(z)$ is decreasing convex in z.*

The result follows because under the stated conditions $\frac{\partial W(p,z)}{\partial p} = (p-z)d'(p) \leq 0$, so social welfare is decreasing in p and its maximum is attained at $p = z$.

The difference $\mathcal{W}(z) - W(p(z), z)$ is known as the *dead weight loss*. It reflects the difference between the optimal social welfare and the social welfare that results when the firm maximizes its profits. As an example, if $d(p) = \lambda e^{-p/\theta}$, then $p(z) = z + \theta$, $\mathcal{W}(z) = \mathcal{S}(z) = \lambda\theta e^{-z/\theta}$, while $W(p(z), z) = 2\lambda\theta e^{-1}e^{-z/\theta}$, so the dead weight loss is equal to $[1 - 2e^{-1}]\mathcal{W}(z)$ or 26% of the maximum social welfare.

Trying to reduce the dead weight loss is difficult because the optimal solution to the social planner's problem is to set $p = z$ and this results in zero profits for the firm with all of the benefits going to the consumers. We will next explore two cases where the dead weight loss can either be eliminated or reduced. The first case requires the use of real options on services when the booking and the consumption of the service are separated by time and consumers are uncertain about their valuations at the time of booking. The second case corresponds to the situation where the consumers and the firm negotiate instead of using a take it or leave it price.

Call Options on Capacity

Consider first the case of a homogeneous group of consumers booking capacity in advance of consumption. Suppose there are λ consumers, each with independent and identically distributed random willingness-to-pay for the service at the time of consumption. We assume that the distribution $H(p) = \mathbb{P}(\Omega \geq p)$ is common knowledge to consumers and the firm, so the aggregate demand function is $d(p) = \lambda\,\mathbb{P}(\Omega \geq p)$. For this model, the surplus function is $\mathcal{S}(p) = \lambda\mathbb{E}[(\Omega - p)^+]$. We further assume that consumers do not learn the realization of demand until the time of consumption. Under these conditions, the firm can benefit from offering call options to consumers. A call option requires an upfront, non-refundable, payment x that gives the customer the non-transferable right to buy one unit of the service at price p at the time of consumption. The special case where $p = 0$ is called advanced selling, and the case $x = 0$ is called spot selling.

Consumers evaluate call options by the surplus they provide. A customer who buys an (x, p) option will exercise his right to purchase one unit of the service at the time of consumption if and only if $\Omega \geq p$. By doing this, an individual consumer obtains expected surplus $s(p) = \mathbb{E}[(\Omega - p)^+] = \mathcal{S}(p)/\lambda$. Since consumers pay x for this right, the consumer receives expected surplus $s(p) - x$ and would find the (x, p) option attractive only if $s(p) - x \geq 0$.

Consider the problem of maximizing the expected profit from selling (x, p) options subject to the participating constraint $s(p) - x \geq \tilde{s}$, where $\tilde{s} \geq 0$ is a lower bound on the individual surplus that needs to be given to consumers to induce them to buy the option. In practice, the firm may set $\tilde{s} = 0$ to extract as much surplus from consumers. Here we will analyze the problem for other values of $\tilde{s}$ to show that it is possible to eliminate the dead weight loss and use $\tilde{s}$ as a mechanism to distribute profits and surplus between the firm and the consumers.

Since the expected profit from selling (x, p) options that satisfy the participating constraint is $x + (p - z)\,\mathbb{P}(\Omega \geq p)$ and there are λ consumers, the expected profits are equal to $\lambda x + (p - z)\lambda\,\mathbb{P}(\Omega \geq p) = \lambda x + R(p, z)$. This is a function of x, and it is optimal to set $x^* = s(p) - \tilde{s}$. This reduces the problem to that of maximizing $\lambda(s(p) - \tilde{s}) + R(p, z) = \mathcal{S}(p) + R(p, z) - \lambda\tilde{s} = W(p, z) - \lambda\tilde{s}$ with respect to p. We already know that $W(p, z)$ is maximized at $p = z$. Thus, the solution to the provider's problem is to set $p = z$ and $x = s(z) - \tilde{s}$, so the provider obtains profits equal to $\lambda x^* = \lambda(s(z) - \tilde{s})$, while consumers receive surplus $\lambda\tilde{s}$. Since the sum of these two quantities is $\mathcal{S}(z) = \lambda s(z)$, the selling of call options eliminates the dead weight loss and $\tilde{s}$ can be used as a mechanism to distribute the dead weight loss.

We now explore the range of values of $\tilde{s}$ that guarantees that both the firm and the consumers are at least as well off as the solution $(x, p) = (0, p(z))$, where price $p(z)$ is offered to consumers after they know their valuations. Under this scheme, the firm makes $\mathcal{R}(z)$ and consumers receive surplus $\mathcal{S}(p(z)) = \lambda s(p(z))$. As a result, consumers are better off whenever $\lambda\tilde{s} \geq \mathcal{S}(p(z))$, while the firm is better off whenever $\mathcal{S}(z) - \lambda\tilde{s} \geq \mathcal{R}(z)$, so a win-win is achieved for any value of $\tilde{s}$ such that $\mathcal{S}(p(z)) \leq \lambda\tilde{s} \leq \mathcal{S}(z) - \mathcal{R}(z)$. Since $\mathcal{S}(z) \geq \mathcal{R}(z) + \mathcal{S}(p(z))$, the win-win interval is non-empty. In practice, absent competition or an external regulator, the provider may simply select $\tilde{s} = 0$, to improve his profits from $\mathcal{R}(z)$ to $\mathcal{W}(z)$ extracting all consumer surplus while also capturing the dead weight loss. The improvement in profits from options can be very significant. Indeed, in the exponential case, $(\mathcal{W}(z) - \mathcal{R}(z))/\mathcal{R}(z) = (e - 1) = 172\%$.

The idea of using call options can be extended to the case where the variable cost Z of providing the service at the time of consumption is random. In this case, the option is designed by setting $x = \mathbb{E}[s(Z)] - \tilde{s}$ and $p = Z$, so that by paying x in advance the option bearer has the right to purchase one unit of the service at the random variable cost Z.

Bargaining Power

Assume again that demand comes from λ homogeneous consumers with willingness to pay Ω, so $d(p) = \lambda\, \mathbb{P}(\Omega \geq p)$. Without negotiation, the firm sets the price at $p(z)$ and consumers make a purchase if $\Omega \geq p(z)$ and leave the system otherwise. In this section, we will show that the dead weight loss can be reduced if the firm and the consumers negotiate instead of using non-negotiable prices. Suppose that consumers know the realization of their willingness to pay, but the firm knows only the distribution of Ω. We will assume that the firm has a reservation price, say p, under which it is not willing to sell. We will assume the firm or an agent for the firm negotiates with each customer. If $\Omega < p$, then no sale takes place, but if $\Omega \geq p$, we will assume that a sale takes place at the Bargaining Nash Equilibrium (BNE) price $\beta\Omega + (1 - \beta)p$, where $\beta \in [0, 1]$ is the negotiating power of the firm and $1 - \beta$ is the negotiating power of the buyers. Notice that if $\Omega \geq p$, then transaction takes place at the reservation price p, when $\beta = 0$, and at Ω, when $\beta = 1$.

The problem for the firm is to select the reservation price, say $p_\beta(z)$, to maximize expected profits taking into account both the unit cost z and the negotiating power β. Let $\delta(\Omega - p)$ be a random variable taking value 1 if $\Omega \geq p$ and 0 otherwise. The firm wants to select the reservation price p to maximize

$$
\begin{aligned}
R_\beta(p, z) &:= \lambda\, \mathbb{E}[(\beta\Omega + (1 - \beta)p - z)\delta(\Omega - p)] \\
&= \lambda\, \mathbb{E}[\beta(\Omega - p)\,\delta(\Omega - p)] + \lambda(p - z)d(p) \\
&= \beta\lambda\, \mathbb{E}[(\Omega - p)^+] + R(p, z) \\
&= \beta\mathcal{S}(p) + R(p, z). \qquad (8.7)
\end{aligned}
$$

Let

$$\mathcal{R}_\beta(z) := \max_p R_\beta(p, z).$$

Let $h(p)$ be the hazard rate of $d(p) = \lambda \mathbb{P}(\Omega \geq p)$. If $ph(p)$ is increasing in p, then the maximizer of $R_\beta(p, z)$, say $p_\beta(z)$, is the unique root of the equation $(p - z)h(p) = 1 - \beta$. It is easy to see that $p_\beta(z)$ is decreasing in β and increasing in z, while $\mathcal{R}_\beta(z)$ is increasing in β and decreasing in z. By substituting $p_\beta(z)$ into the formula for $R_\beta(p, z)$, we obtain

$$\mathcal{R}_\beta(z) = \beta \mathcal{S}(p_\beta(z)) + R(p_\beta(z), z).$$

At $\beta = 0$, we have $p_0(z) = p(z)$ and $\mathcal{R}_0(z) = \mathcal{R}(z)$. Consequently, pricing at $p(z)$ is tantamount to assuming that the firm has no negotiating power, or equivalently relinquishing the negotiating power. This may be done for expediency for relatively inexpensive goods that are sold in high volumes. At $\beta = 1$, $p_1(z) = z$, so $\mathcal{R}_1(z) = \mathcal{W}(z)$, eliminating all of the dead weight loss, with the firm extracting all of the consumers' surplus. In most cases, $\beta \in (0, 1)$, so it makes sense for the firm to negotiate with consumers for goods that are expensive and sold in relatively low volumes. Indeed, prices for real estate, cars, art, and high-end services are often negotiated, while those of groceries are typically not except in economies where people have more time than money.

The consumers' expected surplus is given by

$$\begin{aligned} \mathcal{S}_\beta(p_\beta(z)) &= \lambda \mathbb{E}[(\Omega - \beta\Omega - (1 - \beta)p_\beta(z))\,\delta(\Omega - p_\beta(z))] \\ &= \lambda(1 - \beta)\,\mathbb{E}[(\Omega - p_\beta(z))^+] \\ &= (1 - \beta)\,\mathcal{S}(p_\beta(z)). \end{aligned} \tag{8.8}$$

It is easy to see that the consumers' surplus is decreasing in β, so some of the benefits that the firm derives from negotiation comes from smaller surplus for consumers.

If we now add (8.7) and (8.8), and evaluate it at $p_\beta(z)$, we see that the social welfare that results from negotiation is equal to

$$W_\beta(p_\beta(z), z) := \mathcal{S}_\beta(p_\beta(z)) + \mathcal{R}_\beta(z) = \mathcal{S}(p_\beta(z)) + R(p_\beta(z), z).$$

This quantity is increasing in β. This follows because $W(p, z)$ is decreasing in p and $p_\beta(z)$ is decreasing in β. This implies that the firm makes more than the loss to the consumers when it has negotiating power.

8.5.4 Multiple Market Segments

Suppose that there are multiple market segments with independent demands $d_m(p), m \in \mathcal{M} := \{1, \ldots, M\}$ for $p \in \Re_+$ for a product. We will assume throughout this section that $d_m(p)$ satisfies the conditions of Theorem 8.10 for every $m \in \mathcal{M}$. This guarantees that there exists a $p_m(z)$ increasing in z that maximizes $R_m(p,z) := (p-z)d_m(p)$. If the firm can use direct price discrimination (also known as third degree price discrimination or personalized pricing), then it would use price $p_m(z)$ for market segment $m \in \mathcal{M}$. The possibility to use direct price discrimination arises when it is possible to vary price by time, location, or customer attributes. This is often true for services and less so for physical products as there may be a gray market which creates demand dependencies.

In some cases, we may need to offer the same price for a subset $S \subset \mathcal{M}$ of market segments. This may be due to regulations or if the markets are not sufficiently different. Let $d_S(p) := \sum_{m \in S} d_m(p)$ denote the aggregate demand over market segments in S at price $p \in \Re_+$, and let $R_S(p,z) := (p-z)d_S(p)$ denote the profit function for market segments in S when the variable cost is z. We seek conditions for the existence of a maximizer $p_S(z)$ of $R_S(p,z)$ that is in the convex hull of the set $\{p_m(z) : m \in S\}$.

The following result shows that $d_S(p)$ inherits some desirable properties from the individual market demand functions $d_m(p), m \in S$.

Proposition 8.13 *If $d_m(p)$ satisfies the conditions of Theorem 8.10 for every $m \in \mathcal{M}$, then so does $d_S(p)$. Moreover, there exists a finite price $p_S(z)$, increasing in z, such that $\mathcal{R}_S(z) = R_S(p_S(z), z)$ is decreasing convex in z.*

It may be tempting to conclude that, under the conditions of Proposition 8.13, $p_S(z)$ would lie in the convex hull of $\{p_m(z), m \in S\}$. Example 8.14 shows that this is not true.

Example 8.14 Suppose that $d_1(p) = 1$ for $p \leq 10$ and $d_1(p) = 0$ for $p > 10$. Then $R_1(p,0)$ is maximized at $p_1(0) = 10$ and $\mathcal{R}_1(0) = 10$. Suppose that $d_2(p) = 1$ for $p \leq 9$, $d_2(p) = 0.1$ for $9 < p \leq 99$ and $d_2(p) = 0$ for $p > 99$. Then $R_2(p,0)$ is maximized at $p_2(0) = 99$ resulting in $\mathcal{R}_2(0) = 9.9$. The total profit is equal to 19.9 if each segment is allowed to be priced separately. Let $S = \{1,2\}$, then $R_S(p,0) = R_1(p,0) + R_2(p,0)$ is maximized at $p_S(0) = 9 < \min_{i \in S} p_i(0)$ resulting in $\mathcal{R}_S(0) = 18$.

Since the sum of quasi-concave functions is not, in general, quasi-concave, it should not be surprising that properties of $d_m(p)$ that imply quasi-concavity of $R_m(p,z)$, for each $m \in \mathcal{M}$ are not, in general, inherited by $d_S(p) = \sum_{m \in S} d_m(p)$. Example 8.15 illustrates this.

Example 8.15

(a) Suppose that $d_m(p) = \exp(-p/b_m)$ for $m = 1,2$ with $b_1 < b_2$. Then the hazard rate $h_m(p) = 1/b_m$, is constant, and there is a unique price $p_m(z) =$

$z + b_m$ that maximizes $R_m(p, z)$ for $m = 1, 2$. Let $S = \{1, 2\}$. The hazard rate $h_S(p)$ of $d_S(p)$ is decreasing in p.

(b) Suppose that $d_m(p) = 1/p^{b_m}$ for some $b_m > 1$, then $ph_m(p) = b_m$ and there is a unique price $p_m(z) = b_m z/(b_m - 1)$ that maximizes $R_m(p, z)$ for $m = 1, 2$. Let $S = \{1, 2\}$. The proportional hazard rate $ph_S(p)$ of $d_S(p)$ is decreasing in p.

In both cases in Example 8.15, the profit function $R_S(p, z)$ is actually quasi-concave, even if the aggregate demand function $d_S(p)$ has decreasing hazard rate (Part a) or decreasing proportional hazard rate (Part b). The next result provides sufficient conditions to bound the maximizer of $R_S(p, z)$ to be within the convex hull of $p_m(z), m \in S$.

Proposition 8.16 *Assume that $d_m(p)$ satisfies the conditions of Theorem 8.10 for each $m \in M$, that the hazard rate $h_m(p)$ is continuous in p, and that $ph_m(p)$ is increasing in p for each $m \in M$. Then, $R_S(p, z)$ has a maximizer in the convex hull of $\{p_m(z), m \in S\}$ for all $S \subset \mathcal{M}$.*

Corollary 8.17 *Proposition 8.16 holds if $h_m(p)$ is increasing in p for all $m \in S$.*

We next consider the problem where we are allowed a price menu that consist of at most $J \leq M$ different prices. The limitation to J prices may be managerial in nature, or it may be due to the lack of precise knowledge of the demand parameters for some of the market segments. The extreme cases are $J = 1$, where a single price is used for all the segments (so there is no price discrimination) and $J = M$, where each segment is priced independently (full direct price discrimination). Let $Q_J(z)$ be the maximum profit from using J distinct prices for the M market segments when the marginal cost is z. For $J = 1$, we have $Q_1(z) = R_{\mathcal{M}}(z)$, and for $J = M$, we have $Q_M(z) = \sum_{m \in \mathcal{M}} \mathcal{R}_m(z)$. Clearly $Q_J(z)$ is increasing in J. For $1 < J < M$, the problem is combinatorial in nature, as we need to assign M market segments into J market clusters, with all market segments in a cluster using the same price.

Our aim in this section is to develop a heuristic and a lower bound on the profitability of using J prices. More precisely, we will develop a heuristic with profit $Q_J^h(z)$ such that

$$\frac{Q_J(z)}{Q_M(z)} \geq \frac{Q_J^h(z)}{Q_M(z)} \geq \gamma_J(z),$$

for some function $\gamma_J(z)$ for situations where all of the demand functions $d_m(p), m \in \mathcal{M}$ belong to the same family. As we shall see, it is often possible to obtain most of the potential profits with a relatively small J even if we do not have detailed knowledge of the demand functions.

We will assume that the demand functions $d_m(p), m \in \mathcal{M}$ belong to the same family. By this we mean that $d_m(p) = \lambda_m H_m(p), m \in \mathcal{M}$ and the tail distributions $H_m(p) = P(\Omega_m \geq p), m \in \mathcal{M}$ differ only on their parameters. Examples of families of demand functions include linear, log-linear, CES, logit, among others. We will assume that the profit function $R_m(p, z) = (p - z)d_m(p)$ is quasi-concave for each m and that there is a unique finite maximizer $p_m(z)$ for each $m \in \mathcal{M}$. We will assume that the market segments are ordered so that $p_1(z) \leq \ldots \leq p_M(z)$. Finally, we will assume that for any $S \subset \mathcal{M}$, the profit function $R_S(p, z) = \sum_{m \in S} R_m(p, z)$ has a finite maximizer $p_S(z)$ in the interval $[\min_{m \in S} p_m(z), \max_{m \in S} p_m(z)]$, as guaranteed under the conditions of Proposition 8.16.

Since we will be using heuristic prices, it is convenient to have a measure of how efficient it is to use price p instead of $p_m(z)$ for market segment m. This motivates defining the relative efficiency of using price p instead of price $p_m(z)$ for market segment m as the ratio

$$e_m(p, p_m(z), z) := \frac{R_m(p, z)}{\mathcal{R}_m(z)} \leq 1. \tag{8.9}$$

Notice that $e_m(p, p_m(z), z)$ reaches maximum efficiency at $p = p_m(z)$ and decays on both directions as a result of our quasi-concavity assumption. We will be particularly interested in families of demands for which $e_m(p, p_m(z), z)$ is independent of m. This is true for the linear, the log-linear, and the logit demand functions, among others. It is possible to find closed-form formulas for $e(p, p_m(z), z)$ for many families of demand functions including linear, log-linear, and CES. However, there are distributions that do not admit closed-form expressions for $e(p, p_m(z), z)$ but the results that we will derive here can also be applied, numerically, to distributions that do not admit closed-form expressions. The relative efficiencies of prices will help us deal with situations where we may not know the exact parameters of some of the market segments. On occasions, we will write $e(q, s, z)$ to mean the efficiency of price $q \neq s$ for a (possibly fictitious) market segment for which price s is optimal for z.

We now show how to construct a heuristic that uses $1 < J < M$ prices. The idea is to break down the interval $[p_1(z), p_M(z)]$ into J sub-intervals, which in turn determine market clusters and then to use a common price for all the market segments within a cluster. The precise price used within a cluster will depend on the detailed knowledge of the market segments in a cluster. If only limited information is known, then a *robust* price that maximizes the minimum efficiency will be used, otherwise an *optimal* price all market segments in the cluster will be used.

We start by describing the procedure by showing how to select the break-points and robust prices and later explain how the heuristic can be improved with optimal prices in each cluster.

Consider arbitrary break-points $p_1(z) = s_0 < s_1 < s_2 \ldots < s_{J-1} < s_J = p_M(z)$ and define market clusters $M_j = \{m : p_m(z) \in [s_{j-1}, s_j)\}$ for $j = 1, \ldots, J-1$ and $M_J = \{m : p_m(z) \in [s_{J-1}, s_J]\}$. Let $q_j \in (s_{j-1}, s_j)$ be a common price to be used for all markets in cluster M_j, $j = 1, \ldots, J$. The break-points $s_1, \ldots, s_{J-1}$ and the prices q_j are designed to maximize the minimum efficiency among all of the market segments. More precisely, the s_j's and q_j's are selected so that

$$e(q_j, s_{j-1}, z) = e(q_j, s_j, z) \quad \text{for all} \quad j = 1, \ldots, J \tag{8.10}$$

and

$$e(q_1, s_1, z) = e(q_2, s_2, z) = \ldots = e(q_J, s_J, z). \tag{8.11}$$

Equation (8.10) guarantees that price q_j is just as efficient for s_{j-1} as it is for s_j. Equation (8.11) guarantees that the efficiency of q_j relative to s_j is the same for each market segment. This implies that for any market segment $m \in M_j$, $e(q_j, p_m(z), z) \geq e(q_j, s_j, z)$ for all $j = 1, \ldots, J$.

It is often possible to find the s_j's and the q_j's with very limited information about the market prices $p_m(z)$. Usually, it is sufficient to know the smallest $p_1(z)$ and the largest $p_M(z)$ prices.

Let $Q^h_J(z)$ be the profit obtained by pricing market all market segments in M_j at q_j for all $j = 1, \ldots, J$. Notice that we assign market m to price q_j if j maximizes $e(q_j, p_m(z), z)$, or equivalently $p_m(z)$ is in the interval defined by s_{j-1} and s_j. Thus, relatively little knowledge about the markets is need to implement the heuristic. However, if detailed knowledge is available, then we can improve on the heuristic by using optimal prices $p_{M_j}(z)$ for each market segment m in cluster $M_j, j = 1, \ldots, J$.

By the choice of the break-points s_j and prices q_j, we have

$$\gamma_J(S) := e(q_1, s_1, z) = e(q_2, s_2, z) = \ldots = e(q_J, s_J, z) \leq 1.$$

The next result shows that $Q^h_J(z)/Q_M(z) \geq \gamma_J(z)$. As we shall see $\gamma_J(z)$ can be quite close to one for relatively small values of J. This indicates that we do not need full price discrimination ($J = M$) to obtain most of the potential profits from price discrimination. Put another way, there may be no need to dice the market into tiny segments if the optimal prices for the different segments are not too far apart.

Theorem 8.18 *Assume that the functions $R_m(p, z)$ are quasi-concave and each has a unique finite maximizer $p_m(z)$. Suppose that the market segments are indexed so that $p_m(z)$ is increasing in $m \in \mathcal{M}$. Assume that $e_m(p, p_m(z), z), m \in \mathcal{M}$ is independent of $m \in \mathcal{M}$. Then offering price q_j to all market segment in M_j for $j = 1, \ldots, J$ results in*

$$\frac{Q_J(z)}{Q_M(z)} \geq \frac{Q^h_J(z)}{Q_M(z)} \geq \gamma_J(z).$$

We now illustrate the lower bounds for a variety of demand functions leaving the proofs as exercises. It is important to recall for this purpose that the market segments are ordered so that $p_m(z)$ is increasing in $m \in \mathcal{M}$, so $p_1(z)$ is the lowest price and $p_M(z)$ is the largest price. Let $\Delta_m(z) := p_m(z) - z$ represent the mark-up for market segment m. Clearly $\Delta_1(z) \leq \Delta_m(z) \leq \Delta_M(z)$.

Proposition 8.19 *Consider linear demand functions $d_m(p) = (a_m - b_m p)$, $m \in \mathcal{M}$. Then*

$$e(p, p_m(z), z) = \frac{p - z}{\Delta_m(z)} \left(2 - \frac{p - z}{\Delta_m(z)}\right) \quad \forall \quad m \in \mathcal{M},$$

and

$$\gamma_J(z) = \frac{4\Delta_1(z)^{1/J} \Delta_M(z)^{1/J}}{\left(\Delta_1(z)^{1/J} + \Delta_M(z)^{1/J}\right)^2}. \tag{8.12}$$

To get a feel for this result, suppose that there are M market segments, and the mark-up for market segment M is 4 times the optimal mark-up of segment 1, so $\Delta_M(z) = 4\Delta_1(z)$. Then $\gamma_1(z) = 64.00\%$, $\gamma_2(z) = 88.89\%$, and $\gamma_4(z) = 97.06\%$. These results are independent of the number of market segments. Recall that these are lower bounds assuming a robust price q_j is used for every cluster, so even better results attain if we use optimal prices within each market segment.

We next consider the exponential demand family.

Proposition 8.20 *Consider the exponential demand functions $d_m(p) = a_m \exp(-p/b_m)$, $m \in \mathcal{M}$. Then*

$$e(p, p_m(z), z) = \frac{p - z}{\Delta_m(z)} \exp\left(1 - \frac{p - z}{\Delta_m(z)}\right) \quad \forall \quad m \in \mathcal{M}.$$

Let $u = b_M/b_1$, and let $U_J = \frac{\ln(u)}{J(u^{1/J} - 1)}$. Then

$$\gamma_J(z) = U_J e^{1-U_J}.$$

To get a feel for this result, suppose that there are M market segments and $u = b_M/b_1 = 4$, then $\gamma_1(z) = 79.13\%$, $\gamma_2(z) = 94.21\%$, and $\gamma_4(z) = 98.51\%$. Again, these numbers are independent of the number of market segments. Recall that these are lower bounds assuming a robust price q_j is used for every cluster, so even better results attain if we use optimal prices within each market segment.

In addition to the linear and log-linear demand functions, efficiency functions can be computed for the CES model and for the multinomial logit model. Consequently, pricing heuristics can be computed for those demand functions as well.

So far we have avoided the issue of consumer surplus and total welfare under direct price discrimination. Most of the insights can be obtained from studying what happens with two market segments. As we move from a common optimal price, say $p(z)$, to two prices, say $p_1(z) < p_2(z)$, we typically have $p(z) \in (p_1(z), p_2(z))$ under mild conditions (Example 8.14 shows that this is not always true). In this case, there is a Robin Hood effect that favors the firm and market segment 1 at the expense of market segment 2. The change in total welfare can be either positive or negative. A necessary condition for an increase in total welfare is that the total output increases under direct price discrimination.

8.5.5 Peak Load Pricing

Suppose that a product with variable cost $\alpha > 0$ is sold in different markets or time periods $m \in \mathcal{M}$. We will assume that $d_m(p)$ is continuous in p, and there is a unique price $p_m(\alpha + z_m)$ that maximizes $R_m(p, \alpha + z_m)$ for all $z_m \geq 0$.

Consider the problem of selecting prices to maximize

$$\sum_{m \in \mathcal{M}} R_m(p_m, \alpha) - \beta \max_{m \in \mathcal{M}} d_m(p).$$

We can think of β as the unit cost of serving the peak demand. To tackle this problem, we will assume that the firm will select the prices $p_m, m \in \mathcal{M}$ as well as the installed capacity, say c. The goal is to maximize

$$\sum_{m \in \mathcal{M}} R_m(p_m, \alpha) - \beta c$$

subject to $d_m(p) \leq c$ for all $m \in \mathcal{M}$.

Let z_m be the dual variable associated with the constraint $d_m(p) \leq c$. Then the Lagrangian problem for fixed c is given by

$$L(p, z) := \sum_{m \in \mathcal{M}} R_m(p_m, \alpha + z_m) + \left[\sum_{m \in \mathcal{M}} z_m - \beta \right] c.$$

Maximizing over $p_m, m \in \mathcal{M}$ yields

$$L(z) := \max_p L(p, z) = \sum_{m \in \mathcal{M}} \mathcal{R}_m(\alpha + z_m) + \left[\sum_{m \in \mathcal{M}} z_m - \beta \right] c.$$

Next we consider the convex problem of minimizing $L(z)$ over $z \geq 0$. The solution is to set $z_m = 0$ if $d_m(p_m(\alpha)) \leq c$. If $d_m(p_m(\alpha)) > c$, we select $z_m > 0$ so $d_m(p_m(\alpha + z_m)) = c$. In summary, for fixed c, the solution is given by $z_m(c)$ and $p_m(\alpha + z_m(c))$ for all $m \in \mathcal{M}$ such that $d_m(p_m(\alpha + z_m(c))) \leq c$ is complementary slack with $z_m(c) \geq 0$.

Let

$$L(z(c)) = \sum_{m \in \mathcal{M}} \mathcal{R}_m(\alpha + z_m(c)) + \left[\sum_{m \in \mathcal{M}} z_m(c) - \beta \right] c.$$

At optimality c^* must be selected so that $\sum_{m \in \mathcal{M}} z_m(c^*) = \beta$, as otherwise the objective can be improved by either increasing or decreasing c. Since $\beta > 0$, at least one period has demand equal to capacity. The variable capacity cost β is allocated to the markets in the set $\{m \in \mathcal{M} : z_m(c^*) > 0\}$ with other markets not contributing to the cost of capacity. Peak load pricing has generated its share of controversy, as it is difficult to understand why two markets consuming the peak capacity should pay different prices, and why those consuming less should get a free ride.

8.6 Multi-Product Pricing Problems

For the multiple product cases with $n > 1$, the known conditions for the existence of a finite maximizer $p(z)$ of $R(p, z) = (p - z)'d(p)$ are seldom useful, as they typically require $R(p, z)$ to be concave or quasi-concave over a compact set. The problem is that for $n > 1$, we need to worry about the possibility that at optimality one or more products are priced at infinity. This is equivalent to not offering all of the products, and this makes it is difficult to reduce the domain to a compact set without loss of optimality. Here, we provide some results for substitute products that sometimes allow for the reduction of the optimization problem to a compact set. Let $d(p) = (d_1(p), \ldots, d_n(p))$. We assume that $d_i(p)$ is increasing in $p_j, j \neq i$ to capture the substitution effect (the demand for chicken goes up as the price for beef increases). For convenience, we will write $p = (p_i, p_{-i})$, where p_{-i} represents the price vector of products other than i. By (p_i, ∞) we imply that products $j \neq i$ are not offered. This allows us to define $d_i(p_i) := d_i(p_i, \infty)$, $R_i(p_i, z_i) := (p_i - z_i)d_i(p_i)$, and $\mathcal{R}_i(z_i) := \max_{p_i} R_i(p_i, z_i)$, corresponding to the demand, profit, and optimal profit for product $i \in N := \{1, \ldots, n\}$ that prevail when only product i is offered.

A lower bound on $\mathcal{R}(z)$ can be obtained by selecting the product $i \in N$ with the largest $\mathcal{R}_i(z_i)$, and by setting other prices to infinity. For an upper bound, we have

$$R(p, z) = \sum_{i=1}^{n}(p_i - z_i)d_i(p) \leq \sum_{i=1}^{n}(p_i - z_i)d_i(p_i) = \sum_{i=1}^{n} R_i(p_i, z_i),$$

so

$$\max_{i \in N} \mathcal{R}_i(z_i) \leq \mathcal{R}(z) \leq \sum_{i \in N} \mathcal{R}_i(z_i). \tag{8.13}$$

We are interested in situations where $p(z)$ is bounded when the optimal individual prices $p_i(z_i)$, $i \in N$ are themselves bounded. For this, we will need the concept of super-modularity. We say that $R(p, z)$ is super-modular in $p \in \Re^n_+$, $p \geq z$, for fixed z, if for any two price vectors $p \geq z$ and $\tilde{p} \geq z$

$$R(\max(p, \tilde{p}), z) + R(\min(p, \tilde{p}), z) \geq R(p, z) + R(\tilde{p}, z).$$

If $R(p, z)$ is twice continuously differentiable in p for fixed z, then R is super-modular in p if and only if

$$\frac{\partial^2 R(p, z)}{\partial p_i \partial p_j} \geq 0 \quad \forall i \neq j.$$

One well-known consequence of super-modularity is that if $R(p_i, p_{-i}, z)$ admits a finite maximizer, say $p_i(z \mid p_{-i}) \geq 0$, for fixed p_{-i} and z, then $p_i(z \mid p_{-i})$ can be selected so that it is increasing in p_j for all $j \neq i$. We are now ready to state our next result.

Theorem 8.21 *If $d_i(p)$ is increasing in p_j, $j \neq i$, then* (8.13) *holds. Moreover, if $p(z)$ is a maximizer of $R(p, z)$ and $R(p, z)$ is super-modular in p for all $z \geq 0$, and $p_i(z_i)$ is finite for all $i \in N$, then $p(z)$ is finite and*

$$p_i(z) \leq p_i(z_i) \quad \forall i \in N. \tag{8.14}$$

We now provide sufficient conditions for $p_i(z_i)$, $i \in N$ to be finite and for $R(p, z)$ to be super-modular in p for fixed z.

Corollary 8.22 *A sufficient condition for $p_i(z_i) < \infty$ for all $i \in N$ is that $d_i(p_i)$ is USC and $\bar{d}_i(p_i)$ is $o(1/p_i)$ for all $i \in N$. A sufficient condition for the super-modularity of $R(p, z)$ is that for all $i \in N$, $d_i(p + z)$ is increasing in p_j for all $j \neq i$ and super-modular in p for all $z \geq 0$.*

If $d_i(p)$ is decreasing in p_i and increasing in p_j, $j \neq i$, then $R(p, z)$ is super-modular in (p_i, z_i) for fixed p_{-i} and z_{-i}, and sub-modular in (p_j, z_i) for fixed p_{-j} and z_{-i}. As a result, an optimizer $p_i(z \mid p_{-i})$ of $R(p_i, p_{-i}, z)$ can be selected so that $p_i(z \mid p_{-i})$ is increasing in z_i and an optimizer $p_j(z \mid p_{-j})$ of $R(p_j, p_{-j}, z)$ can be selected so that $p_j(z \mid p_{-j})$ is decreasing in z_i. That $p_i(z \mid p_{-i})$ is increasing in z_i is intuitive as some of the higher costs are passed on to consumers. Less intuitive is that $p_j(z \mid p_{-j})$ is decreasing in z_i. The explanation is that an increase in z_i reduces the profits of product i, and an effort is made to shift demand to other products by reducing their prices.

When an inverse demand function exists, it is possible to write the profit function in terms of sales instead of price. In some cases, the profit function is sub-modular as a function of sales for fixed z. Consequently, an increase in sales of one product leads to a decrease in the optimal sales for other products. This makes intuitive sense as products are substitutes. The sub-modularity of the profit function in terms of sales, together with the super-modularity of the profit function in terms of prices, implies that an increase in the price of one product leads to an increase in optimal prices and optimal sales of all other products. Similarly, a decrease in the price of a product leads to a decrease in optimal prices and optimal sales of all other products. This suggests that a change in price in one product should result in a price change of other products in the same direction, but not to the extent that a change in sales goes in the opposite direction of the change in prices.

8.6.1 Linear Demand Model

Demand functions for substitute products are often justified by looking at consumers who are utility maximizers. Given a vector of prices p, consumers purchase the quantity $q \geq 0$ that maximizes $U(q) - q'p$. It is well known that the quadratic utility $U(q) = w'q - \frac{1}{2}q'Qq$, with $w \in \Re^n_{++}$, Q symmetric and positive definite, leads to linear demand function $d(p) = a - Bp$ over the polyhedral set $\mathcal{P} = \{p \geq 0 : Bp \leq a\}$, where $a := Bw$, and $B := Q^{-1}$; see the proof of Theorem 8.7 for details.

We are interested in finding conditions on a and B that guarantee the existence of a unique, non-negative, profit maximizing price vector $p(z)$ such that $\mathcal{R}(z) = R(p(z), z)$ for all $z \geq 0$ such that $d(z) \geq 0$. This last condition limits the costs z to the polyhedral set where demands are non-negative at z. If one or more products have costs so high that $d(z)$ is negative for one or more products, then these products can be eliminated from consideration and it is necessary to work on the projection of the demand model into the space where demand for all products at cost z is non-negative. Given B, we denote the transpose by B' and form the symmetric matrix $S = B + B'$.

Theorem 8.23 *If S is positive definite, $S_{ij} \leq 0$ for all $i \neq j$, and $a \in \Re^n_{++}$, then*

$$p(z) = S^{-1}(a + B'z) \geq 0, \tag{8.15}$$

maximizes $R(p, z) = (p - z)'d(p)$ for all z such that $d(z) \geq 0$. Moreover,

$$\mathcal{R}(z) = R(p(z), z) = d(z)'Nd(z) \tag{8.16}$$

where $N = S^{-1}BS^{-1}$.

Notice that the requirements of Theorem 8.23 are very mild. The theorem does not even require that $B_{ij} \leq 0$ for all $i \neq j$, but rather the more mild assumption that $S_{ij} = B_{ij} + B_{ji} \leq 0$ for all $i \neq j$. Notice also that $R(p, z)$ is super-modular if and only if $S_{ij} \leq 0$ for all $i \neq j$. The requirement that S is positive definite is also very natural in this setting.

The solution $d(p)$ presented above was the solution to the problem of maximizing $U(q) - q'p$ without any constraints on q. The solution $q = d(p)$ satisfies the non-negativity constraint if $p \in \mathcal{P}$. We now address the problem for cases where $p \geq 0$, but $p \notin \mathcal{P}$, so the demand $d(p)$ is negative for at least one product, so $q = d(p)$ is not a feasible solution to the problem of maximizing $U(q) - q'p$ subject to $q \geq 0$. Considering the non-negativity constraints explicitly in the optimization problem can be shown to be equivalent to solving the linear complementarity problem where some of the prices are reduced resulting in an optimal solution to the constrained problem of the form $q = d(p - y) \geq 0$ where $y \geq 0$ and $y'd(p - y) = 0$, so prices are adjusted downward for products with negative demands.

Suppose for some $p \in \mathcal{P}$, the unconstrained solution $q = d(p)$ does not satisfy a capacity constraint of the form $q \leq c$. The problem of maximizing $U(q) - q'p$ subject to $q \leq c$ can be shown to be equivalent to solving a linear complementarity problem where prices are adjusted upwards by y, so that $q = d(p + y) \leq c$ is an optimal solution with $y \geq 0$ and $y'(c - d(p + y)) = 0$.

A natural extension to the linear demand model is $D(p) = A - Bp$, where the potential demand A is random with $\mathbb{E}[A] = a$. Are profits higher when A is random? The answer is yes if we can observe A before deciding the price $p(z|A) = S^{-1}(A + B'z)$ to offer. From (8.16), we can write the optimal profit function as $\mathcal{R}(z) = (A-Bz)'N(A-Bz)$ which is a convex function of A given that N is positive definitive. By Jensen's inequality $\mathbb{E}_A(A - Bz)'N(A - Bz) \geq (a - Bz)'N(a - Bz)$, which is the revenue if we price at $p(z) = S^{-1}(a + B'z)$. The implication here is that dynamic pricing can also be driven by randomness in the potential demand A even if the variable value of capacity is unchanged.

The inverse demand function is given by $p = d^{-1}(q) = B^{-1}(a - q)$, so the profit function as a function of q is given by $q'(B^{-1}a - B^{-1}q - z)$. This function is sub-modular in q if and only if $B_{ij}^{-1} \geq 0$ for all $i \neq j$. A sufficient condition for this is that B is an m-matrix, i.e., if $B_{ii} > 0$ for all i, $B_{ij} \leq 0$ for all $i \neq j$, and either $\sum_{i \in N} B_{ij} > 0$ for every $j \in N$ or $\sum_{j \in N} B_{ij} > 0$ for every $i \in N$. If B is an m-matrix, then the profit function is sub-modular in q, and if a finite maximizer exits, then it can be selected so that $q_i(z|q_{-i})$ is decreasing in q_j for all $j \neq i$. This is intuitively consistent with the idea of product substitution. If we want to sell more of product j then it is optimal to sell less of product i.

8.6.2 *The Multinomial Logit Model*

The multinomial logit (MNL) demand function is normally derived, as we do in an earlier chapter, from a discrete choice model. Here, we show that the MNL function also arises as a special case of the linear random utility model, where the *indirect utility function*[4] $V(p, y)$ obtained from price vector p and income level y is given by

$$V(p, y) := \mathbb{E}[\max_{i \in N}(y - p_i + a_i + \epsilon_i)],$$

where a_i is a measure of the quality of product i and the ϵ_i's are mean-zero random variables. In this model, it is typically assumed that $y \geq p_i$, so if product i is purchased, then $y - p_i$ is the utility derived from the remaining budget and $a_i + \epsilon_i$ is the utility associated with product i. In this case, $V(p, y) = y + \mathbb{E}\max_{i \in N}(a_i - p_i + \epsilon_i)$. A direct application of the Williams-Daly-Zachary theorem, assuming λ statistically identical consumers, results in

$$d_i(p) = -\lambda \frac{\partial V / \partial p_i}{\partial V / \partial y} = \lambda\, \mathbb{P}(a_i - p_i + \epsilon_i = \max_j(a_j - p_j + \epsilon_j)),$$

[4]The consumer's maximal attainable utility when faced with a vector of prices and income.

so the demand for product i is the expected number of customers that prefer product i over all other alternatives. Notice that the demand is independent of the income level as long as $y \geq p_i$ for all i. The so-called Profit demand function arises if the ϵ_i's are IID normal random variables. The MNL model arises if the ϵ_i's are IID Gumbel random variables. The MNL model results in

$$d_i(p) = \lambda \frac{e^{\alpha_i - \beta p_i}}{1 + \sum_{j \in N} e^{\alpha_j - \beta p_j}} \quad \forall i \in N, \quad \text{and} \quad d_0(p) = \lambda \frac{1}{1 + \sum_{j \in N} e^{\alpha_j - \beta p_j}}$$

for some constants α_j, $j \in N$ and $\beta > 0$, after normalizing the attraction of the no-purchase alternative to 1. One can think of α_i as the quality of product i and β as the sensitivity to price.

For convenience, let $\pi_i(p) = d_i(p)/\lambda$ denote the market share of product $i \in N$. Then

$$\frac{\partial d_i(p)}{\partial p_i} = -\beta d_i(p)(1 - \pi_i(p)) \leq 0 \quad \text{and} \quad \frac{\partial d_k(p)}{\partial p_i} = \beta d_i(p)\pi_k(p) \geq 0 \quad \forall k \neq i.$$

Consequently, the (absolute) elasticity of demand for product i is given by $\beta p_i(1 - \pi_i(p))$ and is proportional to the complement of the market share $\pi_i(p)$ of product i. The cross elasticities of the demand for product k relative to the price of product i are given by $\beta p_i \pi_k(p)$, and it is proportional to the market share of product k. The next theorem characterizes the optimal prices under the MNL model.

Theorem 8.24 *There exists a function $\theta(z)$ independent of i such that*

$$p_i(z) = z_i + \frac{1}{\beta} + \theta(z) \quad \forall i \in N$$

and

$$\mathcal{R}(z) = \lambda \theta(z),$$

where $\theta(z)$ is the root of the Lambert equation

$$\beta \theta e^{\beta \theta} = \sum_{j \in N} e^{\alpha_j - \beta z_j - 1}.$$

It is worth noting that Theorem 8.24 implies that all products should be offered with the same mark-up $p_i(z) - z_i = 1/\beta + \theta(z)$. It is easy to see that the optimal mark-up $1/\beta + \theta(z)$ is equal to the reciprocal of $\beta \pi_0(p(z))$, so

$$p_i(z) - z_i = \frac{1}{\beta \pi_0(p(z))} \quad \forall i \in N.$$

The implication in a competitive setting is that the optimal mark-up is the reciprocal of the product of the price sensitivity and the complement of the market share. Consequently, optimal mark-ups are small if customers are price sensitive and the firm has a small market share.

Let $p_i(z_i) = z_i + 1/\beta + \theta_i$ be the optimal price for the set when the set of finite prices is $F = \{i\}$, corresponding to the case $p_j = \infty$ for all $j \neq i$. Then, from the proof of Theorem 8.24, we see that

$$p_i(z_i) = z_i + \frac{1}{\beta} + \theta_i \leq z_i + \frac{1}{\beta} + \theta = p_i(z).$$

This inequality goes in the opposite direction to that of the linear demand model. This may suggest to the reader that $R(p, z)$ may be sub-modular, but this is not the case.

The analysis can be extended to the case where the demand function is of the form

$$d_i(p) = \lambda \frac{e^{\alpha_i - \beta_i p_i}}{1 + \sum_{j \in N} e^{\alpha_j - \beta_j p_j}} \quad \forall i \in N,$$

with $d_0(p) = 1 - \sum_{i \in N} d_i(p)$, so that the sensitivity to price is now product dependent. In this case, it is also optimal to offer all products, and there is a function $\theta(z)$, independent of i, such that

$$p_i(z) = z_i + \frac{1}{\beta_i} + \theta(z) \quad \forall i \in N,$$

and $\mathcal{R}(z) = \lambda\theta(z)$. However, θ is no longer the root of a Lambert equation, but the root of a slightly more complicated function.

8.6.3 The Nested Logit Model

In this section, we consider pricing under the nested logit (NL) model, which is a popular generalization of the standard MNL model. For a certain range of parameters, the NL model is an example of a random utility model where the random component of the utilities of products within a nest are positively correlated and independent of the utilities of products outside the nest. The probability of selecting a product with the largest utility can then be viewed as a sequential decision: At the upper level, customers select a nest of products; at the lower level, they select a product within the nest. Suppose that the substitutable products constitute n nests and nest i has m_i products. Let $p_i = (p_{i1}, p_{i2}, \ldots, p_{i,m_i})$ be the price vector corresponding to nest $i = 1, \ldots, n$, and let $p = (p_1, \ldots, p_n)$ be the price vector for all the products in all the nests. Let $Q_i(p_1, \ldots, p_n)$ be the probability that a customer selects nest i at the upper level; and let $q_{k|i}(p_i)$ denote the probability that product k of nest i is selected at the lower level, given that the customer selects nest i. Under the NL model, the quantities $Q_i(p_1, \ldots, p_n)$ and $q_{k|i}(p_i)$ are given by

$$Q_i(p_1, \ldots, p_n) = \frac{e^{\gamma_i I_i}}{1 + \sum_{l=1}^{n} e^{\gamma_l I_l}}$$

$$q_{j|i}(p_i) = \frac{e^{\alpha_{ij} - \beta_{ij} p_{ij}}}{\sum_{s=1}^{m_i} e^{\alpha_{is} - \beta_{is} p_{is}}},$$

where α_{is} can be interpreted as the "quality" of product s in nest i, $\beta_{is} \geq 0$ is the product-specific price sensitivity for that product, $I_l = \log \sum_{s=1}^{m_l} e^{\alpha_{ls} - \beta_{ls} p_{ls}}$ represents the attractiveness of nest l, which is the expected value of the maximum of the utilities of all the products in nest l, and nest coefficient γ_i can be viewed as the degree of inter-nest heterogeneity and is a measure of the correlation among the utilities of the products in nest i. When $\gamma_i = 1$ for all i, the model reduces to the MNL model. The case $\gamma_i \in (0, 1]$ is consistent with random utility theory.

The probability that a customer will select product j of nest i, which can also be considered as the market share of that product, is

$$\pi_{ij}(p_1, \ldots, p_n) = Q_i(p_1, \ldots, p_n) q_{j|i}(p_i). \tag{8.17}$$

The monopolist's problem is to determine the price vectors $(p_1, \ldots, p_n)$ to maximize the total expected profit

$$R(p, z) := \sum_{i=1}^{n} \sum_{j=1}^{m_i} \lambda (p_{ij} - z_{ij}) \pi_{ij}(p_1, \ldots, p_n), \tag{8.18}$$

where $z = (z_1, \ldots, z_n)$, and z_i is the vector of unit costs for nest i, and λ is the market size. Let $\mathcal{R}(z) := \max_{(p_1, \ldots, p_n)} R(p, z)$. The objective function $R(p, z)$ fails to be quasi-concave in prices. When the objective function is rewritten with market shares as decision variables, then the objective function can be shown to be concave if the price sensitivity parameters $\beta_{ij} = \beta_i$ are product independent in each nest and $\gamma_i \in (0, 1]$ for all i. However, the objective function fails to be concave in the market shares in the more general case where the price sensitivities are product dependent or some of the parameters γ_i are allowed to exceed one.

The results of Theorem 8.24 extend to the NL model, where the optimal price $p_{ij}(z)$ for product j in nest i as a function of the vector of unit costs z is of the form $p_{ij}(z) = z_{ij} + 1/\beta_{ij} + \theta_i$. Also, the nest dependent constants $\theta_i, i = 1, \ldots, n$ are linked to a single parameter as explained in the following theorem.

Theorem 8.25 *If* $\gamma_i \geq 1$ *or* $\frac{\max_s \beta_{is}}{\min_s \beta_{is}} \leq \frac{1}{1-\gamma_i}$, *then there exists a unique constant* ϕ *such that*

$$\theta_i + \left(1 - \frac{1}{\gamma_i}\right) w_i(\theta_i) = \phi,$$

and

$$p_{ij}(z) = z_{ij} + \frac{1}{\beta_{ij}} + \theta_i,$$

where $w_i(\theta) = \sum_{k=1}^{m_i} \frac{1}{\beta_{ik}} \cdot q_{k|i}(\theta_i)$ *and* $q_{k|i}(\theta_i) = \frac{e^{\tilde{\alpha}_{ik}-\beta_{ik}\theta_i}}{\sum_{s=1}^{m_i} e^{\tilde{\alpha}_{is}-\beta_{is}\theta_i}}$, *and* $\tilde{\alpha}_{is} = \alpha_{is} - \beta_{is}z_{is} - 1$ *for all* i *and all* s. *Moreover,*

$$\mathcal{R}(z) = \lambda\phi.$$

Theorem 8.25 is interesting because a non-concave optimization problem over $\sum_{i=1}^{n} m_i$ variables can be reduced, under mild conditions, to a root finding problem over the single variable ϕ. Notice that each value of ϕ gives a set of θ_i's dictated by the first equation in the theorem. For these θ_i's, the second equality in the theorem gives the prices. If $\gamma_i \geq 1$ or $\frac{\max_s \beta_{is}}{\min_s \beta_{is}} \leq \frac{1}{1-\gamma_i}$ fails to hold, then the problem reduces to a single variable maximization problem of a continuous function over a bounded interval, so the problem can be easily solved numerically. Also, if different firms control different nests, then the pricing problem under competition is strictly log-super-modular in the nest mark-up constants, so the equilibrium set is nonempty with the largest equilibrium preferred by all the firms.

8.7 End of Chapter Problems

1. Show that if $d(p) = 1$ for $p \in [0, 10)$ and $d(p) = 0$ for $p \geq 10$, then $\mathcal{R}(z) = (10 - z)^+$ but the maximum is not attained.
2. Show that Theorem 8.10 applies to the demand function $d(p) = a\exp(-bp)\sin^2(p)$ by showing that $\bar{d}(p) \leq a\exp(-bp)$ and $p\bar{d}(p) \to 0$ as $p \to \infty$. Find a formula for $p(z)$.
3. Determine the form of $p(z)$ for $d(p) = \lambda\exp(-p/\theta)$ for $\lambda, \theta > 0$.
4. Determine the form of $p(z)$ for $d(p) = ((a - bp)^+)^c$ for $a, b, c > 0$.
5. Determine the form of $p(z)$ for $d(p) = (a + bp)^{-c}$ for $a, b > 0, c > 1$.
6. Show that if $1/h(p)$ is concave in p, then $p(z)$ is increasing concave in z.
7. Consider a single product where c units are available for sale. Let $d_c(p) = \min(d(p), c)$ and consider the following two formulations: (i) $\max_p p\, d_c(p)$, and (ii) $\max p\, d(p)$ subject to $d(p) \leq c$. Solve each problem for $c = 2$, if $d(p) = 3$ for $p \leq 10$ and $d(p) = 0$ for $p > 10$. What happens if $d(p)$ is continuous?
8. Consider a single product problem with a strictly decreasing demand function $q = d(p)$. Let $\tilde{p}(q)$ be the inverse demand function. Assume that $\tilde{r}(q) = q\,\tilde{p}(q)$ is concave with a bounded maximizer, say q^*. Suppose $c < q^*$. Show that $q = c$ solves the problem of maximizing $\tilde{r}(q)$ subject to $q \leq c$ and that there is a $z \geq 0$ such that $q = c$ maximizes $\tilde{r}(q) - qz$. Is z unique? What

if $\tilde{r}(q)$ is differentiable? Is the concavity of $\tilde{r}(q)$ a sufficient condition for the existence of a unique maximizer $p(z)$ of $R(p, z)$? What else may you need? Consider the problem $\Gamma(c) = \min_{z \geq 0}[\mathcal{R}(z) + cz]$. Is the concavity of $\tilde{r}(q)$ a sufficient condition for $\Gamma(c) = \tilde{r}(c)$?

9. Consider a single product, single resource formulation (8.4) with sales horizon $[0, T]$, $T = 1$, $d_t(p) = a_1 - bp$ for $t \in (0, 1/2]$, and $d_t(p) = a_2 - bp$ for $t \in (1/2, 1]$. Find the solution $p_t(0)$ for $t \in [0, T]$. For what values of c is this solution optimal? Show that for such values of c, $\bar{V}(T, c) = (a_1^2+a_2^2)/8b$. Now solve formulation (8.6) and show that $\bar{V}_f(T, c) = (a_1 + a_2)^2/16b$. Show that the optimal profits are 25% higher under dynamic pricing when $a_1 = 50$ and $a_2 = 150$.
10. Consider the demand function $d(p) = \lambda P(\Omega \geq p)$ and assume that Ω has a gamma distribution with parameters α and β, so $\mu = \mathbb{E}[\Omega] = \alpha\beta$ and $\sigma^2 = \text{Var}[\Omega] = \alpha\beta^2$. We can fit any mean μ and variance σ^2 by setting the parameter $\beta = \sigma^2/\mu$ and $\alpha = \mu^2/\sigma^2$. One may wonder how $p(z)$ and $\mathcal{R}(z)$ behave as a function of σ^2 for fixed μ. Does more variance lead to higher or lower prices and profits? Construct a table of $p(z)$ and $\mathcal{R}(z)$ for $z = 400$, $\lambda = 1$ and $\mathbb{E}[\Omega] = 500$ for values of $\sigma/\mu \in \{k/8 : k = 0, 1, \ldots, 16\}$. What happens to $p(z)$ as σ increases? What happens to $\mathcal{R}(z)$ as σ increases?
11. Consider the problem of maximizing $R(p, z) = (p - z)d(p)$ when $d(p) = \lambda e^{-p/\mu}$ subject to the constraint $d(p) \leq c$. Let $L(p, z) = R(p, z) - \gamma(d(p) - c) = R(p, z + \gamma) + \gamma c$. Argue that $\min_{\gamma \geq 0} L(p, z)$ is equivalent to $\min_{\gamma \geq 0} [\mathcal{R}(z + \gamma) + \gamma c]$. Show that this is a convex program in γ and that if γ_c is an unconstrained minimizer of $R(p + \gamma) + \gamma c$, then $\gamma_c^* = \max(\gamma_c, 0)$ solves the Lagrangian problem, and that $p(z + \gamma_c^*) = \max(p(z), p_{\min}(c))$ is the optimal price, where $p_{min}(c)$ is the root of $d(p) = c$.
12. Consider the multiple market segment problem and show that the total welfare can go up when we move from an optimal common price to direct price discrimination only if the total output goes up. Hint: Use the fact that the surplus function is convex.
13. For the linear function $d(p) = a - Bp$ and for $p \in \mathcal{P} = \{p \geq 0 : d(p) \geq 0\}$, $q = d(p)$ is the solution to the problem $\max_{q \geq 0}[U(q) - p'q]$ as presented in Sect. 8.6.1. Suppose that $p \geq 0$, but $p \notin \mathcal{P}$ and $a - Bp$ has both positive and negative components. To find the demand at p, we need to solve $\max_{q \geq 0}[U(q) - p'q]$ without ignoring the non-negativity constraints. Let $y \geq 0$ be the vector of dual variables to the constraint $q \geq 0$. Show that the optimal solution is given by $q = d(p) + By$, where $y \geq 0$ minimizes $y'[d(p) + By]$. Notice that this is a linear complementarity problem. Solve the linear complementarity problem for the case of $n = 2$, when $d_1(p) > 0$ and $d_2(p) < 0$ to see how the demand for product one is reduced by $b_{12}\, y_2$.
14. Consider again the linear demand model $d(p) = a - Bp$, but assume now that there are n firms with firm i selecting the price of product $i = 1, \ldots, n$. More precisely, assume that firm i maximizes $R_i(p, z) = (p_i - z_i)d_i(p)$ over $p_i \geq z_i$. This results in the best response price $p_i(p_{-i})$ for each firm i, and what we seek

is the equilibrium price vector, so that the prices constitute a Nash Equilibrium. Show that first-order conditions can be written in matrix form as

$$a + \mathrm{diag}(B)z - (B + \mathrm{diag}(B))p = 0.$$

Assume that $M = B + \mathrm{diag}(B)$ is an m-matrix, so that the inverse of M exists and is non-negative and show that the equilibrium prices are given by

$$\tilde{p}(z) = M^{-1}(a + \mathrm{diag}(B)z) = z + M^{-1}d(z).$$

Show also, that $d(\tilde{p}(z)) = \mathrm{diag}(B)M^{-1}d(z)$ and that

$$\sum_{i=1}^{n} R_i(\tilde{p}(z), z) = d(z)'M'^{-1}\mathrm{diag}(B)M^{-1}d(z)$$

$$= (\tilde{p}(z) - z)'\mathrm{diag}(B)(\tilde{p}(z) - z)),$$

and that the monopolist formulas (8.15) and (8.16) coincide with the competition formulas if $\mathrm{diag}(B) = B$, i.e. if there are no cross elasticities. Otherwise, we expect competitive prices to be lower, demand to be higher and aggregate profits to be lower under competition, with more surplus going to the consumers. The results can be extended to the case where each competitor controls the prices of a subset of the products.

8.8 Bibliographical Remarks

Theorems 8.1, 8.5, and 8.7 show that the firm prefers randomness in z, the consumer's prefer randomness in p, and under mild conditions both the firm and the representative consumer prefer prices that respond to randomness in variable costs. Theorem 8.10 allows for demand functions that are not necessarily decreasing or eventually decreasing. Theorem 8.11 provides bounds on optimal prices. The analysis of consumer surplus is due to Chen and Gallego (2019). The reader is directed to van den Berg (2007) and references therein for earlier efforts to characterize the existence or uniqueness of global maximizers. The reader is also referred to Larriviere and Porteus (2001) for an equivalent assumption where the absolute value of the price elasticity of demand $|e(p)| = ph(p)$ is called the generalized hazard rate. Caplin and Nalebuff (1991) have some interesting conditions on the inverse demand function for an optimal price $p(z)$ to exist. Ziya et al. (2004) discuss the relationship between several assumptions used to ensure that the expected revenue function is well behaved. The results on options are due to Gallego and Sahin (2010), Png (1989), Shugan and Xie (2000), and Xie and Shugan (2001). The section on priority pricing is based on the work of McAfee (2004), where he considers the social benefit of coarse matching. See Johnson (1970) for a discussion of positive definite matrices. The pricing results for the NL model can be found in Li and Huh (2011), Gallego and Wang (2014), and Rayfield et al. (2015). The development and discussion of the NL model can be found in McFadden (1974) and Carrasco and de Ortuzar (2002). The MNL and NL models are special cases of a broader class of choice models, called the generalized extreme value models. Zhang et al. (2018) work on pricing problems under generalized extreme value models.

Keller et al. (2014) give mathematical programming formulations for pricing problems under generalizations of the MNL model. Du et al. (2016), Wang and Wang (2017) and Du et al. (2018) study pricing problems under a variant of the MNL model, where the attraction value of a product depends on the market size it garners. Yan et al. (2017) study a joint parameter estimation and pricing problem when the marginal distribution information is available on the utilities. Wang (2018b) studies a pricing problem under the MNL model, where customers form a reference price by using the prices of the offered products and adjust their reactions accordingly. Cui et al. (2018) and Wang et al. (2019) study multi-product pricing problems when products are sold as ancillary to others. Amornpetchkul et al. (2018) examine promotion models, when the amount of promotion depends on the quantity purchased by a customer.

Maglaras and Zeevi (2005) study pricing problems when the firm offers services with different levels of quality using a common pool of capacity. Besbes et al. (2010) design tests to check the validity of a fitted price-demand curve not from the perspective of statistical goodness of fit but from the perspective of operational performance. Eren and Maglaras (2010) consider pricing problems when the price-demand curve is unknown to the firm. Cachon and Feldman (2011) study pricing models to understand the tradeoff between charging on a per-use basis or selling subscriptions. Kostami et al. (2017) give a pricing model when the utility of a customer depends on the presence of the other customers in the system. Cohen et al. (2017b) give performance bounds when only partial information about the price-demand relationship is available. Similarly, Chen et al. (2017a) study pricing problems with only limited information about the price-demand relationship. Cachon et al. (2017) study a stylized pricing model for a two-sided platform where the demand and the supply are both endogenous. Hu and Nasiry (2018) demonstrate that a price-demand model that is obtained by aggregating the behavior of individual customers may not reflect the individual customers anymore. Elmachtoub et al. (2018) bound the relative expected revenue gain when a firm knows the exact willingness to pay of a customer rather than the distribution of willingness to pay, providing insights on the effectiveness of personalized pricing. Boyaci and Akcay (2018) study pricing models when customers cannot fully evaluate the quality of a product. Ho et al. (1998) study a model to understand the reaction of consumers to different pricing strategies. Petruzzi and Dada (1999) and Lu and Simchi-Levi (2013) study incorporating pricing decisions into the newsvendor problem. Tang et al. (2004) analyze the benefits from providing advance booking discounts to reduce demand uncertainty. Tang and Yin (2007) develop a joint procurement and pricing model under deterministic demand.

Rusmevichientong et al. (2006) work on a nonparametric pricing problem, where each customer is represented by a budget and a preference list of products. Caldentey and Wein (2006) give fluid approximations for a joint pricing and admission control problem. Hu et al. (2013a) study a pricing problem with a gray market, which acts as an authorized channel to sell the authentic products of a supplier. Phillips (2013) describes a host of practical issues in pricing credit and gives a mathematical model. In an opaque product, a feature of a product, such as color for a piece of apparel or departure time for a flight, is hidden from the customer until the purchase occurs. Elmachtoub and Wei (2015) and Elmachtoub and Hamilton (2017) study pricing problems for opaque products. Belkaid and Martinez-de-Albeniz (2017) estimate the effect of weather conditions on demand and study the effectiveness of weather-dependent pricing strategies. Courty and Nasiry (2018) observe that certain products with different observable qualities are sold at uniform price and develop a model to resolve this paradox. Ma and Simchi-Levi (2018) develop a model that exploits the information extracted from bundled products to estimate individual price sensitivities.

The reader is referred to Anderson et al. (1992) for more on discrete choice theory of product differentiation and Vives (2001) for comparative static tools and oligopoly pricing.

Appendix

Proof of Theorem 8.1 It is clear that $R(p, z)$ is decreasing in z and that this implies that $\mathcal{R}(z)$ is decreasing in z. To verify convexity, let $\alpha \in (0, 1)$. Then for any $z, \tilde{z}$,

$$
\begin{aligned}
\mathcal{R}(\alpha z + (1-\alpha)\tilde{z}) &= \max_{p \in X} R(p, \alpha z + (1-\alpha)\tilde{z}) \\
&= \max_{p \in X} R(\alpha p + (1-\alpha)p, \alpha z + (1-\alpha)\tilde{z})) \\
&= \max_{p \in X} \left[\alpha(p-z)' + (1-\alpha)(p-\tilde{z})'\right] d(p) \\
&= \max_{p \in X} \left[\alpha R(p, z) + (1-\alpha)R(p, \tilde{z})\right] \\
&\le \alpha \max_{p \in X} R(p, z) + (1-\alpha) \max_{p \in X} R(p, \tilde{z}) \\
&= \alpha\mathcal{R}(z) + (1-\alpha)\mathcal{R}(\tilde{z}).
\end{aligned}
$$

□

Proof of Proposition 8.2 This follows from a direct application of the Taylor's expansion around $\mathcal{R}(\mathbb{E}[Z])$. □

Proof of Corollary 8.4 The proof of the first part is left as an exercise. From the concavity of g we have $g(\alpha z + (1-\alpha)\tilde{z}) \geq \alpha g(z) + (1-\alpha)g(\tilde{z})$ for any $z, \tilde{z} \in \Re^m$ and any $\alpha \in [0,1]$. Since $\mathcal{R}$ is decreasing, it follows that $\mathcal{R}(g(\alpha z + (1-\alpha)\tilde{z}) \leq \mathcal{R}(\alpha g(z) + (1-\alpha)g(\tilde{z}))$. From the convexity of $\mathcal{R}$, we have $\mathcal{R}(\alpha g(z) + (1-\alpha)g(\tilde{z})) \leq \alpha\mathcal{R}(g(z)) + (1-\alpha)\mathcal{R}(g(\tilde{z}))$. Consequently, $\mathcal{R}(g(\alpha z + (1-\alpha)\tilde{z})) \leq \alpha\mathcal{R}(g(z)) + (1-\alpha)\mathcal{R}(g(\tilde{z}))$, showing that $\mathcal{R}(g(z))$ is convex in z. From Jensen's inequality, it follows that $\mathbb{E}[\mathcal{R}(g(Z))] \geq \mathcal{R}(g(\mathbb{E}[Z]))$. □

Proof of Theorem 8.5 That $\mathcal{S}(p)$ is decreasing follows directly from the fact that $S(q,p)$ is decreasing in p. To verify convexity, let $\alpha \in (0,1)$. Then for any $p, \tilde{p}$

$$\begin{aligned}
\mathcal{S}(\alpha p + (1-\alpha)\tilde{p}) &= \max_{q\geq 0} S(\alpha p + (1-\alpha)\hat{p}, q) \\
&= \max_{q\geq 0} \left[U(q) - (\alpha p + (1-\alpha)\hat{p})'q\right] \\
&= \max_{q\geq 0} \left[\alpha(U(q) - p'q) + (1-\alpha)(U(q) - \tilde{p}'q)\right] \\
&\leq \alpha \max_{q\geq 0} S(p,q) + (1-\alpha)\max_{q\geq 0} S(\tilde{p}, q) \\
&= \alpha\mathcal{S}(p) + (1-\alpha)\mathcal{S}(\tilde{p}).
\end{aligned}$$

Notice that $\mathcal{S}(p) = U(d(p)) - p'd(p)$, so $\nabla\mathcal{S}(p) = \nabla d(p)\nabla U(d(p)) - d(p) - \nabla d(p)p = -d(p)$ on account of $\nabla U(d(p)) = p$ for all $p \in \mathcal{P}$.[5]

□

Proof of Theorem 8.7 Let $w'q - \frac{1}{2}q'Qq$ be the quadratic approximation to an increasing concave utility function U, where w is a vector of positive components, and Q is symmetric positive definite matrix.[6] Let $B = Q^{-1}$ and write $d(p) = B(w-p)$ over the set $\mathcal{P} = \{p : p \geq 0, B(w-p) \geq 0\}$. Then

$$\mathcal{S}(p) = U(d(p)) - d(p)'p = \frac{1}{2}(w-p)'B(w-p) \quad \text{over} \quad p \in \mathcal{P},$$

which is decreasing convex in $p \in \mathcal{P}$ since B is positive definite. The firm's problem is to find $p = p(z)$ that maximizes

$$R(p,z) = (p-z)'B(w-p).$$

The optimizer is given by $p(z) = (w+z)/2$, which is an increasing linear function of $z \in \mathcal{P}$. The composite function $\mathcal{S}(p(z))$ is therefore convex. □

---

[5]Notice here that $\nabla d(p)$ is the Jacobian of $d(p)$, i.e., the matrix of partial derivatives $\partial d_i(p)/\partial p_j$.

[6]If Q is not symmetric we can transform $Q \leftarrow (Q+Q')/2$ to make it symmetric without changing the utility function.

Proof of Theorem 8.10 Since $d(p)$ is USC and the product of non-negative USC functions is also USC, it follows that $R(p,z)$ is USC. The USC of $d(p)$ implies the USC of $\bar{d}(p)$ for if $\bar{d}(p)$ is not USC at p_0, then there exist a $p_1 > p_0$ at which $d(p_1) = \bar{d}(p_o)$ fails to be USC. As a result $\bar{R}(p,z)$ is also USC in $p \in [z, \infty)$. If $d(p) = 0$ for all $p \geq z$, then $p(z) = z$ and $\mathcal{R}(z) = R(z,z) = 0$ and there is nothing to prove. Otherwise there exists a price $\hat{p} > z$ such that $0 < \bar{d}(\hat{p}) < \infty$, for if $\bar{d}(p) = \infty$ for all $p > z$, then $\bar{d}(p)$ is not $o(1/p)$. Let $\epsilon = \bar{R}(\hat{p}, z) > 0$. We will show that there is a price $\tilde{p} > \hat{p}$ such that $\bar{R}(p,z) \leq \epsilon$ for all $p > \tilde{p}$, for if not, then for any $\tilde{p} > z$, we can find a $p > \tilde{p}$ such that $\bar{R}(p,z) > \epsilon$, or equivalently, $p\bar{d}(p) > p\epsilon/(p-z)$, contradicting the fact that $p\bar{d}(p) \to 0$ as $p \to \infty$. Given that $\bar{R}(p,z) \leq \epsilon$ for all $p \geq \tilde{p}$, we can restrict the optimization of $\bar{R}(p,z)$ without loss of optimality to the compact set $[z, \tilde{p}]$. The extreme value theorem guarantees the existence of a finite price, say $\bar{p}(z) \in [z, \tilde{p}]$, that maximizes $\bar{R}(p,z)$. We will now show that $p(z) = \bar{p}(z)$ also maximizes $R(p,z)$ so $\mathcal{R}(z) = \bar{\mathcal{R}}(z)$. Assume for a contradiction that $\bar{p}(z)$ is not a maximizer of $R(p,z)$. Then

$$(\bar{p}(z) - z)\bar{d}(\bar{p}(z)) = \bar{\mathcal{R}}(z) \geq \mathcal{R}(z) > (\bar{p}(z) - z)d(\bar{p}(z))$$

implies that $d(\bar{p}(z)) < \bar{d}(\bar{p}(z)) = \sup_{p \geq \bar{p}(z)} d(p)$. Then there exists a $p' > \bar{p}(z)$ such that $d(p') = \bar{d}(p(z))$, but then $\bar{R}(p',z) > \bar{R}(\bar{p}(z), z) = \bar{\mathcal{R}}(z)$ contradicting the optimality of $\bar{p}(z)$. □

Proof of Theorem 8.11 First, we show Part a. If $h(p)$ is continuous and increasing in p, then $f(p)$ is continuous and strictly decreasing in $p \geq z$. Equivalently, $(p - z)h(p)$ is continuous and strictly increasing in p. Now $f(z) = 1 > 0$ implies that $p(z) > z$, while $f(z + 1/h(z)) = 1 - h(z + 1/h(z))/h(z) \leq 0$ on account of $h(z + 1/h(z)) \geq h(z) > 0$ implies that $p(z) \leq z + 1/h(z)$. Because $(p - z)h(p)$ is continuous and strictly increasing in p, there exist a unique $p(z)$ satisfying $p(z) = \sup\{p : f(p) \geq 0\}$ that is bounded below by z and above by $z + 1/h(z)$. Suppose that $z' > z$, then $(p(z) - z')h(p(z)) < 1$, so $p(z') > p(z)$ showing that $p(z)$ is strictly increasing in z. To show that $\Delta(z) = p(z) - z$ is decreasing in z, let $p' = z' + \Delta(z)$ and notice that $(p' - z')h(p') = \Delta(z)h(p') \geq \Delta(z)h(p(z)) = 1$, so $p(z') = z' + q(z') \leq p' = z' + \Delta(z)$ implying that $\Delta(z') \leq \Delta(z)$. For the exponential demand function $d(p) = \lambda e^{-p/\theta}$, we have $h(z) = 1/\theta$ and $p(z) = z + \theta = z + 1/h(z)$, so the upper bound is attained.

Next, we show Part b. If $ph(p)$ is continuous and strictly increasing in p and $\tilde{z}h(\tilde{z}) > 1$, then $f(p)$ is continuous in $p > z$ and the equation $f(p) = 0$ can be written as $ph(p) = p/(p - z)$ with the left hand side increasing in p and the right hand side strictly decreasing to one for $p > z$. Since $zh(z) < \infty$ it follows that $p(z) > z$. Notice that $z/(1 - \tilde{z}h(\tilde{z}))$ is the root of $\tilde{z}h(\tilde{z}) = p/(p - z)$. Since $ph(p) \geq \tilde{z}h(\tilde{z}) \geq p/(p - z)$ for all $p \geq z/(1 - \tilde{z}h(\tilde{z}))$, it follows that $p(z)$ is unique and bounded above by $z/(1 - 1/\tilde{z}h(\tilde{z}))$. Suppose that $z' > z$, then $p(z') > z'$, so if $z' \geq p(z)$ it follows immediately that $p(z') \geq p(z)$. Suppose now that $z < z' < p(z)$, then at $p = p(z)$ we have $ph(p) < p/(p - z')$ implying that $p(z') > p(z)$. For $d(p) = \lambda p^{-b}$, with $b > 1$, we have $ph(p) = b$ for all p, and

$p(z) = bz/(b-1) = z/(1-1/b) = z/(1-1/\tilde{z}h(\tilde{z}))$, so the upper bound is attained. Let $m(z) := 1/h(z)$. Then, using the implicit function theorem on $f(p, z) = 0$, we can find the first and second derivatives of $p(z)$ in terms of $m(z)$. It is easy to see that the first derivative is given by $p'(z) = (1-m'(p(z))^{-1}$, so the second derivative is given by

$$p''(z) = \frac{m''(p(z)p'(z))}{(1-m'(p(z)))^2} \leq 0,$$

since $m''(z) \leq 0$ and $p'(z) > 0$.

Finally, we show Part c. Clearly $\tilde{f}(p) \leq f(p)$ so $\tilde{p}(z) \leq p(z)$. □

Proof of Proposition 8.13 Since the sum of USC is USC it follows that $d_S(p)$ is USC. Moreover $\bar{d}_m(p) = o(1/p)$ for all $m \in \mathcal{M}$ implies that $\bar{d}_S(p) = o(1/p)$. As a result $d_S(p)$ satisfies the conditions of Theorem 8.10 so there exists a finite price $p_S(z)$, increasing in z, such that $\mathcal{R}_S(z) = R_S(p_S(z), z)$ is decreasing convex in z. □

Proof of Proposition 8.16 It is easy to see that $p_m(z) > z$ is the root of $p/(p-z) = ph_m(p)$. Since the left hand side is decreasing in p and $ph_m(p)$ is increasing in p, it follows that there is a unique root $p > z$. This observation implies that $f_m(p) > 0$ on $p < p_m(z)$ and $f_m(p) < 0$ on $p > p_m(z)$. Let $f_S(p) = 1-(p-z)h_S(p)$ where $h_S(p)$ is the hazard rate of $d_S(p)$. Since $f_S(p)$ is a convex combination of $f_m(p) = 1-(p-z)h_m(p)$ with weights $\theta_m(p) = d_m(p)/d_S(p)$, it follows that $f_S(p) > 0$ for all $p < \min_{m\in S} p_m(z)$ because over that interval $f_m(p) > 0$ for all $m \in S$. Also $f_S(p) < 0$ for all $p > \max_{m\in S} p_m(z)$ because over that interval $f_m(p) < 0$ for all $m \in S$. Since the derivative of $R_S(p, z)$ is proportional to $f_S(p)$ it follows that $R_S(p, z)$ is increasing over $p < \min_{m\in S} p_m(z)$ and decreasing over $p > \max_{m\in S} p_m(z)$. Moreover, since $R_S(p, z)$ is continuous over the closed and bounded interval $[\min_{m\in S} p_m(z), \max_{m\in S} p_m(z)]$, appealing to the EVT yields the existence of a global maximizer $p_S(z)$ of $R_S(p, z)$. □

Proof of Theorem 8.18 Clearly

$$\begin{aligned}
\frac{Q_J(z)}{Q_M(z)} \geq \frac{Q_J^h(z)}{Q_M(z)} &= \frac{\sum_{j=1}^{J}\sum_{m\in M_j} R_m(q_j, z)}{Q_M(z)} \\
&= \sum_{j=1}^{J}\sum_{m\in M_j} e(q_j, p_m(z), z)\frac{\mathcal{R}_m(z)}{Q_M(z)} \\
&\geq \gamma_J(z)\frac{\sum_{m\in\mathcal{M}} \mathcal{R}_m(z)}{Q_M(z)} \\
&= \gamma_J(z).
\end{aligned}$$

□

Proof of Theorem 8.21 We have already shown that inequalities in (8.13). To show (8.14), notice that the super-modularity of $R(p, z)$ in p for fixed z, allows us to select $p_i(z|p_{-i})$ so that it is increasing in p_{-i}. Consequently, $p_i(z|p_{-i}) \leq p_i(z_i|\infty) = p_i(z_i)$ for all i. In particular, $p_i(z) = p_i(z|p_{-i}(z)) \leq p_i(z_i)$ for all $i \in N$. □

Proof of Theorem 8.23 Maximizing $R(p, z) = (p - z)'d(p)$ with respect to p is equivalent to minimizing $\frac{1}{2}p'Sp - (a + B'z)'p + a'z$ which is quadratic function. A sufficient condition for this function to be convex is that S is positive definitive. It is known that S is positive definitive, if and only if B is, see Johnson (1970). If B is positive definitive then S is invertible and since S is symmetric, so it is inverse S^{-1}. If B is positive definitive then the maximizer of $R(p, z)$ is given by (8.15). A sufficient condition for $p(0) = S^{-1}a \geq 0$ is for $S^{-1} \geq 0$, since $a > 0$. However, this is true because S is an s-matrix, i.e. a real symmetric, positive definitive matrix with non-positive off-diagonal elements. It is known that an s-matrix has a non-negative inverse implying that $S^{-1} \geq 0$, and consequently that $p(0) = S^{-1}a \geq 0$. Since $p(z)$ is non-decreasing in z by Theorem 8.1, it follows that $p(z) \geq p(0) \geq 0$ for all $z \geq 0$ such that $d(z) \geq 0$.

By adding and subtracting Bz to the expression in parenthesis on the right hand side of (8.15) we can write $p(z) - z = S^{-1}d(z)$, where $d(z)$ is the demand at $p = z$. It is also possible to write $d(p(z)) = a - Bp(z) = a - B(p(z) + z - z) = a - Bz - B(p(z) - z) = (I - BS^{-1})d(z)$ and then use the fact that $I - BS^{-1} = B'S^{-1}$ to obtain $d(p(z)) = B'S^{-1}d(z)$. This allows us to write $\mathcal{R}(z) = (p(z) - z)'d(p(z)) = d(z)'S^{-1}B'S^{-1}d(z) = d(z)'S^{-1}B'S^{-1}d(z)$ resulting in (8.16). □

Proof of Theorem 8.24 The first-order conditions are of the form

$$\frac{\partial R(p, z)}{\partial p_i} = d_i(p)[1 + \beta R(p, z)/\lambda - \beta(p_i - z_i)] = 0 \quad \forall i \in N.$$

For every subset $F \subseteq N$, let $p^F(z)$ be the solution to the first-order conditions obtained by setting the expression in brackets equal to zero for all $i \in F$ and by setting $d_i(p) = 0$ for all $i \notin F$. Then,

$$p_i = z_i + 1/\beta + R(p, z)/\lambda \quad \forall i \in F \quad \text{and} \quad p_i = \infty \quad \forall i \notin F.$$

For each $F \subseteq N$, there exists a constant, $\theta_F = R(p^F(z), z)/\lambda$, given by the root of the Lambert type equation

$$\beta\theta e^{\beta\theta} = \sum_{j \in F} e^{\alpha_j - \beta z_j - 1},$$

such that $p_i^F(z) = z_i + 1/\beta + \theta_F$ for all $i \in F$, and $R(p^F(z), z) = \lambda\theta_F$, so θ_F represents the optimal profit per customer when we are allowed to offer only products in F. Since the root θ_F is increasing in F, it follows that among all the 2^n

solutions to the first-order conditions, the one with the highest profit corresponds to $F = N$. Thus, at optimality, we have

$$p_i(z) = z_i + \frac{1}{\beta} + \theta,$$

where θ is the root of the Lambert equation for $F = N$. Moreover, $\mathcal{R}(z) = \lambda\theta$ is the optimal profit. □

Chapter 9
Dynamic Pricing Over Finite Horizons

9.1 Introduction

In this chapter, we first consider the problem of dynamically pricing one or more products that consume a single resource. Sales take place over a finite selling horizon, and the objective is to maximize the expected revenue that can be obtained from a finite inventory of the resource. We will assume that inventories cannot be replenished during the sales horizon. This problem setup holds for hotels, airlines, and seasonal merchandise including fashion retailing that have long procurement lead times. In this chapter, we focus on models that explicitly consider the stochastic and dynamic nature of demand. We use dynamic programming formulations to compute an optimal policy as a function of the remaining inventory and the time-to-go. In some cases, we are able to give closed-form solutions for the value function and the optimal pricing policy. In other cases, we resort to numerical solutions and to heuristic policies.

In Sect. 9.2, we provide a dynamic programming formulation of the dynamic pricing problem for a single resource. We also provide structural properties for the value function and the optimal pricing policy. In Sect. 9.3, we give several extensions of the basic formulation to allow for inventory replenishments, discounted cash flows, holding costs, multiple market segments, bargaining power, batch arrivals, nonlinear pricing, and strategic consumers. In Sect. 9.4, we show conditions for a fixed pricing policy to be asymptotically optimal, as the initial inventory of the resource and the demand are scaled up at the same rate. In Sect. 9.5, we establish conditions under which a bid-price policy is asymptotically optimal. In Sect. 9.7, we study the customer surplus process, and in Sect. 9.8, we study dynamic pricing for multiple products consuming multiple resources.

G. Gallego, H. Topaloglu, *Revenue Management and Pricing Analytics*, International Series in Operations Research & Management Science 279, https://doi.org/10.1007/978-1-4939-9606-3_9

9.2 Single Product Dynamic Pricing

Sales take place over a finite sales horizon of length T. There is a single resource with c units of capacity at the beginning of the sales horizon. We assume that inventories cannot be replenished during the sales horizon and that the salvage value at the end of the horizon is zero. If there is a positive salvage value, then the objective is to maximize the expected revenue in excess of salvage value. We will measure time backwards so that $t \in [0, T]$ is the time-to-go until the end of the sales horizon. Customers arrive as a time heterogeneous Poisson or compound Poisson process. Expositionally, it helps to introduce the basic formulation for the Poisson case and later take care of the changes needed to deal with the compound Poisson case. It is also helpful to initially work with a single product and then show that under mild conditions the same formulation works for multiple products consuming a single resource.

Let $d_t(p)$ be the Poisson arrival rate of customers willing to buy at price $p \in \Re_+$ at time t. We assume that customers unwilling to buy at price p leave the system, so customers are not inter-temporally strategic. We will address the issue of strategic customers in a later section. Let $R_t(p, z) := (p - z)d_t(p)$. We know from Chap. 8 that $\mathcal{R}_t(z) := \sup_{p \geq 0} R_t(p, z) \geq 0$ exists and is decreasing convex in z. Let $\bar{r}_t$ be the choke-off price of $d_t(p)$. This is defined as the smallest price (possibly infinity) such that $d_t(p) = 0$ for all $p \geq \bar{r}_t$. Notice that $\mathcal{R}_t(z) > 0$ as long as $z < \bar{r}_t$ because there must be a price $p \in (z, \bar{r}_t)$ such that $d_t(p) > 0$, so $\mathcal{R}_t(z) \geq R_t(p, z) = (p - z)d_t(p) > 0$. Furthermore, if $d_t(p)$ is continuous and $o(1/p)$, then there exists a finite price $p_t(z)$, increasing in z, such that $\mathcal{R}_t(z) = \max_p R_t(p, z) = R_t(p_t(z), z)$, thus avoiding the use of the supremum. As always, the terms increasing and decreasing should be interpreted in the weak sense unless stated otherwise.

Let $V(t, x)$ be the maximum expected revenue when the time-to-go is t, and the remaining inventory is $x \geq 1$. We will refer to (t, x) as the state of the system, with (T, c) being the initial state. Our goal is to find $V(T, c)$ and an optimal dynamic pricing policy that results in expected revenue $V(T, c)$.

Suppose that the current state is (t, x), and consider a time increment δt small enough to approximate the probability of a request for one unit at price p by $d_t(p)\delta t$. Then, with probability $d_t(p)\delta t$ a unit is sold at price p during the interval $(t - \delta t, t]$ and the state changes to $(t - \delta t, x - 1)$. With probability $1 - d_t(p)\delta t$, there are no requests during the time interval $(t - \delta t, t]$ and the state changes to $(t - \delta t, x)$. We have the dynamic programming formulation

$$
\begin{aligned}
V(t, x) = \max_p \Big\{ & d_t(p)\, \delta t\, [p + V(t - \delta t, x - 1)] \\
& + (1 - d_t(p)\, \delta t)\, V(t - \delta t, x) \Big\} + o(\delta t). \qquad (9.1)
\end{aligned}
$$

The term $o(\delta t)$ captures the remote possibility of more than one request during the interval $(t - \delta t, t)$ and is a function that goes to zero faster than δt. If $d_t(p)$ is continuous in t, we can subtract $V(t - \delta t, x)$, divide by δt, and take the limit as δt goes to zero to obtain the Hamilton–Jacobi–Bellman (HJB) partial differential equation

$$\frac{\partial V(t,x)}{\partial t} = \max_p R_t(p, \Delta V(t,x)) = \mathcal{R}_t(\Delta V(t,x)), \tag{9.2}$$

where $\Delta V(t,x) = V(t,x) - V(t,x-1)$ is the marginal value of the x-th unit of capacity for integer $x \geq 1$. The boundary conditions are $V(0,x) = 0$ for $x > 0$ and $V(t,0) = 0$. The HJB equation (9.2) has a unique solution which is equal to the value function $V(t,x)$. If $d_t(p)$ is piecewise continuous, then the HJB equation (9.2) holds over each subinterval where $d_t(p)$ is continuous where the boundary condition is modified to be the value function over the remaining time horizon. Let $p_t(z) = \arg\max_p R_t(p,z)$. Define $P(t,x) := p_t(\Delta V(t,x))$ as an optimal solution to the maximization problem on the right side of (9.2).

9.2.1 Examples with Closed Form Solution

We now present two examples for which we can solve the HJB equation (9.2) without resorting to numerical solutions. These examples illustrate the behavior of $V(t,x)$ and $P(t,x)$ as functions of t and x.

Example 9.1 Suppose that willingness to pay is a random variable Θ that is exponentially distributed with mean θ, so $H(p) = P(\Omega \geq p) = \exp(-p/\theta)$. The demand function $d_t(p) = \lambda H(p)$ for all $t \in [0,T]$ corresponds to a time homogeneous arrival rate λ and a time homogenous exponential willingness to pay Ω. It is possible to show that the value function is given by

$$\begin{aligned} V(t,x) &= \theta\lambda^* t + \theta \ln(\mathbb{P}\{N^*(t) \leq x\}) \\ &= \theta \ln\left(\sum_{j=0}^{x} \frac{(\lambda^* t)^j}{j!}\right), \end{aligned} \tag{9.3}$$

where $N^*(t)$ is the time homogeneous Poisson process with rate $\lambda^* = \lambda/e$. One can verify that (9.3) satisfies (9.2) by taking the partial derivative with respect to t. The corresponding optimal price policy is given by

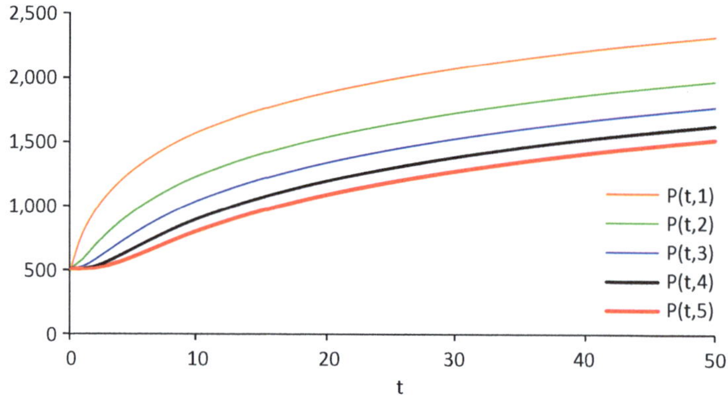

Fig. 9.1 Optimal pricing policy in Example 9.1

$$\begin{aligned} P(t,x) &= \theta + \Delta V(t,x) \\ &= \theta\left(1+\ln\left(\frac{\mathbb{P}\{N^*(t)\le x\}}{\mathbb{P}\{N^*(t)\le x-1\}}\right)\right). \end{aligned} \tag{9.4}$$

To illustrate the solution, suppose that the arrival rate is $\lambda = 2$, $T = 50$ and the willingness to pay is exponentially distributed with mean $\theta = 500$. Figure 9.1 shows the price paths $P(t,x)$, $x \in \{1,\ldots,5\}$ for $t \in \{5,10,\ldots,50\}$. The figure confirms that $P(t,x)$ increases with t and decreases with x. Since t is the time-to-go, prices decrease as time elapses and there is a price increase $P(t,x-1) - P(t,x)$ when a sale occurs at state (t,x). Notice also that the price paths are neither convex nor concave in t for fixed x.

Example 9.2 Suppose that $d_t(p) = \lambda_t\, p^{-b}$ for some $b > 1$. Let $\Lambda_t = \int_0^t \lambda_s ds$ be the expected number of sales at price $p = 1$, and let k_x be a sequence defined by $k_0 = 0$ and for integer $x \ge 1$, $k_x = \left(\frac{b-1}{b}\right)^{b-1}(k_x - k_{x-1})^{1-b}$. It is possible to show that the value function and the optimal pricing policy satisfy

$$\begin{aligned} V(t,x) &= \Lambda_t^{1/b} k_x \\ P(t,x) &= \Lambda_t^{1/b} k_x^{-1/(b-1)}. \end{aligned}$$

Furthermore, for large x, one can build the approximations $V(t,x) \approx (\Lambda_t)^{1/b} x^{1-1/b}$ and $P(t,x) \approx (\Lambda_t/x)^{1/b}$.

To illustrate the solution, suppose that $\lambda = 2$, $T = 50$ and $b = 1.5$. For $c = 30$, $V(50,30) = 65.44$. The price paths $P(t,x)$ for $x \in \{1,\ldots,5\}$ and $t \in \{5,10,\ldots,50\}$ are given in Fig. 9.2.

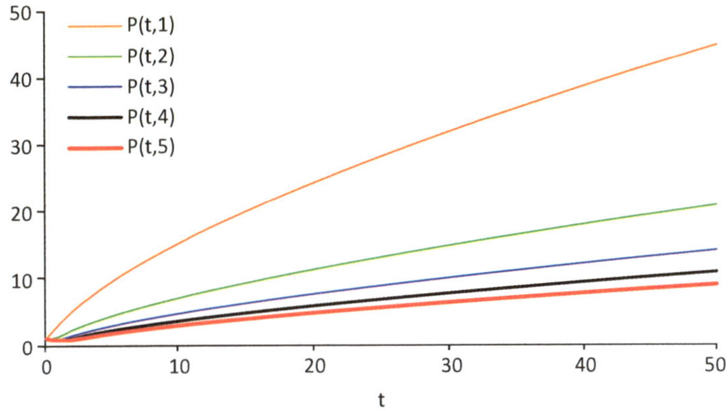

Fig. 9.2 Optimal pricing policy in Example 9.2

9.2.2 *Structural Results*

In this section, we present structural results for the value function $V(t, x)$ and the optimal pricing policy $P(t, x)$.

Theorem 9.3 *$V(t, x)$ is increasing in t and in x. $\Delta V(t, x)$ is increasing in t and decreasing in x. Moreover $P(t, x)$ is decreasing in x. If $p_t(z)$ is increasing in t, then $V(t, x)$ is strictly increasing in t and $P(t, x)$ is increasing in t. A sufficient condition for $p_t(z)$ to be increasing is that $d_t(p) = d(p)$ is independent of t. Another sufficient condition is that the hazard rate $h_t(p)$ is decreasing in t. Moreover, if $\mathcal{R}_t(z)$ is differentiable in z, and $\Delta V(t, x) < \bar{r}_t$ for all t, then $V(t, x)$ is concave in t.*

Remark 9.4 It is possible for $V(t, x)$ to fail to be strictly increasing and concave in t. This may happen if there is an intermediate period of low willingness to pay where it is optimal to use the choke-off price, so $V(t, x)$ is flat in t over this interval.

9.2.3 *Factors Affecting Dynamic Pricing*

A solution to the dynamic pricing problem (9.2) is to price at

$$P(t, x) = p_t(\Delta V(t, x)). \tag{9.5}$$

There are two factors that cause price variations in dynamic pricing:

1. Changes in the marginal value of capacity $\Delta V(t, x)$ as time elapses and inventory is depleted.
2. Changes in the demands over time that lead to changes in the optimal price $p_t(z)$ as a function of $t \in (0, T]$.

Consider first the case where a time-invariant price $p(z) = p_t(z)$ is optimal for all $z \in \Re_+$ for all $t \in (0, T]$. In this case, $P(t, x) = p(\Delta V(t, x))$ and all price changes are due only to changes in state dynamics. By Theorem 9.3, $P(t, x)$ is increasing in t and decreasing in x. Suppose now that $p_t(z)$ depends on t, then $P(t, x) = p_t(\Delta V(t, x))$ is still decreasing in x as shown in Theorem 9.3, but $P(t, x)$ may be either increasing or decreasing in t. A sufficient condition for $P(t, x)$ to increase in t is that $p_t(z)$ is increasing in t and a sufficient condition for this is that $h_t(p)$ is decreasing in t.

9.2.4 Discrete Time Formulation and Numerical Solutions

Except for a few selected cases, it is virtually impossible to solve the HJB equation (9.2) in closed form. One can, of course, resort to numerical computations. One approach is to use numerical methods to locally solve the partial differential equation (9.2) and to use smooth pasting techniques to put together an approximate solution. A more common approach, however, is to discretize the dynamic program and then solve it numerically. This is typically done by rescaling time and setting $\delta t = 1$. To see how this works, select a real number $a > 1$, so that $T \leftarrow aT$ is a large integer and rescale demand so that $d_t(p) \leftarrow \frac{1}{a} d_{t/a}(p) < \epsilon$, for all p and all $t \in [0, T]$. Now set $\delta t = 1$ in (9.1) and drop the term $o(\delta t)$. Reorganizing terms leads to

$$V(t, x) = V(t - 1, x) + \mathcal{R}_t(\Delta V(t - 1, x)) \tag{9.6}$$

with the same boundary conditions. The value of ϵ governs the accuracy of the discrete time dynamic program (9.6), and the smaller the value of ϵ the better the accuracy. We suggest using values $\epsilon \leq 0.1$. One can then compute $V(t, x)$ and the corresponding optimal price $P(t, x)$ for all $(t, x) \in S = \{(t, x) : t \in \{0, 1, \ldots, T\}, x \in \{0, 1, \ldots, c\}\}$. The resulting policy is *optimal* up to the small error introduced by the discrete-time formulation. Both of these numerical approaches yield approximate value functions and approximately optimal pricing policies. For a given time increment δt, the level of accuracy that can be obtained by using smooth pasting techniques is typically higher than using discrete-time dynamic programming. Nevertheless, coding the discrete time approximation is easier, so from a practical point of view we recommend the discrete time approximation.

Example 9.5 We reconsider Example 9.1, with $c = 50$, $\theta = 500$, $\lambda = 2$, and $T = 50$, but use a rescaling factor of $a = 200$, to obtain $\lambda = 0.01$ and $T = 10000$. For this example, we have $V(T, c) = \$18,386.41$, which is quite close to the true value $\$18,386.31$. Scaling by a factor of $a = 2000$ results in $V(T, c) = \$18,386.32$. As we can see, the accuracy of the discrete time approximation improves as we use a more aggressive scaling.

9.3 Extensions of Basic Model

In this section, we present extensions to replenishments, holding costs, discounted cash flows, multiple market segments, compound Poisson demands, and to situations where multiple products use the same scarce resource. We also present an extension to the case where customers negotiate.

9.3.1 Inventory Replenishments

Suppose that it is possible to replenish inventories at unit cost w. Then the differential equation is of the same form, except that at $x = 0$ the differential equation is given by

$$\frac{\partial V(t,0)}{\partial t} = \mathcal{R}_t(w).$$

9.3.2 Holding Costs

In some cases, there may be a holding cost rate, say $h(x)$, charged on inventories. Let $\Delta h(x) = h(x) - h(x-1)$ for $x \geq 1$. Then,

$$\frac{\partial V(t,x)}{\partial t} = \mathcal{R}_t(\Delta V(t,x) - \Delta h(x)) - h(x).$$

9.3.3 Discounted Cash Flows

Suppose that cash flows are discounted continuously, at rate α, to the beginning of the sales horizon T. Thus, a dollar received at time-to-go $t \in [0, T]$ is worth $e^{-\alpha(T-t)}$. Following the same logic used to find the differential equation (9.2) for the case without discounting, we obtain the following HJB equation for the value function $V(t,x)$ with discounting:

$$\frac{\partial V(t,x)}{\partial t} = \mathcal{R}_t(\Delta V(t,x)) - \alpha V(t,x).$$

Since this equation has $V(t,x)$ explicitly, it is often written as

$$V(t,x) = \frac{1}{\alpha}\left\{\mathcal{R}_t(\Delta V(t,x)) - \frac{\partial V(t,x)}{\partial t}\right\}.$$

9.3.4 Multiple Market Segments

Suppose there is a single product or resource but there is demand from different market segments. We will assume that markets are segmented by customer attributes such as age, gender, occupation, time-of-purchase, geography, or club membership. If markets are segmented by verifiable customer attributes, then the capacity provider can dynamically change the price charged to each market. Let $d_{mt}(p)$ be the Poisson arrival rate of customers of market segment $m \in \mathcal{M} := \{1, \ldots, M\}$ willing to purchase at price p. The value function $V(t, x)$ is governed by the HJB equation

$$\frac{\partial V(t,x)}{\partial t} = \sum_{m=1}^{M} \mathcal{R}_{mt}(\Delta V(t,x)) \tag{9.7}$$

with boundary conditions $V(t, 0) = V(0, x) = 0$, where $\mathcal{R}_{mt}(z) := \max_{p \geq z} R_{mt}(p, z)$ and $R_{mt}(p, z) := (p - z)\, d_{mt}(p)$. If $p_m(z)$ is a maximizer of $R_m(p, z)$ for each $m \in \mathcal{M}$, then it is optimal to set prices $P_m(t, x) = p_{mt}(\Delta V(t, x)), m \in \mathcal{M}$ at state (t, x). Notice that all markets use the same marginal value of capacity to set prices.

Example 9.6 Suppose there are $M = 4$ market clusters with exponential willingness to pay distributions with respective means \$100, \$150, \$250, \$300 and time homogeneous arrival rates 0.25, 0.5, 0.5, 0.25, and $T = 100$. Solving (9.7) numerically for capacity of $c = 50$ results in $V(100, 50) = \$10,801.65$, $\Delta V(100, 50) = \$31.93$ and initial prices \$131.93, \$181.93, \$281.93, \$331.93. Recall that the marginal value of capacity $\Delta V(t, x)$ is increasing in t and decreasing with x so prices change continuously (going down between sales and up after each sale).

9.3.5 Dynamic Pricing when Customers Negotiate

Suppose that the Poisson arrival rate is of the form $d_t(p) = \lambda_t P(\Omega_t \geq p)$ where Ω_t is the willingness to pay for customers arriving at time-to-go t. Suppose that instead of posting a price $P(t, x)$ as a take it or leave it price, the seller is willing to negotiate with each customer as they arrive. We will assume that the seller will dynamically select a reservation price, say $P^\beta(t, x)$, below which he will not sell the product. This price is not made public, but helps the seller negotiate differently with each customer depending on the state of the system (t, x). We will assume that the negotiation results in a bargaining Nash equilibrium (BNE) equal to $(1 - \beta)P^\beta(t, x) + \beta\Omega$ when $\Omega \geq P^\beta(t, x)$ with no transaction occurring if $\Omega < P^\beta(t, x)$. Here $\beta \in [0, 1]$ reflects the negotiation power of the seller and $1 - \beta$ the negotiating power of the buyer. Suppose z is the marginal cost at state (t, x).

Let

$$R_t^\beta(p,z) := \lambda_t \, \mathbb{E}\{[\beta\Omega_t + (1-\beta)p - z]1(\Omega_t \geq p)\} = \lambda_t \, \beta \, \mathbb{E}[(\Omega_t - p)^+] + R_t(p,z)$$

be the expected revenue that can be obtained by negotiating with reservation price p when the marginal cost is z. Let

$$\mathcal{R}_t^\beta(z) := \max_p R_t^\beta(p,z).$$

It can be shown that $\mathcal{R}_t^\beta(z)$ is decreasing convex in z, and increasing in β. Let $h_t(p) := -d_t'(p)/d_t(p)$ be the hazard rate corresponding to $d_t(p)$, and consider the equation

$$(p - z)h_t(p) = 1 - \beta. \tag{9.8}$$

Under mild conditions on $h_t(p)$ (e.g. $h_t(p)$ or $ph_t(p)$ strictly increasing in p), the root of (9.8) is the unique maximizer, say $p^\beta(z)$, of $R_t^\beta(p,z)$. From (9.8), it is clear that the reservation price $p^\beta(z)$ is *increasing* in z and *decreasing* in β.

The HJB equation for the case of negotiation results in

$$\frac{\partial V^\beta(t,x)}{\partial t} = \mathcal{R}_t^\beta(\Delta V^\beta(t,x)), \tag{9.9}$$

where $z = \Delta V^\beta(t,x) = V^\beta(t,x) - V^\beta(t,x-1)$ is the marginal value of capacity. The boundary conditions are $V^\beta(t,0) = V^\beta(0,x) = 0$. At state (t,x), it is optimal to use reservation price $P^\beta(t,x) = p^\beta(\Delta V^\beta(t,x))$.

If $\beta = 0$, then transactions occur at $P^0(t,x)$ if $\Omega \geq P^0(t,x)$, so this is equivalent to dynamic pricing without negotiation, where the price $P^0(t,x)$ is announced as a take it or leave it price. Consequently, the dynamic pricing formulation (9.2) implicitly assumes that the seller has *no* negotiating power. $V^\beta(t,x) > V(t,x)$ holds for all (t,x) and all $\beta > 0$ on account of $\mathcal{R}_t^\beta(z) > \mathcal{R}_t(z)$ for all relevant values of z. The case $\beta = 1$, results in $P^1(t,x) = \Delta V^1(t,x)$, resulting in sales at Ω whenever $\Omega > \Delta V^1(t,x)$. Thus, the seller extracts all the consumer surplus from consumers willing to pay $\Delta V^1(t,x)$, resulting in

$$\frac{\partial V^1(t,x)}{\partial t} = \mathcal{R}_t^\beta(\Delta V^1(t,x)) = \lambda_t \, \mathbb{E}[(\Omega_t - \Delta V^1(t,x))^+].$$

For the exponential case, $h_t(p) = \theta$ results in $p^\beta(z) = z + (1-\beta)\theta$, and $\mathcal{R}_t^\beta(z) = \mathcal{R}_t(z)e^\beta > \mathcal{R}_t(z) = \theta \exp(-z/\theta)\exp(-1)$.

If only a fraction of customers negotiate, then the provider can post a take it or leave it price p and use a reservation price $\underline{p}$ to use as a lower bound for customers that do negotiate.

9.3.6 Compound Poisson

The dynamic program (9.2) is written under the implicit assumption that requests arrive for a single unit. Here, we extend the analysis to allow for multiple purchases. We start with a model where customers arriving at time-to-go t request a random number of units, say Z_t. The random variables Z_t are independent of anything else. To formulate the problem, we need to make assumptions of how to price multiple units. Under dynamic linear pricing, the price for k units is kp for some $p \in \Re_+$ that varies dynamically with the state of the system. Several formulations are possible under this framework. As an example, the seller may honor the posted price for request up to capacity. This is unsatisfactory for two reasons. First, a consumer requesting $k > x$ units of capacity may or may not be interested in taking the x units available. Second, the firm may lose money even if $k < x$ when $pk < \Delta_k V(t,x) = V(t,x) - V(t,x-k)$. A better formulation is to accept a request of size k at unit price p only if $kp \geq \Delta_k V(t,x)$ and assume that rejected request disappears from the system. Under these assumptions, the formulation for the compound Poisson case is given by

$$\frac{\partial V(t,x)}{\partial t} = \max_{p \geq 0} \sum_{k=1}^{\infty} \mathbb{P}\{Z_t = k\}\, d_t(p)(kp - \Delta_k V(t,x))^+, \tag{9.10}$$

with boundary conditions $V(0,x) = V(t,0) = 0$. Thus, an arriving customer requesting up to x units will buy them at rate $d_t(p)$.

9.3.7 Dynamic Nonlinear Pricing

In dynamic nonlinear pricing, the seller is not constrained to offer price kp for a request of size k. We consider two models. Let $d_{kt}(p)$ be the k-unit demand rate at *aggregate* price p at time-to-go t. If the demands for different request sizes are independent, then the formulation can be written as

$$\frac{\partial V(t,x)}{\partial t} = \sum_{k=1}^{x} \mathcal{R}_{kt}(\Delta_k V(t,x)), \tag{9.11}$$

where $R_{kt}(p,z) := (p - z)d_{kt}(p)$ and $\mathcal{R}_{kt}(z) := \max_{p \geq 0} R_{kt}(p,z)$. Here, $P_k(t,x) = p_{kt}(\Delta_k V(t,x))$ is the optimal price for a request of size k at time-to-go t when the remaining inventory is x. Of course, if $k > x$, $\Delta_k V(t,x) = \infty$, and $P_k(t,x) = \infty$ reflecting the fact that request of size $k > x$ cannot be satisfied.

As a special case, assume that consumers arrive at rate λ_t, and have willingness to pay Ω per unit of capacity, and request k units with probability $\mathbb{P}(Z_t = k)$, $k = 1, \ldots, K$. Then $d_{kt}(p) = \lambda_t \mathbb{P}(Z_t = k)\,\mathbb{P}(\Omega \geq p/k)$. If Ω is exponential with mean

θ, then

$$P_k(t,x) = k\theta + \Delta_k V(t,x) \quad k \le x,$$

is the total price for a request of size k which is clearly nonlinear.

Example 9.7 We reconsider Example 9.5, with $c = 50$, $\theta = 500$, $\lambda = 2$, and $T = 50$, with time rescaled by a factor $a = 200$, to obtain $\lambda = 0.01$ and $T = 10000$. We assume that the distribution of request sizes is time homogenous with $\mathbb{P}(Z = 1) = 0.7, \mathbb{P}(Z = 2) = 0.15, \mathbb{P}(Z = 3) = 0.1, \mathbb{P}(Z = 4) = 0.05$. For this example, $V(T, c) = \$18,364.21$. We can also see that at state (T, c) the prices for requests of size $k = 1, 2, 3,$ and 4 are, respectively, \$508.77, \$1,019.62, \$1,532.92, and \$2,049.081. The unit prices are nearly uniform in this case, increasing from \$508.77 for a single unit to \$512.27 for four units. However, more significant differences are seen at states where c is tight relative to capacity. As an example, at state $(T, c/10)$, the unit prices range from \$1,266.82 for one unit, to \$1,906.60 for four units.

Formulation (9.11) assumes that the markets for different demand sizes are independent. In practice, consumers may adjust their demand sizes to the prevailing prices. Thus, a more general model would involve the consumer choice for the best number of units to purchase at any price. A simple, but useful model is to assume that the utility of purchasing k units is $u_k(\theta)$, where θ is a parameter reflecting the taste for the product. Let $F(\theta)$ be the cumulative distribution of the consumers' tastes. If we let $\rho_k(\theta) := u_k(\theta) - u_{k-1}(\theta)$ be the marginal utility of the k-th unit, then it is reasonable to assume that $\rho_k(\theta)$ is increasing in θ and decreasing in k. Let p_j be the marginal price for the j-th unit. A consumer would buy the k-th unit if $\theta \ge \theta_j(p_j) := \rho_j^{-1}(p_j)$ for all $j \le k$. If the $\theta_j(p_j)$'s are monotone increasing, then a consumer would buy the k-th unit if $\theta \ge \rho_k^{-1}(p_k)$, and this happens with probability $\bar{F}(\theta_k(p_k))$. Let $R_k(p, z_k) := (p_k - z_k)\bar{F}(\theta_k(p_k))$. Let $p_k(z_k)$ be the price that maximizes $R_k(p, z_k)$ and $\mathcal{R}_k(z_k) := R_k(p_k(z_k), z_k)$. Consider the following dynamic nonlinear pricing formulation that separately selects the marginal prices:

$$\frac{\partial V(t,x)}{\partial t} = \sum_{k=1}^{x} \mathcal{R}_k(\Delta_k V(t,x)).$$

The resulting optimal marginal prices are given by $p_k(\Delta_k V(t, x))$. The model, as stated, is correct only if $\theta_k(p_k(\Delta_k(t, x)))$ are monotone increasing in $k \in \{1, \dots, x\}$ for all (t, x). It is possible to show that this is indeed true for a variety of models including $\rho_k(\theta) = a_k\theta$ where the a_k's are monotonically decreasing. It is possible to show that $\Delta V(t, x)$ is increasing concave in x for this nonlinear pricing model, a property that may fail to hold under dynamic linear pricing.

9.3.8 Strategic Customers and Monotone Pricing Policies

All of the formulations so far allow prices to vary dynamically, and this pricing flexibility helps maximize expected revenues as long as customers are not strategic. If prices exhibit strong fluctuations, then some customers may decide to wait for better prices and if a significant number of them act this way the strategy of allowing price flexibility may backfire resulting in losses rather than gains. To avoid this, capacity providers may limit price flexibility. An extreme case would be to insist on monotone prices. Such formulations are a little more difficult to solve because the solution must implicitly price the cost of making irreversible decisions, i.e. price increases. We can generalize the formulation (9.10) by expanding the state space to (t, x, p_t) where p_t is the current price vector and by imposing lower bound pricing constraints, say $p \geq p_t$ to ensure that price paths are monotone. In many cases, the loss in potential revenue from the restriction to monotonic prices is relatively small and it imposes more discipline in the price process at the same time it discourages speculators who want to arbitrage inter-temporal price differences.

9.4 Fixed Price Policies for Time Independent Demands

We return here to continue our study of the base case in (9.2). Fixed price policies are expected to do poorly if $R_t(p, z)$ has significantly different optimizers, $p_t(z)$, over $t \in [0, T]$. When $p_t(z)$ is time invariant for all $z \geq 0$, there is no advantage in changing prices over time in the deterministic case. The hope is that fixed pricing policies may be near-optimal in some stochastic settings, and our goal here is to understand when this is true.

If $p_t(z)$ is time invariant, then changes in the marginal value of capacity is the only source of price variation in dynamic pricing. Let $N_t(p)$ be a Poisson random variable with parameter $D_t(p) := \int_0^t d_s(p)ds$. Notice that $D_t(p)$ is the expected number of customers that purchase at price p over the sales horizon $[0, t]$. Let $V^f(t, x) = \max_p p\,\mathbb{E}\min(x, N_t(p))$ be the maximum expected profit that can be obtained by using a fixed pricing policy over the horizon $[0, t]$ starting with x units of inventory. Under mild conditions, there is a finite fixed price maximizer, say $p_f(t, x)$, of $p\,\mathbb{E}\min(x, N_t(p))$. It is important to understand when the difference $V(t, x) - V^f(t, x)$ is significant enough, relative to $V(t, x)$, to warrant using a dynamic pricing policy over a fixed pricing policy. The following examples indicate the performance of fixed pricing policies for the exponential and constant elasticity demand functions.

Example 9.8 Consider again the exponential demand function of Example 9.1. $V(50, 50) = \$18,386.31$ is only 0.06% larger than $V^f(50, 50) = \$18,374.49$. In contrast, $V(50, 5) = \$10,625.94$ is 5.19% larger than $V^f(50, 5) = \$10,101,51$. Similarly, $V(200, 1)$ is 7.19% larger than $V^f(200, 1)$, while $V(200, 20)$ is less than 2% larger than $V^f(200, 20)$.

Example 9.9 Consider again the constant elasticity demand model $d_t(p) = \lambda_t p^{-b}$ with $\lambda_t = 2$ and $b = 1.5$ of Example 9.2. $V(50, 30) = \$65.44$ is 4.8% higher than $V^f(50, 30) = \$62.44$. $V(50, 10) = \$43.82$ is 6.5% larger than $V^f(50, 10) = 40.98$, while $V(10, 1) = \$5.11$ is 7.8% higher than $V^f(10, 1) = 4.70$.

These examples give us some indications of when the benefits of dynamic pricing are or are not significant relative to a fixed pricing policy in the case that $p_t(z)$ is time invariant. Consider first the situation with ample capacity. Since the revenue rate $\mathcal{R}_t(0) \geq \mathcal{R}_t(z)$ for all $z > 0$, integrating (9.2) we see that $V(t, x) \leq \int_0^t \mathcal{R}_s(0)ds$ for all x and all t. Consider now using the fixed price $p(0)$ over $[0, t]$ as long as capacity is available. If $x \gg D_t(p(0))$, there is little risk of running out of capacity at the fixed price $p(0)$. As a result, the revenues $V^f(t, x)$ will be close to $\int_0^t \mathcal{R}_s(0)ds$ and dynamic pricing should be of little benefit. We can see this at work in some of the examples above. For instance, in Example 9.8, the expected demand over the horizon $[0, 50]$ at price $p(0) = 500$ is $100/e = 36.78 \ll 50$, and the advantage of dynamic of static pricing is only 0.06%. At the other extreme, state (t, x) is of severe shortage if $x \ll D_t(p(0))$ and in this case we expect that the potential benefits of dynamic pricing can be significant. This is illustrated by the state $(T, c) = (50, 5)$ and more extremely by the state $(T, c) = (200, 1)$ for the exponential demand in Example 9.8.

On the other hand, we expect the benefits to be small when $N_T(p)$ has a large mean and consequently, a small coefficient of variation, and capacity is moderately large so that $\Delta V(t, x)$ is fairly stable over the sales horizon. This is illustrated by the state $(T, c) = (2000, 50)$ for the exponential demand distribution. Similar observations hold for the constant elasticity of substitution demand model, except that in this case the benefits of dynamic pricing tend to be more significant than in the case of the exponential demand function.

We now present a simple heuristic to approximate the optimal fixed price. Let $p_{mkt}(c)$ be the market clearing price, so that $\mathbb{E}N_T(p) = D_T(p) = c$ at $p_{mkt}(c)$. The fixed price heuristic is given by:

$$p_F = \max(p_{mkt}(c), p(0)). \tag{9.12}$$

It turns out that using the fixed pricing policy (9.12) over the entire selling horizon $[0, T]$ has good asymptotic properties when $p_t(z)$ is time invariant. Consequently, we expect the fixed pricing policy to do well for large systems. To be more precise, let $V_b(T, c)$ be the optimal expected revenue from dynamically pricing a system indexed by $b \geq 1$, with capacity bc and demand rates $bd_t(p)$ for $0 \leq t \leq T$. Also, let $V_b^F(T, c)$ be the expected revenue from using price p_F. Notice that $p_{mkt}(c)$, $p(0)$, and consequently p_F are all independent of b.

Theorem 9.10 *If $N_T^b(p)$ is Poisson, then*

$$\lim_{b \to \infty} \frac{V_b^F(T, c)}{V_b(T, c)} = 1.$$

The theorem above actually holds more generally as we will see in the next section. The theorem exploits the fact that the coefficient of variation of $N_T^b(p)$ is $\sqrt{bD_T(p)}/bD_T(p) = 1/\sqrt{bD_T(p)} \to 0$ as $b \to \infty$. This means that the system behaves as a deterministic system when b is large. It is easy to verify that p_F is the optimal price for the deterministic system, and this essentially explains the asymptotic optimality of the fixed pricing policy (9.12). In practice, it does not take a very large b for the asymptotic results to kick in.

9.5 Bid-Price Heuristics

The main disadvantage of fixed price policies is that they do not react to changes in $p_t(z)$ over $t \in [0, T]$. In this section, we present a heuristic that reacts to changes in $p_t(z)$ over $t \in [0, T]$, but ignores changes in the marginal value $\Delta V(t, x)$ of capacity due to random changes in the state space. More specifically, the heuristic is of the form $P^h(t, x) = p_t(z(T, c)),\ 0 \leq t \leq T$. The quantity $z(T, c)$ is chosen to capture the marginal value of capacity of a fluid model where demands are replaced by their expectations. For the basic problem (9.2), $z(T, c)$ can be obtained by solving the one-dimensional optimization problem

$$\bar{V}(T, c) = \min_{z \geq 0} \left[\int_0^T \mathcal{R}_t(z)dt + cz \right]. \tag{9.13}$$

In this formulation, z is the dual variable of the capacity constraint of the problem of maximizing $\int_0^T R_t(p, 0)dt$ subject to $\int_0^T d_t(p)dt \leq c$. From the chapter on basic pricing theory, we know that $\mathcal{R}_t(z)$ is convex in z. Consequently, the minimization problem in (9.13) is a convex program in a single variable, so if z_c is the unconstrained minimizer of (9.13), then $z(T, c) = \max(z_c, 0)$ is an optimal solution to the constrained problem. It is possible to show that $\bar{V}(T, c) \geq V(T, c)$ is an upper bound on the optimal expected revenue at state (T, c).

If $\mathcal{R}_t(z)$ is differentiable with respect to z, then the derivative of $\int_0^T \mathcal{R}_t(z)dt + cz$ is $c - \int_0^T d_t(p_t(z))dt$, which is the difference between capacity and demand at price path $p_t(z), t \in [0, T]$. If the derivative is positive at $z = 0$, then $\int_0^T \mathcal{R}_t(z)dt + cz$ is increasing in $z \geq 0$ and consequently $z = z(T, c) = 0$ is optimal. Otherwise, $z(T, c)$ is given by

$$z(T, c) = \sup \left\{ z \geq 0 : c - \int_0^T d_t(p_t(z))dt \leq 0 \right\}.$$

If $\mathcal{R}_t(z)$ is continuously differentiable in z for all $t \in [0, T]$, and $\int_0^T d_t(p_t(0))dt > c$, then the market clears at $p_t(z(T, c)), t \in [0, T]$ for some $z(T, c) > 0$, in the sense that the expected demand at $p_t(z(T, c)), t \in [0, T]$ is exactly c. If $d_t(p)$ is discontinuous then it is possible to have $c <$

$\int_0^T d_t(p_t(z(T, c)))dt$. To see this, consider a problem with demand rate $d_t(p) = 3$ if $p \leq 10$ and $d_t(p) = 0$ if $p > 10$ over the horizon $[0, T]$ with $T = 1$ and verify that $z(T, c) = 10$ and $\int_0^T d_t(p_t(z(T, c)))d_t = 3 > c$ for $c \in \{1, 2\}$.

We can think of $z(T, c)$ as an approximation to the marginal value of capacity. In practice, the marginal value of capacity is known as the bid-price. Using $z(T, c)$, as an approximate bid-price, we can use the deterministic price path $P^h(t, x) = p_t(z(T, c))$, $t \in (0, T]$, $x > 0$ as a bid-price heuristic for the stochastic problem. We emphasize, that given a bid-price $z(T, c)$, we use the optimal price $p_t(z(T, c))$ for that bid-price for all (t, x) such that $t > 0$ and $x > 0$. The bid-price heuristic is modified at boundary points (t, x), $t > 0$, and $x = 0$, by issuing the choke price $\bar{r}_t$.

Crude as it may sound, the bid-price heuristic is asymptotically optimal, in the sense of Theorem 9.10, as we will soon show. The bid-price heuristic can be made more dynamic by resolving (9.13) at reading dates $T = t_1 > t_2 > \ldots > t_K > t_{K+1} = 0$, and this typically results in a significant improvement in performance. More precisely, at state (t_k, x_k) we compute $z(t_k, x_k)$ and use the price path $P^h(s, x) = p_s(z(t_k, x_k))$ over $s \in (t_{k+1}, t_k]$ for $k = 1, \ldots, K$. Starting from a large state (T, c), the bid-price heuristic makes many good decisions, and the corrections made at reading dates help when sales deviate significantly from expectations. However, towards the end of the horizon, the coefficient of variation of the remaining demand is large, and optimal policies can be significantly different from those based on the bid-price heuristic. Consequently, resolving towards the end of the horizon can cause revenue deteriorations that are small in the aggregate. If the system is updated continuously, we get a feedback policy $p_t(z(t, x))$, $t \in [0, T]$. It is possible to show that under mild conditions the feedback policy is asymptotically optimal.

We previously justified (9.13) via the fluid model presented in an earlier chapter. We can also justify (9.13) by looking at a HJB equation similar to (9.2) where $V(t, x)$ is replaced by $G(t, x)$ and $\Delta V(t, x)$ is replaced by $\partial G(t, x)/\partial x$ resulting in:

$$\frac{\partial G(t, x)}{\partial t} = \mathcal{R}_t\left(\frac{\partial G(t, x)}{\partial x}\right) \tag{9.14}$$

with boundary $G(0, x) = G(t, 0) = 0$ for all $x \geq 0$ and $t \geq 0$. This formulation is the limit of speeding up the clock but making the request smaller and smaller. We have the following result.

Proposition 9.11 *If $\mathcal{R}_t(z)$ is continuously differentiable in z for all $t \in [0, T]$, then $\bar{V}(t, x) = G(t, x)$ for all (t, x).*

One implication of this is that if we continuously adjust the bid-price heuristic, we obtain $z(t, x) = \partial G(t, x)/\partial t$, which is an approximation of the marginal capacity $\Delta V(t, x)$ at state (t, x). This same approximation is the one proposed when the bid-price heuristic is updated at reading date k, as then $z(t_k, x_k)$ is approximately $\partial G(t, x)/\partial t$ evaluated at $(t, x) = (t_k, x_k)$.

It is possible to generalize the upper bound $\bar{V}(T,c)$ to multiple market segments and compound Poisson. In this case, the upper bound problem is given by

$$\bar{V}(T,c) = \min_{z \geq 0} \left[\sum_{m \in \mathcal{M}} \mathbb{E}[Z_m] \int_0^T \mathcal{R}_{mt}(z)dt + cz \right]. \tag{9.15}$$

Here, Z_m is the random demand from market segment m and $\mathcal{R}_{mt}(z) = \max_p (p - z)d_{mt}(p)$ is the profit function for market segment m. Notice that the objective function is convex in z as cz is linear and each $\mathcal{R}_{mt}(z)$ is convex in z. Let z_c be the unconstrained maximizer of $cz + \sum_{m \in \mathcal{M}} \mathbb{E}[Z_m] \int_0^T \mathcal{R}_{mt}(z)dt$. Then $z(T,c) = \max(z_c, 0)$.

The prices the heuristic uses are given by

$$P_m^h(t,x) = p_{mt}(z(T,c)), \quad m = 1, \ldots, M, \quad \forall t > 0, x > 0.$$

9.6 Asymptotic Optimality of the Bid-Price Heuristic

Just as the fixed pricing policy p_F is asymptotically optimal for a single market with time-invariant optimal price functions $p_t(z)$, the bid-price heuristic is asymptotically optimal for multiple market segments with compound Poisson demands. To make this statement more precise, we let $V^h(T,c)$ be the total expected revenue of using the bid-price heuristic until the end of the horizon or until capacity is exhausted. Clearly $V^h(T,c) \leq V(T,c)$. Consider a sequence of dynamic pricing problems indexed by $b > 1$ where the demand rates $d_{mt}^b(p) = bd_{mt}(p)$ and capacity $c^b = bc$. Let $V_b^h(T,c)$ and $V_b(T,c)$ denote, respectively, the performance of the bid-price heuristic and the optimal dynamic pricing policy for any $b \geq 0$.

Theorem 9.12 *Suppose that for every market segment $m \in \mathcal{M}$, there is an optimal price function $p_{mt}(z)$ for all $t \in [0,T]$, that the demand sizes Z_m have finite first and second moments, and that $\mathcal{R}_{mt}(z)$ is continuously differentiable in z for each market segment $m \in \mathcal{M}$ and for each $t \in [0,T]$. Then,*

$$\lim_{b \to \infty} \frac{V_b^h(T,c)}{V_b(T,c)} = 1$$

The proof of this result can be found in the appendix. Theorem 9.12 tells us that for large systems, the bid-price heuristic is almost as good as the optimal dynamic pricing policy, in the sense that it captures almost all the revenue relative to the optimal solution. This suggest that the bid-price heuristic can be used effectively for moderately large systems, but even large systems can benefit from refinements, i.e. by resolving problem (9.15) at reading dates $T = t_1 > t_2 > \ldots > t_K > t_{K+1} = 0$ as described before. One additional advantage of the bid-price heuristic is that prices are not as nervous as optimal dynamic prices that react instantaneously to state dynamics, e.g. decreasing prices between sales and increasing them after each sale. This is an important advantage in practice as bid-price heuristics are easier to implement. In practice, limits are often imposed on allowable prices, so the optimization is restricted to $p \in X_t$ where X_t may be a finite price menu. The design of the price menu may be itself considered part of the problem, where the goal is to design a small enough menu that allows for near-optimal expected revenues.

We remark that the assumption that $\mathcal{R}_t(z)$ is continuously differentiable cannot be easily relaxed as the following example shows. Assume there are two markets, and market i can only use price p_i with $p_1 > p_2$ or a choke-off price. Suppose further that $\Lambda_i = \int_0^T d_t(p_i)dt$ for $i = 1, 2$, with $\Lambda_1 = (1-\epsilon)c$ and $\Lambda_2 \gg c$. Then $z(T, c) = p_2$, $p_{1t}(z(T, c)) = p_1$, and $p_{2t}(z(T, c)) = p_2$ for $t \in [0, T]$, so the bid-price heuristic accepts all requests until capacity is exhausted. If arrivals are low-to-high, then the entire capacity will be consumed by low-fare customers resulting in expected revenue $V^h(T, c) \leq cp_2$ that can be far below the expected revenue $(c(1-\epsilon)p_1)$ from accepting only fare 1 customers. One may try to modify the bid-price heuristic to accept only requests from market 1, but then things go wrong if $\Lambda_1 = \epsilon c$ as most of the capacity would get spoiled. Notice that neither version of the heuristic improves as demand and capacity are scaled by $b > 1$. As a result, bid-price heuristics are not, in general, asymptotically optimal. However, even in this example, things improve dramatically if the bid-price heuristic is updated frequently during the horizon, resulting in revenue that is close to the upper bound $p_1c(1-\epsilon)+p_2c\epsilon$.

We end this section with two results that bound below the performance of bid-price heuristics. The first result shows that the simple bid-price $z = 0.5\bar{V}(T, c)/c$ guarantees at least half of the optimal revenue. This heuristic bid-price is simply half the per-unit expected revenue for the upper bound $\bar{V}(T, c)$. The second heuristic shows that the bid-price $z = z(T, c)$ explained above guarantees at least $1 - 1/e$ of the optimal revenue when the demands are time-invariant.

Theorem 9.13 *The modified bid-price heuristic, that uses $z = \bar{V}(T, c)/2c$ as the bid-price, has revenue $V^m(T, c)$ that satisfies*

$$\frac{V(T, c)}{\bar{V}(T, c)} \geq \frac{V^m(T, c)}{\bar{V}(T, c)} \geq \frac{1}{2}.$$

Theorem 9.14 *If $d_t(p) = d(p)$ for all $t \in [0, T]$ is continuous and $o(1/p)$, then the heuristic $z = z(T, c)$ satisfies*

$$\frac{V^h(T, c)}{V(T, c)} \geq \frac{V^h(T, c)}{\bar{V}(T, c)} \geq 1 - 1/e.$$

9.7 The Surplus Process

We next consider the impact of dynamic pricing on consumers' surplus. We will denote by $S(t, x)$ the expected consumers' surplus starting from state (t, x). More precisely, $S(t, x)$ measures the surplus from the customers that arrive over the interval $[0, t]$, when the inventory at time-to-go t is x, and the vendor uses an optimal dynamic pricing policy. Using similar logic to the one that led to the HJB equation for $V(t, x)$, we can derive the corresponding equations for $S(t, x)$:

$$\frac{S(t,x)}{\partial t} = \mathcal{S}(P(t,x)) - \Delta S(t,x)d_t(P(t,x)), \tag{9.16}$$

with boundary conditions $S(t, 0) = S(0, x) = 0$, where $\mathcal{S}(p) = \int_p^\infty d(y)dy$ is the surplus at price p, and $P(t, x)$ is the price at (t, x) used by the dynamic pricing policy.

Our main interest is to compare expected surplus $S(T, c)$ to the expected surplus, say $S^f(T, c)$, from using the fixed pricing policy p that maximizes expected revenues. Let $V^f(t, x)$ be the expected revenue associated with the optimal fixed price policy. The process $V^f(t, x)$ and the surplus process $S^f(t, x)$ are given by

$$\frac{\partial V^f(t,x)}{\partial t} = R(p_f, \Delta V^f(t,x)), \tag{9.17}$$

$$\frac{\partial S^f(t,x)}{\partial t} = \mathcal{S}(p_f) - \Delta S^f(t,x)d_t(p_f). \tag{9.18}$$

with boundary conditions $V^f(t, 0) = V^f(0, x) = S^f(t, 0) = S^f(0, x) = 0$, where p_f is the optimal fixed price for the system (T, c).

In general, it is very difficult to find closed-form solutions for S and S^f, so it is difficult to compare S and S^f. Numerical studies indicate that $S(T, c) \geq S^f(T, c)$ for most demand models, so consumers benefit in expectation from dynamic pricing.

For the constant elasticity of substitution (CES) demand model $d(p) = ap^{-b}, b > 1$, the expected surplus at price p is given by $\mathcal{S}(p) = (b - 1)pd(p)$. For the CES model it is easy to show that $S(t, x) = (b - 1)V(t, x)$ and that $S^f(t, x) = (b - 1)V^f(t, x)$. Then $V(t, x) \geq V^f(t, x)$ implies that $S(T, c) \geq S^f(T, c)$, so customers obtain more surplus under dynamic pricing than under the best fixed price policy. This is also true for the exponential demand function, and for the linear demand function except for cases where c/T is very small.

9.8 Multi-Product Dynamic Pricing Problems

Let $N := \{1, \ldots, n\}$ denote the set of potential products that can be offered, and let $M := \{1, \ldots, m\}$ be a set of resources utilized by the products. Let A be an $m \times n$ matrix where $A_{ij} \geq 0$ is the number of units of resource $i \in M$ consumed by product $j \in N$. For example, $A_{ij} = 1$ if product j uses one unit of resource i and $A_{ij} = 0$ otherwise. Let $z \in \Re_+^m$ be a vector of marginal cost of the resources, Then $A'z$ is a vector in $\Re_+^n$ representing the marginal cost of the products. If we sell one unit of product j, then we collect a revenue of p_j and incur a marginal cost $A_j'z$ where A_j is the jth column vector of A. Let $c = (c_1, \ldots, c_m)$ be the vector of initial inventories. There is a sales horizon of length $T > 0$ over which it is not possible to replenish inventories. We assume that consumers arrive as a Poisson process with λ_t, and upon arrival, a consumer sees the current price vector, say $p = (p_j)\ j \in N$, and selects product j with probability $\pi_j(p)$, with $\sum_{j\in N} \pi_j(p) \leq 1$ and $\pi_0(p) = 1 - \sum_{j\in N} \pi_j(p)$ represents the no-purchase probability. For convenience, we let $d_t(p) = \lambda_t \pi(p)$ be the expected demand vector.

Let

$$R_t(p, A'z) := \sum_{j\in N} (p_j - A_j' z)\, d_{tj}(p)$$

be the profit at price vector p when the marginal cost is z, and let

$$\mathcal{R}_t(A'z) := \max_{p\in\Re_+^n} R_t(p, A'z). \tag{9.19}$$

Let $V(t, x)$ be the maximum expected revenue that can be obtained from state (t, x), where $t \in [0, T]$ is the time-to-go, and $x \in [0, c]$ is the vector of remaining inventories. An argument similar to that used to establish (9.2) results in the HJB equation:

$$\frac{\partial V(t,x)}{\partial t} = \max_{p\in\Re_+^n} R_t(p, \Delta V(t,x)) = \mathcal{R}_t(\Delta V(t,x)), \tag{9.20}$$

and boundary conditions $V(t, 0) = V(0, x) = 0$, where $\Delta V(t, x)$ is a vector with components $\Delta_j V(t, x) = V(t, x) - V(t, x - A_j)$ for x such that $x \geq A_j$ and $\Delta_j V(t, x) = \infty$ otherwise. This serves to make the cost of product j equal to infinity when there is not enough capacity to produce a unit of product $j \in N$. Notice that similarity between (9.2) and (9.20). The only difference is that x and $\Delta V(t, x)$ are vectors of dimension m, and p is a vector of dimension n in (9.20), whereas $n = m = 1$ in (9.2).

Many, but not all of the results for the single product problem carry through. $V(t, x)$ is increasing in t and in x, and $\Delta V(t, x)$ is increasing in t and decreasing in x_j for each j, but $V(t, x)$ is not necessarily concave in x. It is also possible to show that

$$\bar{V}(T, c) = \min_{z \in \Re_+^m} \left[\int_0^T \mathcal{R}_t(A'z)dt + c'z \right]. \tag{9.21}$$

is an upper bound on $V(T, c)$. If $p_t(A'z)$ is an optimal solution to (9.19) for all $t \in [0, T]$, and $z(T, c)$ is a solution to problem (9.21), then it is possible to show that if $\mathcal{R}(A'z)$ is continuously differentiable, then the price path $p_t(A'z(T, c)), t \in [0, T]$ is asymptotically optimal as capacity and demand are scaled up.

9.9 End of Chapter Problems

1. Consider the problem of selling a house within T time periods. At each time period, we get an offer say p_j with probability λ_j. The possible price offers for the house are in the set $\{p_1, p_2, \ldots, p_n\}$, and one of the prices may be 0. If we have not sold the house by the beginning of the last time period, we have to accept whatever offer we get at this time period. The objective is to maximize the expected revenue from selling the house.

   (a) Formulate the problem as a dynamic program.

   (b) Use the structural properties of the value function, show that if it is optimal to accept the price offer p_j at time-to-go t, then it is optimal to accept the price offer p_j at time-to-go $t - 1$.

   (c) How would you modify the problem if at the end of the horizon you can salvage the house for a certain price p_0?

2. Consider a dynamic pricing problem, where we are also allowed to replenish the inventory of the product. The problem takes place over a discrete sales horizon of length T. The purchasing cost for the product at time-to-go t is c_t. There is at most one unit of demand for the product at each time period. If we charge the price p_t at time period t, then we observe a demand with probability $\lambda_t(p_t)$. Assume that $\lambda_t(\cdot)$ is strictly decreasing and we denote its inverse by $P_t(\cdot)$. Thus, $\lambda_t(p_t) = a_t$ if and only if $P_t(a_t) = p_t$. At each time period, we decide how much inventory to purchase, if any. We immediately receive the purchased inventory. Following this decision, we decide what price to charge. Assume that there exists a prohibitively large price p^∞ so that $\lambda_t(p^\infty) = 0$ for all $t = 1, \ldots, T$ and the one-period expected revenue function $a_t \times P_t(a_t)$ is a differentiable and concave function of a_t for all $t = 1, \ldots, T$. The objective is to maximize the total expected profit over T time periods.

   (a) Show that the optimal policy is of the following structure. For each time period t, there exist a critical inventory level x_t^* and a critical price level p_t^* so that if the inventory level x_t at the beginning of time period t is less than x_t^*, then we replenish up to the critical level x_t^* and charge the critical price p_t^*. Otherwise, we do not purchase any inventory at all and charge a price that depends on the inventory level at the beginning of the time period.

(b) Show that the optimal price to charge at each time period is a decreasing function of the inventory level at the beginning of the time period.

3. The price of a product changes randomly over time. A firm is purchasing this product over time periods $\{1, \ldots, T\}$ to satisfy the random demand that occurs at time period $T + 1$. Here, we assume that we count time forwards. Time period 1 is the first procurement period, and T is the last procurement period. At the beginning of each time period, the firm observes the total purchased inventory x_t and the random price P_t of the product. Depending on the inventory and price, the firm decides how much more product to purchase. We assume that the prices at different time periods are independent of each other. At the end of the purchasing horizon, the firm observes the random demand D for the product. If the total purchased inventory is not enough to cover all of the demand, then the firm incurs a shortage cost of θ for each unit short. The goal is to minimize the total expected cost of purchasing the product and the shortage.

   (a) Formulate the problem as a dynamic program.
   (b) Let $V(t, x_t)$ be the optimal total expected cost over time periods $\{t, \ldots, T + 1\}$ when we start at time period t with a total purchased inventory of x_t. Show that $V(t, \cdot)$ is convex for all $t = 1, \ldots, T + 1$.
   (c) Show that $V(t, x + 1) - V(t, x) \geq V(t + 1, x + 1) - V(t + 1, x)$, so the marginal value of inventory decreases in t.

4. Consider the problem of controlling the inventory of a product over T time periods. At each time period t, the following sequence of events take place. The inventory position x_t is observed, a quantity of u_t is ordered if necessary, the order quantity is immediately received, and a random demand of D_t is realized. If there is excess inventory after covering the demand, then a holding cost of h per unit per time period is incurred. The demand is backlogged in the sense that if we do not have enough inventory to cover the demand, then the inventory position can take a negative value indicating that we are behind. For each unit we are behind, we incur a backlogging cost of b per time period. The purchasing cost for the product is c per unit. Assume that $b > c$.

   (a) Write down a dynamic programming formulation to find the optimal policy.
   (b) Assume that the value functions are convex functions of the inventory position. Under this assumption, come up with a succinct characterization of the optimal policy similar to the protection level policies.
   (c) Show that the value functions are indeed convex functions of the inventory position.

5. Consider the time-homogeneous version of the HJB equation (9.6) and argue from the fundamental theorem of calculus that for any (T, c), we can write the value function as:

$$V(T, c) = \int_0^T \mathcal{R}(\Delta V(t, c))dt.$$

Let U be a continuous uniform random variable with support $[0, T]$, and consider the random variable $\Delta V(U, c)$.

(a) Argue that $V(T, c) = T\mathbb{E}[\mathcal{R}(\Delta V(U, c))]$.
(b) Use the convexity of $\mathcal{R}$ to show that

$$V(T, c) \geq T\mathcal{R}(\mathbb{E}[\Delta V(U, c)]).$$

(c) Use the fact that $\mathcal{R}(z)$ is decreasing in z to show that

$$V(T, c) \geq T\mathcal{R}(z)$$

for all $z \geq \mathbb{E}[\Delta V(U, c)]$.
(d) Consider the time homogeneous version of Eq. (9.13)

$$\bar{V}(T, c) = \min_{z \geq 0}[T\mathcal{R}(z) + cz]$$

and show that $\bar{V}(T, c)$ is an upper bound on $V(T, c)$.
(e) Conclude that

$$T\mathcal{R}(z) \leq V(T, c) \leq T\mathcal{R}(z) + cz$$

for all $z \geq \mathbb{E}[\Delta V(U, c)]$.
(f) Argue that if c is large then $z = \Delta V(T, c) \simeq 0$ so the lower bound and the upper bound are almost identical.

6. Consider the dynamic pricing problem for $T = 500$, $c = 50$, when the arrival rate is $\lambda_t = 0.02$ and the willingness to pay Ω_t is exponential with mean equal to 500 for all $t \in [0, T]$. Suppose the bargaining power of the firm is $\beta \in [0, 1]$. Plot the value functions $V^{\beta}(T, c)$ for $\beta \in \{0, 0.1, \ldots, 0.9, 1.0\}$. What is the maximal relative gain for $\beta = 1$ relative to $\beta = 0$?
7. One of your salesmen claims to be a good negotiator and tells you that are leaving money in the table by posting a take it or leave it price $P(t, x)$. You set $\beta = 0.4$ and give him a table of values $P_{\beta}(t, x)$ and allow him to negotiate for a price that is at least $P_{\beta}(t, x)$ at stage (t, x). You monitor his performance for a while and compare it to $V^{\beta}(T, c)$.

(a) What should you do if the revenues he generates exceed $V^{\beta}(T, c)$?
(b) What if he generates revenues between $V(t, x)$ and $V^{\beta}(T, c)$?
(c) What if he generates revenues below $V(t, x)$?
(d) How would you share the additional revenues with a seller with negotiating power $\beta > 0$?

9.10 Bibliographical Remarks

Reviews of dynamic pricing models include Bitran and Caldentey (2003) and Elmahraby and Keskinocak (2003). Gallego and van Ryzin (1994) study the canonical single-product dynamic pricing problem. We refer the reader to Gihman and Skorohod (1979) for a discussion of the fact that the HJB equation in (9.2) has a unique solution that gives the value function $V(t, x)$. The closed-form solutions in Sect. 9.2.1 are from Gallego and van Ryzin (1994) and McAfee and te Velde (2008). Theorem 9.10 is due to Gallego and van Ryzin (1994). Maglaras and Meissner (2006) show that the feedback policy discussed in Sect. 9.5 is asymptotically optimal when the arrival rate is time invariant: $d_t(p) = d(p)$ for all $t \in [0, T]$. Feng and Gallego (1995) provide a discussion of different approaches for numerically solving the HJB equation. Feng and Gallego (2000) consider formulations of the dynamic pricing problem with monotone prices. Such a formulation avoids the possibility of strategic consumers waiting for lower prices. Kuo et al. (2011) study a model where a fraction of the customers negotiate for the price. The results on nonlinear pricing are due to Gallego et al. (2018), while the results concerning consumer surplus are due to Chen and Gallego (2019). See also McAfee and te Velde (2008). Theorems 9.13 and 9.14 are due to Gallego et al. (2016c). Hu and Hopp (2009) study sub- and super-martingale properties of the optimal price path.

Federgruen and Heching (1999), Chen and Simchi-Levi (2004a), Chen and Simchi-Levi (2004b) Chen and Simchi-Levi (2006), Song et al. (2009), Chen et al. (2011), Chen and Simchi-Levi (2012), Chen and Hu (2012) and Chen et al. (2014) focus on dynamic pricing problems when the inventory of the product can be replenished. Boyaci and Gallego (2002) focus on similar joint inventory and price optimization problems, but over a network of locations. Aydin and Ziya (2008) and Aydin and Ziya (2009) study dynamic pricing models when the customers provide a signal to the firm in terms of their willingness to pay amount. Li and Huh (2012) develop a model that determines price change times a priori when the customer displays different price sensitivities over the selling horizon. Chen and Farias (2013) show that simple policies perform well when there is uncertainty in the size of the market but the demand-price relationship is known. Ceryan et al. (2013) and Ceryan et al. (2018) discuss pricing models when the demand for one product can be shifted to another in case of demand-supply mismatch. Cohen et al. (2018b) develop a dynamic pricing model that uses samples of the demand quantities and give performance guarantees. Cohen et al. (2016), Hu et al. (2016b), Chen et al. (2016c) and Chen et al. (2017b) study dynamic pricing model that incorporates reference price effects. Cohen et al. (2017a), Cohen et al. (2017c) and Baardman et al. (2019) give integer programming formulations to choose the timing and magnitude of promotions, as well as the timing of promotion activities such as flyers and commercials.

Levin et al. (2007), Levin et al. (2008) and Levin et al. (2014) study dynamic pricing problems, respectively, with price guarantees, risk concerns, and quantity discounts. Zhang and Kallesen (2008) study a dynamic pricing model that incorporates the prices of the competitors. Besbes and Maglaras (2009) give a fluid approximation for the dynamic pricing problem in a queueing system, where the customers make a decision based on the price charged and the expected waiting time in queue. Besbes and Maglaras (2012) study a dynamic pricing problem, where certain number of sales and a certain amount of revenue must be achieved at certain points in the selling horizon. Akan et al. (2013) work on a dynamic pricing and inventory problem when a product can be sold as new or remanufactured. Besbes and Lobel (2015) and Lobel (2017) focus on pricing models where the customers are patient and make a purchase only when the price goes down their willingness to pay. Wang (2016) discusses a pricing model with a reference price that is determined by the prices of the products offered to the customer. Chen and Nasiry (2019) work on a dynamic pricing problem with heterogeneous loss-averse customers and characterize when a cyclic discounting policy is optimal. Sogomonian and Tang (1993) and Chan et al. (2006) develop models to coordinate pricing and production decisions. Besbes et al. (2018) consider a dynamic pricing model, where the firm is expected to cover a debt at the end of the selling horizon. Bitran et al. (1998), Heching et al. (2002), Caro and Gallien (2012) and Ferreira et al. (2016) discuss practical applications of dynamic pricing models in retail.

There is emerging literature on strategic customers. In the models that we discussed so far, customers do not anticipate the pricing decisions of the firm. In reality, however, customers are strategic and they time their purchases anticipating the pricing decisions of the firm, often resulting in a game between the firm and the customers. For a sample of papers that involve strategic customers, we refer the reader to Gallego et al. (2008b), Zhang and Cooper (2008), Liu and van Ryzin (2008b), Aviv et al. (2009), Yin et al. (2009), Bansal and Maglaras (2009), Cachon and Swinney (2009), Levina et al. (2009a), Levin et al. (2010), Osadchiy and Vulcano (2010), Cachon and Swinney (2011), Liu and van Ryzin (2011), Nasiry and Popescu (2012), Mersereau and Zhang (2012), Lim and Tang (2013), Borgs et al. (2014), Akan et al. (2015), Ata and Dana (2015), Cachon and Feldman (2015), Levina et al. (2015), Liu and Cooper (2015), Wang et al. (2015), Ozer and Zheng (2016), Surasvadi et al. (2017), Caldentey et al. (2017), Chen and Farias (2017), Golrezaei et al. (2017), Adida and Ozer (2019), Chen and Farias (2018) and Chen and Jasin (2018), Chen et al. (2018a) and Chen and Shi (2019). Also, the models that we discussed work under the assumption that the prices are posted. An alternative to posted prices is to run an auction, in which case, the buyers compete with each other. The papers by Vulcano et al. (2002), van Ryzin and Vulcano (2004), Gallien (2006), Lim and Tang (2006), Caldentey and Vulcano (2007), Bertsimas et al. (2009), Kakade et al. (2013), Kanoria and Nazerzadeh (2014), Celis et al. (2014) and Golrezaei and Nazerzadeh (2017) study auctions in the revenue management context.

Gallego and van Ryzin (1997) show the asymptotic optimality of the prices derived from a deterministic nonlinear programming formulation for the multi-product case. Bertsimas and de Boer (2004), Zhang and Cooper (2009), Kunnumkal and Topaloglu (2010b), Zhang and Lu (2013), Wang and Ye (2016), Zhang and Weatherford (2017) also study dynamic pricing models with multiple ODF's. Adida and Perakis (2006), Adida and Perakis (2007) and Adida and Perakis (2010b) work on multi-product joint inventory and pricing problems. Chen et al. (2016b) study multi-product pricing problems when the products can serve as substitutes of each other and the price adjustments need to be infrequent. Motivated by e-retail, Lei et al. (2018) work on a dynamic pricing problem where the products are available at multiple warehouses and the retailer needs to decide which warehouse to use to satisfy a customer. Besbes and Zeevi (2012) consider the case where the price-demand relationship is not known and has to be learned over time.

Appendix

Proof of Theorem 9.3 Notice that

$$\frac{\partial V(t,x)}{\partial t} = \mathcal{R}_t(\Delta V(t,x)) \geq 0,$$

with the inequality strict as long as $d_t(p) > 0$ for some $p > \Delta V(t,x)$, or equivalently if $\Delta V(t,x) < \bar{r}_t$. This shows that $V(t,x)$ is increasing in t. To show that $V(t,x+1) \geq V(t,x)$, consider a sample path argument where the system with $x+1$ units of inventory uses the optimal policy for the system with x units of inventory until either the system with x units runs out of stock or time runs out. If the system with x units of inventory runs out at time s, then the system with $x+1$ units of inventory can still collect $V(s,1) \geq 0$. On the other hand, if time runs out, the two systems collect the same revenue. Consequently, the system with $x+1$ units of inventory makes at least as much revenue resulting in $V(t,x+1) \geq V(t,x)$.

Clearly $\Delta V(t,1) \leq \Delta V(t,0) = \infty$. Assume as the inductive hypothesis that $\Delta V(t,y)$ is decreasing in $y \leq x$ for all $t \geq 0$. We want to show that $\Delta V(t,x+1) \leq \Delta V(t,x)$, or equivalently that

$$V(t,x+1) + V(t,x-1) \leq V(t,x) + V(t,x). \tag{9.22}$$

We will use a sample path argument to construct the establish inequality (9.22). Consider four systems, one with $x+1$ units of inventory, one with $x-1$ units of inventory, and two with x units of inventory. Assume that we follow the optimal policy for the system with $x+1$ and for the system with $x-1$ that are on the left hand side of inequality (9.22). For the two systems on the right, we use the sub-optimal policies designed for $x+1$ and $x-1$ units of inventory, respectively. We follow these policies until one of the following events occurs: time runs out, the difference in inventories for the systems on the left drops to 1, or the inventory of the system with $x-1$ units drops to zero. After that time, we follow optimal policies for all four systems.

To establish inequality (9.22), we will show that the revenues obtained for the systems in the right are at least as large as for the systems on the left, even though sub-optimal policies are used for the systems in the right. This is obviously true if we run out of time since the realized revenues of the two systems on the right are exactly equal to the realized revenues from the two systems on the left. Assume now that at time $s \in (0, t)$, the difference in inventories of the two systems on the left hand side drops to 1, so that the states are $(s, y+1)$ and (s, y) for some $y < x$. This means that system on the left with $x+1$ units of inventory had $x-y$ units of sale and the system with $x-1$ units of inventory had $x-1-y$ units of sale. This implies that the system on the right that was following the policy designed for $x+1$ reaches state (s, y), while the system that was using the policy designed for $x-1$ reaches state $(s, y+1)$. Clearly, the additional optimal expected revenues over $[0, s]$ for each pair of systems are $V(s, y+1) + V(s, y) = V(s, y) + V(s, y+1)$, showing that the system on the right gets as much revenue as the system on the left even if sub-optimal policies are used for part of the horizon. Finally, if the inventory of the system with $x-1$ units of inventory drops to 0 at some time $s \in [0, t)$, so that state of the systems on the left are, respectively, (s, y) and $(s, 0)$ for some y, such that $1 < y \leq x$, while the systems on the right are $(s, y-1)$ and $(s, 1)$. From the inductive hypothesis, we know that $\Delta V(s, y) \leq \Delta V(s, 1)$ for all $y \leq x$ and all $s \leq t$. Consequently,

$$V(s, y) + V(s, 0) \leq V(s, y-1) + V(s, 1),$$

and once again the pair of systems on the right result in at least as much revenue even though sub-optimal policies are used for part of the sales horizon.

We now show that $\Delta V(t, x)$ is increasing in t. This is equivalent to

$$\frac{\partial V(t, x)}{\partial t} = \mathcal{R}_t(\Delta V(t, x)) \geq \mathcal{R}_t(\Delta V(t, x-1)) = \frac{\partial V(t, x-1)}{\partial t},$$

but this is true on account of $\mathcal{R}_t(z)$ being decreasing in z and $\Delta V(t, x)$ being decreasing in x. We now show that $P(t, x) = p_t(\Delta V(t, x))$ is decreasing in x. This follows because $p_t(z)$ is increasing in z and $\Delta V(t, x)$ is decreasing in x.

Clearly, $P(t, x) = p_t(\Delta V(t, x))$ is also increasing in t provided that $p_t(z)$ is increasing in t since $\Delta V(t, x)$ is also increasing in t and $p_t(z)$ is increasing in z. Clearly, $p_t(z) = p(z)$ holds if $d_t(p) = d(p)$ for all t. Moreover, if $h_t(p)$ is decreasing in t, then $p_t(z)$ is the root of the equation $(p - z)h_t(p) = 1$, and this is increasing in t.

We now show that $V(t, x)$ is strictly increasing in t if $\Delta V(t, x) < \bar{r}_t$. This follows because

$$\frac{\partial V(t, x)}{\partial t} = \mathcal{R}_t(\Delta V(t, x)) > 0,$$

as any price $p \in (\Delta V(t, x), \bar{r}_t)$ returns a positive profit. Notice that this is automatically true if $p_t(z)$ is increasing in z because

$$\frac{\partial V(t, x)}{\partial t} = \mathcal{R}_t(\Delta V(t, x)) \geq \mathcal{R}_t(\Delta V(t, 1)) = \mathcal{R}_t(V(t, 1)) > 0,$$

where the first inequality follows because $\mathcal{R}_t$ is decreasing and $V(t,1) = \Delta V(t,1) \geq \Delta V(t,x)$ for all $x \geq 1$. The strict inequality follows because $V(t,1) < \bar{r}_t$, as otherwise the single unit must sell with probability one at the choke-off point, but by definition the demand at the choke-off point is zero. Consequently, there is a $p_t > V(t,1)$ with $d_t(p) > 0$ implying that $\mathcal{R}_t(V(t,1)) > 0$.

To show that $V(t,x)$ is concave, notice that if $\mathcal{R}_t(z)$ is differentiable, then

$$\frac{\partial^2 V(t,x)}{\partial t^2} = \mathcal{R}_t'(\Delta V(t,x))\frac{\partial \Delta V(t,x)}{\partial t} \leq 0,$$

follows since $\mathcal{R}_t'(z) \leq 0$, on account of $\mathcal{R}_t(z)$ being decreasing in z, and from the fact that $\Delta V(t,x)$ is increasing in t. □

Proof of Proposition 9.11 Recall that $\bar{V}(t,x) = \int_0^t \mathcal{R}_s(z(t,x))ds + xz(t,x)$. Since $\mathcal{R}_t(z)$ is continuously differentiable, either $z(t,x) = 0$ or $z(t,x) > 0$ is the root of $\int_0^t d_s(p_s(z(t.x)))ds = x$. If $z(t,x) = 0$, then $\bar{V}(t,x) = \int_0^t \mathcal{R}_s(0)ds$, so

$$\frac{\partial \bar{V}(t,x)}{\partial t} = \mathcal{R}_t(0).$$

On the other hand, if $z(t,x) > 0$, then

$$\frac{\partial \bar{V}(t,x)}{\partial t} = \mathcal{R}_t(z(t,x)) + [x - \int_0^t d_s(p_s(z(t,x))dt]\frac{\partial z(t,x)}{\partial t} = \mathcal{R}_t(z(t,x))$$

since the last term cancels on account of $\int_0^t d_s(p_s(z(t,x))dt = x$. Let $\partial G(t,x)/\partial x = z(t,x)$. Then

$$\frac{\partial \bar{V}(t,x)}{\partial t} = \mathcal{R}_t\left(\frac{\partial G(t,x)}{\partial t}\right) = \frac{\partial G(t,x)}{\partial t}.$$

Since the boundary conditions $\bar{V}(t,0) = G(t,0) = \bar{V}(0,x) = G(0,x) = 0$, it follows that $\bar{V}(t,x) = G(t,x)$ as claimed. □

Proof of Theorem 9.12 For ease of exposition, we will give the proof of Theorem 9.12 for the case of a single market with Poisson, rather than compound Poisson demands. However, the proof holds as stated. Clearly $V_b^h(T,c) \leq V_b(T,c) \leq \overline{V}_b(T,c)$. The idea of the proof is to show that the difference $\overline{V}_b(T,c) - V_b^h(T,c)$ becomes negligible relative to $V_b^h(T,c)$. First, notice that scaling arrival rates and demands does not change $z(T,c)$ or the bid-price policy $P(t,x) = p_t(Z(T,c))$. To evaluate $V_b^h(T,c)$ let $t = T - s$ be the time-to-go for each $s \in [0,T]$. The bid-price at time s is $q_s = P_{T-s}(z(T,c))$. The demand intensity at s is $\gamma_s = d_{T-s}(P_{T-s}(z(T,c))$. Let N_s be a Poisson process with intensity $\Gamma_s = \int_0^s \gamma_s ds$, and let τ_c be the first time the process reaches c, i.e. $\tau_c = \inf\{s \geq 0 : N_s = c\}$. The bid-price policy earns revenues at rate $q_s dN_s$ until time $\min(T, \tau_c)$, so

$$V_b^f(T,c) = \mathbb{E}\left\{\int_0^{\min(T,\tau_c)} q_s dN_s\right\} = \mathbb{E}\left\{\int_0^{\min(T,\tau_c)} q_s \gamma_s ds\right\}$$

where the equality follows from Watanabe's characterization of Poisson processes, see Bremaud (1980). Let $\Theta_s = \int_0^s \theta_u du$, where $\theta_u = q_u \gamma_u$, also let $\bar{\theta} = \max_{0 \leq u \leq T} \theta_u$. Then

$$V_b^f(T,c) = \Theta_T - \mathbb{E}[\Theta_T - \Theta_{\min(T,\tau_c)}] \geq \Theta_T - \bar{\theta}\, \mathbb{E}(T - \tau_c)^+.$$

On the other hand, we can write

$$\bar{V}(T,c) = \Theta_T - (\Gamma_T - c)z(T,c).$$

The result follows since the gap between $\bar{\theta}\mathbb{E}(T - \tau_c)^+ - (\Gamma_T - c)z(T,c)$ grows as the square root of b while the lower bound grows linearly with b. □

Proof of Theorem 9.13 Our plan is to construct a heuristic with expected revenue $V^m(T,c)$ such that

$$\frac{V(T,c)}{\bar{V}(T,c)} \geq \frac{V^m(T,c)}{\bar{V}(T,c)} \geq \frac{1}{2}.$$

Let

$$\bar{\pi} = \bar{V}(T,c) = \min_{z \geq 0}\left(\int_0^T \mathcal{R}_t(z)dt + zc\right),$$

and set $z = 0.5\bar{\pi}/c$. Clearly

$$\bar{\pi} = \bar{V}(T,c) \leq \int_0^T \mathcal{R}_t(0.5\bar{\pi}/c)dt + 0.5\bar{\pi},$$

so

$$0.5\bar{\pi} \leq \int_0^T \mathcal{R}_t(0.5\bar{\pi}/c)dt.$$

Consider a heuristic policy that prices at $p_s(0.5\bar{\pi}/c), 0 \leq s \leq T$, and let $V^m(T,x)$ be its expected revenue. We will show that $V^m(T,c) \geq 0.5\bar{\pi}$. Suppose for a contradiction that $V^m(T,c) < 0.5\bar{\pi}$. Then,

$$V^m(T,c) < 0.5\bar{\pi} \leq \int_0^T \mathcal{R}_t(0.5\bar{\pi}/c)dt.$$

We will show that in fact $V^m(T,c) \geq \int_0^T \mathcal{R}_t(0.5\bar{\pi}/c)dt$. Notice that

$$0.5\bar{\pi} > V^m(T,c) = \sum_{x=1}^{c} \Delta V^m(T,x),$$

and since $\Delta V^m(T, x)$ is decreasing in x, it follows that

$$\Delta V^m(T, c) \leq V^m(T, c)/c < 0.5\bar{\pi}/c.$$

But then

$$\begin{aligned}
V^m(T, c) &= \int_0^T R_t(p_t(0.5\bar{\pi}/c), \Delta V^m(t, c))dt \pm 0.5\bar{\pi}/c \int_0^T d_t(p_t(0.5\bar{\pi}/c))dt \\
&= \int_0^T \mathcal{R}_t(0.5\bar{\pi}/c)dt + \int_0^T [0.5\bar{\pi}/c - \Delta V^m(t, c)]d_t(p_t(0.5\bar{\pi}/c))dt \\
&\geq \int_0^T \mathcal{R}_t(0.5\bar{\pi}/c)dt,
\end{aligned}$$

where the inequality follows because $0.5\bar{\pi}/c > \Delta V^h(t, c)$ for all t, since $\Delta V^h(t, c)$ is increasing in t and the result holds at T. This shows that the assumption that $V^m(T, c) < 0.5\bar{V}(T, c)$ leads to a contradiction so it must be that $V^m(T, c) \geq 0.5\bar{V}(T, c)$. □

Proof of Theorem 9.14 Because $d_t(p) = d(p)$ is continuous and $o(1/p)$, there exists a $p_t(z) = p(z)$ for all $z \geq 0$ that maximizes $R_t(p, z) = (p - z)d_t(p)$ for all $t \in [0, T]$. Let $y = \int_0^T d_t(p(0))dt = Td(p(0))$. If $y < c$, then $z(T, c) = 0$ is optimal and the upper bound is $T\mathcal{R}(0) = yp(0)$. The bid-price heuristic would price at $p_t = p(0)$ for all $t \in (0, T]$ and would face Poisson demand with parameter y over $[0, T]$, resulting in expected revenue $\min(c, \text{Poisson}(y))p(0)$. The ratio is therefore

$$\frac{V^h(T, c)}{V(T, c)} \geq \frac{V^h(T, c)}{\bar{V}(T, c)} = \frac{\mathbb{E}\min(c, \text{Poisson}(y))}{y}$$

with $y < c$. On the other hand, if $c \geq y$, then the assumed properties of $d(p)$ guarantee the existence of a $z = z(T, c) > 0$ such that $Td(p(z)) = c$, so the upper bound yields $cp(z)$ while the lower bound yields $\mathbb{E}\min(c, \text{Poisson}(c))p(z)$. Thus, the ratio is given by

$$\frac{V^h(T, c)}{V(T, c)} \geq \frac{V^h(T, c)}{\bar{V}(T, c)} = \frac{\mathbb{E}\min(c, \text{Poisson}(c))}{c}.$$

It is well known that these ratios are bounded below by $1 - 1/e$. □

Chapter 10
Online Learning

10.1 Introduction

In the models that we have studied so far, we have assumed that the demand model and its parameters are all known. In practice, demand models need to be estimated before dynamic pricing, assortment optimization, and revenue management can be effectively done. In some instances, there is enough data over a long period of time to calibrate different demand models, do model selection, and update parameter estimates. At the other extreme, we may be pricing for products for which we have little or no information. In this case, demand learning needs to be done on the fly. This is particularly true for online retailing of new products. In this chapter, we address the problem of online demand learning. We study the expected loss in revenue of a learning-and-earning policy relative to an optimal clairvoyant policy that knows the expected demand function. We consider both the case of ample and constrained capacity and measure how the regret grows as the length of the sales horizon increases. We present only the strongest available results for both the case of ample and the case of constrained capacity. In Sect. 10.2, we consider the case with ample capacity, whereas in Sect. 10.3, we consider the case with constrained capacity.

10.2 Ample Inventory Model

Let $D_t(p) \in \{0, 1\}$ be the random demand in period t at price p. We will assume that $D_t(p), t \in \{1, \ldots, T\}$ are independent and identically distributed Bernoulli random variables with mean $d(p) = \mathbb{E}[D_t(p)]$ for all $p \in [l, u]$ where l and u are non-negative constants. Without loss of generality we rescale prices if necessary so that $[l, u] = [0, 1]$. We assume that the $d(p)$ is unknown and that the goal of the

G. Gallego, H. Topaloglu, *Revenue Management and Pricing Analytics*, International Series in Operations Research & Management Science 279,
https://doi.org/10.1007/978-1-4939-9606-3_10

seller is to maximize revenues over the selling horizon via a learning and earning algorithm. We assume that capacity c is ample. In the context of the Bernoulli model just described, this means that $c \geq T$, because there is at most one unit of sale at each period.

We will study both parametric and non-parametric models. In parametric models, the seller knows the form of the expected demand, say $d(p) = \lambda e^{-p/\theta}$, but needs to estimate the unknown parameters (in this case λ and θ). The non-parametric case makes no assumptions about the form of $d(p)$. In the parametric case, the parameters may be updated over time by following several techniques including the Bayesian approach, maximum likelihood, or least squares. For non-parametric models, the exploration does not attempt to estimate the demand function itself as its main concern is to obtain prices that work well empirically.

Let $R(p) = pd(p)$ be the revenue at price p. We assume there exist a unique maximizer of $R(p)$, say $p^* \in [0, 1]$. Over the selling horizon, the expected revenue obtained by the clairvoyant policy is $TR(p^*)$. The objective is to design a non-anticipating pricing policy π_t to maximize the total reward $\sum_{t=1}^{T} \mathbb{E}[R(\pi_t)]$. The information structure of π_t requires that the decision for period t, π_t, only relies on the history of the process until time $t - 1$. This is similar to the multi-armed bandit problem, but here the decision variable is continuous rather than a finite set.

10.2.1 Regret

A standard measure used in the literature for the performance of a policy is the regret incurred compared to the clairvoyant policy. More formally, we define the regret of a policy π_t to be the expected gap in revenue from the clairvoyant policy, given by

$$r_\pi(T) := \sum_{t=1}^{T} \mathbb{E}\left[R(p^*) - R(\pi_t)\right].$$

This is a learning and earning problem where the demand is learned on the fly with a tradeoff between exploration and exploitation, whose goal is to design a policy π_t for which $r_\pi(T)$ scaling gracefully as $T \to \infty$. Since $r_\pi(T)$ depends on the unknown function $d(p)$, we require the designed policy to perform well for a wide class $\mathcal{C}$ of functions, i.e., we seek for optimal policies in terms of the minimax regret

$$\inf_{\pi_t} \sup_{d \in \mathcal{C}} r_\pi(T).$$

Although it is usually impossible to find the exact policy that achieves the minimax regret, most authors focus on policies whose regret is at least comparable to (of the same order as) the minimax regret as $T \to \infty$. To state this more formally, we will use the big O notation. We say that $f(T) = O(g(T))$ as $T \to \infty$ if there are constants C and t_0 such that $f(t) \leq Cg(t)$ for all $t \geq t_0$. We say that $f(T) = \Omega(g(T))$ as $T \to \infty$ if there are constants C and t_0 such that $f(t) \geq Cg(t)$ for all $t \geq t_0$. We say that $f(T) = O^*(g(T))$ if there constants C and t_0 such that $f(t) \leq Cg(t)p(t)$ for all $t \geq t_0$ for some poly-logarithmic factor $p(t)$ of order lower than $g(t)$, so the big O^* notation neglects multiplicative terms of lower order.

For the ample capacity case, it is possible to show under mild conditions that there is a pricing policy for parametric models based on the maximum likelihood framework that achieves regret $O(\sqrt{T})$. For the non-parametric case, it is possible to show that there is a policy that achieves regret $O(\log(T)^{1/2}\sqrt{T})$, and for both cases, the regret is at least $\Omega(\sqrt{T})$. In summary, under mild assumptions, there are policies that have regret $O^*(\sqrt{T})$ for both the parametric and non-parametric cases. These regret bounds are for models that ignore customer characteristics that are crucial for personalized pricing.

In this section, we describe a non-parametric model that allows for personalized pricing. In this framework, each consumer arrives with a vector x of covariates in a bounded d-dimensional hypercube which we take without loss of generality to be $[0, 1)^d$. The expected demand function $\mathbb{E}[D(p, x)] = d(p, x)$ depends both on the price p and on the covariate vector x. A clairvoyant policy would observe x and return $p(x) = \arg\max R(p, x)$, where $R(p, x) = pd(p, x)$ is the revenue at price p when the covariate vector is x.

The main result of this section is that under mild assumptions there is an algorithm that returns a policy with regret at most $O(\log(T)^2 T^{(2+d)/(4+d)})$ where d is the dimension of the covariate vector. We also show that all policies have regret at least $\Omega(T^{(2+d)/(4+d)})$, so there exist a policy that is $O^*(T^{(2+d)/(4+d}))$. Without covariates, $d = 0$, the regret is $O^*(\sqrt{T})$, matching the performance of earlier algorithms. As d increases the lower bound deteriorates and becomes nearly linear in T. This suggests that only the most salient covariates should be included in personalized pricing, perhaps after applying a dimension-reduction algorithm. Thus, there is a tradeoff between trying to exploit covariate information and minimizing the regret, particularly as d gets large.

10.2.2 Assumptions

For any convex subset $B \subset [0, 1)^d$, let $R_B(p) := \mathbb{E}[r(X, p)|X \in B]$, where the expectation is taken over the distribution of the covariate space.

Assumption 1 $D(p, x)$ is a Bernoulli random variable with mean $d(p, x) := \mathbb{E}[D(p, x)] \in [0, 1]$ for all $p \in [0, 1]$ and all $x \in [0, 1)^d$.

Assumption 2 The expected revenue function $R(p, x) = pd(p, x)$ is Lipschitz continuous, i.e., there exists $M_1 > 0$ such that $|R(x_1, p_1) - R(x_2, p_2)| \leq M_1(\|x_1 - x_2\|_2 + |p_1 - p_2|)$ for all $x_i \in [0, 1)^d$ and $p_i \in [0, 1]$ with $i = 1, 2$.

Assumption 3

3.1 The function $R_B(p)$ has a unique maximizer $p^*(B) \in [0, 1]$. Moreover, there exist uniform constants $M_2, M_3 > 0$ such that for all $p \in [0, 1]$, $M_2(p^*(B) - p)^2 > R_B(p^*(B)) - R_B(p) > M_3(p^*(B) - p)^2$.
3.2 The maximizer $p^*(B)$ of $R_B(p)$ is inside the interval

$$[\inf\{p^*(x) : x \in B\}, \sup\{p^*(x) : x \in B\}].$$

3.3 Let δ_B be the diameter of B. Then there exists a uniform constant $M_4 > 0$ such that $\sup\{p^*(x) : x \in B\} - \inf\{p^*(x) : x \in B\} \leq M_4\delta_B$.

Assumption 1 is very mild. Lipchitz continuity or similar smoothness conditions are common in the literature and are needed for past experiments to be informative. The intuition behind the third assumption is to consider a learning problem associated with B without covariates. Indeed, if we only know that $X \in B$, the learning objective would be $R_B(p)$, so the clairvoyant policy would set price $p^*(B)$ in each period where $X \in B$. Assumption 3 is satisfied by many of the parametric families studied in the literature. For example, in the linear case, we have $d(p, x) = \alpha' x - \beta p$, so $R_B(p) = p(\alpha' \mathbb{E}[X|X \in B] - \beta p)$ and $p^*(B) = \alpha' \mathbb{E}[X|X \in B]/2\beta$.

10.2.3 Preliminary Concepts

We start by defining a bin and its children.

Definition 10.1 A bin is a hyper-rectangle in the covariate space. More precisely, a bin is of the form

$$B = \{x : a_i \leq x_i < b_i,\ i = 1, \ldots, d\}$$

for $0 \leq a_i < b_i \leq 1, i = 1, \ldots, d$.

We can *split* a bin B by bisecting it in all the d dimensions to obtain 2^d *child* bins of B, all of equal size. For a bin B with boundaries a_i and b_i for $i = 1, \ldots, d$, its children are indexed by the 2^d vectors in $\{0, 1\}^d$. Indeed, for any $w \in \{0, 1\}^d$, we have the child

$$B_w = \left\{x : a_i \leq x_i < \frac{a_i + b_i}{2} \text{ if } w_i = 0, \right.$$
$$\left. \frac{a_i + b_i}{2} \leq x_i < b_i \text{ if } w_i = 1,\ i = 1, \ldots, d\right\}$$

that choses the first half of the range of component i if $w_i = 0$ and the second half if $w_i = 1$ for each $i = 1, \ldots, d$.

Denote the set of all child bins of B by $C(B) = \{B_w : w \in \{0,1\}^d\}$. Notice that $C(B)$ is a mutually exclusive and collectively exhaustive partition of B into 2^d child bins. For any $B' \in C(B)$, we refer to B as the *parent* bin of B', denoted by $P(B') = B$.

The adaptive binning and exploration (ABE) algorithm given below starts with bin $B_\emptyset = [0,1)^d$ and successively splits it as data is collected. Any bin B produced during the process is the *offspring* of $B_\emptyset$. Therefore, one can use a sequence of vectors in $\{0,1\}^d$, $w^1, w^2, \ldots, w^k$ to represent a bin that is build during the algorithm. The bin $B_{w^1,w^2,\ldots,w^k}$ refers to a bin that is obtained by k split operations of $B_\emptyset$. After the first split, we obtain B_{w^1} from $B_\emptyset$. When B_{w^1} is split, we obtain its child $B_{w^1w^2}$ and so on. In the last operation, when $B_{w^1\ldots w^{k-1}}$ is split, we obtain its child $B_{w^1\ldots w^k}$. For such a bin, we define its *level* to be k, denoted by $l(B) = k$, with $l(B_\emptyset) = 0$.

At the end of the ABE algorithm, there is a partition, say $\mathcal{P}$, of the covariate space, and for each $B \in \mathcal{P}$, the function $R_B(p)$ is estimated from data for values of p in a grid partition of an interval $[p_B^l, p_B^h] \subset [0,1]$ produced by the algorithm. The algorithm then selects the price in this grid that maximizes the approximation of $R_B(p)$, which should be close to $p^*(B)$, and in turn close to $p(x)$ for $x \in B$ given the Lipchitz continuity assumptions. The intuition is that for large T, we should be able to get reliable estimates for fairly small bins, and the approximation should be very accurate. The algorithm tries to do this learning efficiently by judiciously deciding when to split bins. This is done by a set of discrete decisions (referred to as the *decision set* hereafter) for each bin in the partition.

The algorithm keeps a dynamic partition $\mathcal{P}_t$ of the covariate space consisting of offspring of $B_\emptyset$ in each period t, starting with $\mathcal{P}_0 = \{B_\emptyset\}$. Each bin in $\mathcal{P}_{t+1}$ has an ancestor (or itself) in $\mathcal{P}_t$. Each time a bin is partitioned, that bin is removed and replaced by all of its children. The process can also be interpreted as the sequential splitting of a *branching process* and relates to decision trees in statistical learning.

The decision set consists of equally spaced grid points of an interval associated with the bin. When a covariate X_t is generated inside a bin B, a price is chosen successively in a grid and is applied to X_t. The realized reward for this decision is recorded. When sufficient covariates are observed in B, the average reward for each price in the grid is recorded as an estimate of $R_B(p)$. The best price in the grid is the *empirically-optimal* decision and is close to $p^*(B)$ with high confidence.

Adaptive Binning and Exploration (ABE)

Step 1. Initialization

(A) Input: T, d

(B) Constants: $M_1, M_2, M_3, M_4, \sigma$

(C) Parameters: K and Δ_k, n_k, N_k for $k = 0, \ldots, K$

(D) Set partition: $\mathcal{P} \leftarrow \{B_\emptyset\}$, $p_l^{B_\emptyset} \leftarrow 0$, $p_u^{B_\emptyset} \leftarrow 1$, $\delta_{B_\emptyset} \leftarrow 1/(N_0 - 1)$, $\bar{Y}_{B,j}, N_{B_\emptyset,j} \leftarrow 0$ for $j = 0, \ldots, N_0 - 1$, $N(B_0) = 0, l(B_0) = 0$

Step 2. Learning and Earning

(A) For $t = 1$ to T do
(B) Observe X_t
(C) $B \leftarrow \{B \in \mathcal{P} : X_t \in B\}$
(D) $k \leftarrow l(B)$, $N(B) \leftarrow N(B) + 1$
(E) If $k < K$ then

   (a) If $N(B) < n_k$ then
   (b) $j \leftarrow N(B) - 1 \pmod{N_k}$
   (c) $\pi_t \leftarrow p_l^B + j\delta_B$; apply π_t and observe revenue Z_t
   (d) $\bar{Y}_{B,j} \leftarrow \frac{1}{N_{B,j}+1}(N_{B,j}\bar{Y}_{B,j} + Z_t)$, $N_{B,j} \leftarrow N_{B,j} + 1$
   (e) Else
   (f) $j^* \in \arg\max_{j\in\{0,1,\ldots,N_k-1\}}\{\bar{Y}_{B,j}\}$, $p^* \leftarrow p_l^B + j^*\delta_B$
   (g) $\mathcal{P} \leftarrow (\mathcal{P} \setminus B) \cup C(B)$
   (h) For $B' \in C(B)$

      - $N(B') \leftarrow 0$
      - $p_l^{B'} \leftarrow \max\{0, p^* - \Delta_{k+1}/2\}$; $p_u^{B'} \leftarrow \min\{1, p^* + \Delta_{k+1}/2\}$
      - $\delta_{B'} \leftarrow (p_u^{B'} - p_l^{B'})/(N_{k+1} - 1)$
      - $N_{B',j}, \bar{Y}_{B',j} \leftarrow 0$, for $j = 0, \ldots, N_{k+1} - 1$
      - End For

   (i) End If

(F) Else $\pi_t \leftarrow (p_l^B + p_u^B)/2$
(G) End If
(H) End For

The parameters for the algorithm include K, the maximal level of the bins, Δ_k the length of the interval for level-k bins, n_k the maximum number of covariates observed in a level-k bin, N_k the number of decisions to explore in the decision set of level-k bins (consisting in N_k evenly spaced points in the interval $[p_l^B, p_u^B]$ specified by the algorithm). We initialize with the root bin $\mathcal{P}_0 = \{B_\emptyset\}$. Its decision set spans the whole interval [0, 1] with N_0 equally spaced grid points. That is, the j-th decision is $j\delta_{B_\emptyset} := j/(N_0-1)$ for $j = 0, \ldots, N_0-1$. The initial average reward and the number of explorations already applied to the j-th decision are set to $\bar{Y}_{B_\emptyset,j} = N_{B_\emptyset,j} = 0$. We set $K = \lfloor \frac{\log(T)}{(d+4)\log(2)} \rfloor$, $\Delta_k = 2^{-k}\log(T)$, $N_k = \lceil \log(T) \rceil$, and

$$n_k = \max\left\{0, \left\lceil \frac{2^{4k+18}\sigma}{M_2^2 \log^3(T)}(\log(T) + \log(\log(T)) - (d+2)k\log(2)) \right\rceil\right\}.$$

To give a sense of their magnitudes, the maximal level of bins is $K \approx \log(T)/(d+4)$. The range of the decision set (Δ_k) is proportional to the edge length of the bin (2^{-k}). The number of decisions in a decision set is approximately $\log(T)$. Therefore, the grid size $\delta_B \approx 2^{-k}$ for a level-k bin B. The number of covariates to collect in a level-k bin B is roughly $n_k \approx 2^{4k}/\log(T)^2$. When k

is small, n_k can be zero according to the expression. In this case, the algorithm immediately splits the bin without collecting any covariate in it.

Suppose the partition is $\mathcal{P}_t$ at t and a covariate X_t is generated (Step B). The algorithm determines the bin $B \in \mathcal{P}_t$ which the covariate falls into. The counter $N(B)$ records the number of covariates already observed in B up to t when B is in the partition (Step C). If the level of B is $l(B) = k < K$ (i.e., B is not at the maximal level) and the number of covariates observed in B is not sufficient, then the algorithm further explores and test prices in the decision set $\{p_l^B + j\delta_B\}$ for $j = 0, \ldots, N_k - 1$. There are N_k decisions in the set and they are equally spaced in the interval $[p_l^B, p_u^B]$. They are explored sequentially as new covariates are observed in B. The algorithm applies decision $\pi_t = p_l^B + j\delta_B$ where $j = N(B) - 1 \pmod{N_k}$ to the $N(B)$-th covariate observed in B (Step b). Step d updates the average reward and the number of explorations for the j-th decision.

If the level of B is $l(B) = k < K$ and we have observed sufficient covariates in B (Step e), then the algorithm splits B and replaces it by its 2^d child bins in the partition (Step g). For each child bin, Step h initializes the counter, the interval that encloses the decision set, the grid size of the decision set, and the average reward/number of explorations that have been conducted for each decision in the decision set. In particular, to construct the decision set of a child bin, the algorithm first computes the empirically optimal decision in the decision set of the parent bin B; that is, $j^* \in \arg\max_{j\in\{0,1,\ldots,N_k-1\}}\{\bar{Y}_{B,j}\}$ in Step f. Then, the algorithm creates an interval centered at this optimal decision with width Δ_{k+1}, properly cut off by the boundaries $[0, 1]$. The decision set is then an equally spaced grid of the above interval. If the level of B is already K, then the algorithm simply applies a single decision inherited from its parent (Step F) repeatedly without further exploration. For such a bin, its size is sufficiently small and the algorithm has narrowed the range of the decision set K times. The applied decision, which is the middle point of the interval, is close enough to all $p^*(x)$, $x \in B$, with high probability.

The following result provides upper and lower bounds on the regret.

Theorem 10.2 *For any function $R(p, x)$ satisfying Assumptions 1–3, the regret incurred by the ABE algorithm is bounded by*

$$r_{\pi_{ABE}}(T) \leq KT^{\frac{2+d}{4+d}} \log(T)^2$$

for a constant $K > 0$ that is independent of T. For all non-anticipating polices, we have

$$\inf_{\pi} \sup_{f\in C} R_\pi \geq kT^{\frac{2+d}{4+d}}$$

for a constant $k > 0$ that is independent of T.

We illustrate the key steps of the algorithm by an example with $d = 2$. Figure 10.1 illustrates a possible outcome of the algorithm in periods $t_1 < t_2 < t_3$ (top panel, mid panel, and bottom panel, respectively). Up until period t_1, there is

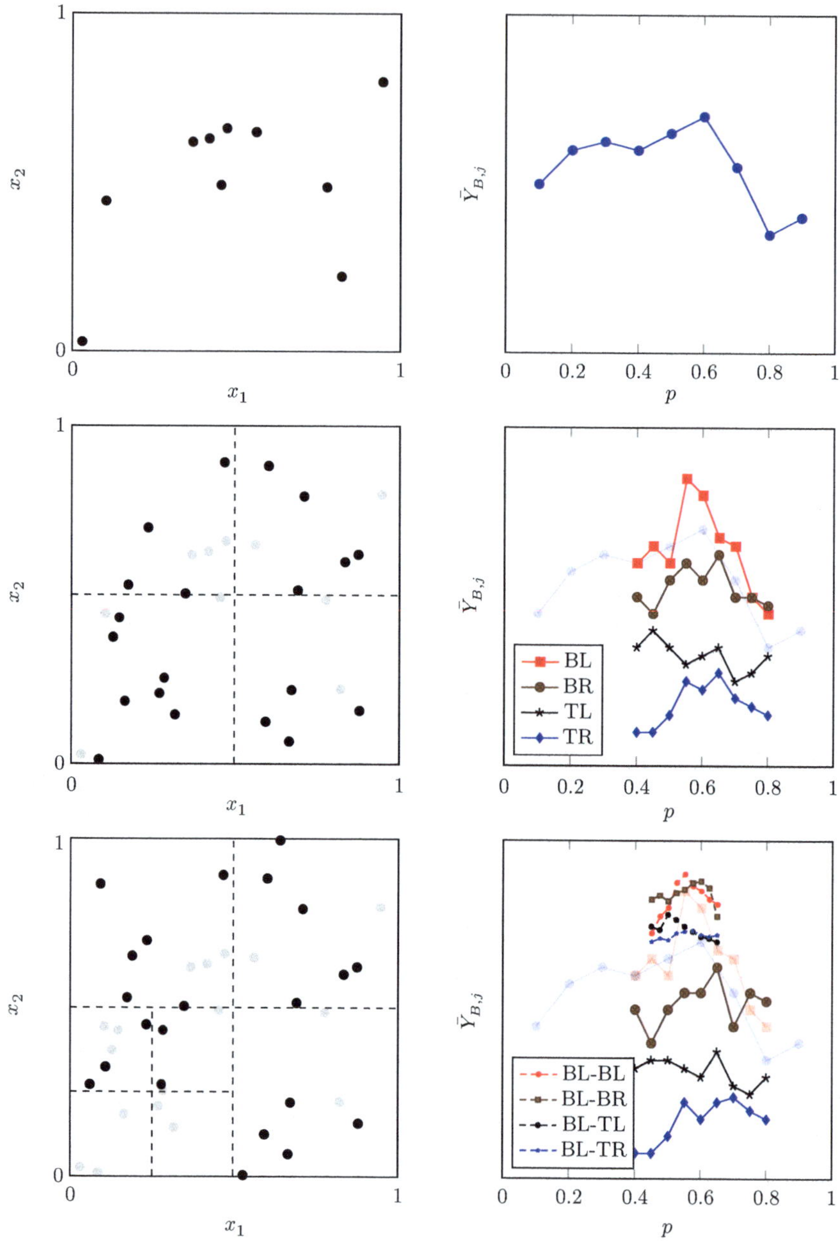

Fig. 10.1 A schematic illustration of the ABE algorithm

a single bin and the observed values X_t for $t \leq t_1$ are illustrated in the top left panel. The algorithm has explored the objective in the decision set, in this case $p \in \{0.1, 0.2, \ldots, 0.9\}$, and recorded the average reward $\bar{Y}_{B,j}$, illustrated by the top right panel. At $t_1 + 1$, sufficient observations are collected and Step e is triggered in the algorithm. Therefore, the bin is split into four child bins.

From period $t_1 + 1$ to t_2, new covariates are observed in each child bin (mid left panel). Note that the covariates generated before t_1 in the parent bin are no longer used and colored in gray. For each child bin (the bottom-left bin is abbreviated as BL and similarly for others), the average reward for the decision set is demonstrated in the mid right panel. The decision sets are centered at the empirically optimal decision of their parent bin, in this case $p^* = 0.6$ from the top right panel. They have narrower ranges and finer grids than that of the parent bin. At $t_2 + 1$, sufficient observations are collected for BL, and it is split into four child bins.

From period $t_2 + 1$ to t_3, the partition consists of seven bins, as shown in the bottom left panel. The BR, TL, and TR bins keep collecting covariates and updating the average reward, because they have not collected sufficient data. Their status at t_3 is shown in the bottom panels. In the four newly created child bins of BL (the bottom-left bin of BL is abbreviated as BL-BL and similarly for others), the decisions in the decision sets are applied successively and their average rewards are illustrated in the bottom right panel.

10.3 Constrained Inventory Model

Constrained inventory models require a slightly more careful analysis. Most of the models assume that the demand is either a Bernoulli or a Poisson process. In the Bernoulli process case, the inventory is ample whenever $c \geq T$ since in this case capacity exceeds potential demand. In the Poisson case, things are more subtle as the demand over the sales horizon at price is $Td(p)$ where $d(p)$ is the sales rate at price p, and $d(p)$ may be larger than one. If $d(p)$ is bounded above, a change of variables $T \leftarrow aT$ and $d(p) \leftarrow d(p)/a$ for a sufficiently large a results in $d(p) \ll 1$ for all p. In this case, the Poisson process can be closely approximated by a Bernoulli process. With this scaling, the ample inventory model corresponds to the case $\rho := c/T \geq 1$, and the constrained inventory model to the case $\rho < 1$. For a fixed ρ, we are interested in the learning and earning problem as $T \to \infty$, which means that initial inventory $c = \rho T < T$ also grows at the same rate. If we want to keep c integer (and this is important for the formulation of the dynamic pricing problem with finite capacity), we can restrict T to be of the form $T = nc/\rho$, where n is an integer so $\rho T = nc$. In the literature, we often see a different but equivalent scaling mechanism for the Poisson case, where T is held fixed (often normalized to one) and c and $d(p)$ are scaled up by a factor n, so the initial inventory is nc and the demand rate is $nd(p)$. Most authors use the second scaling method $(nc, nd(p))$, but it should be clear to the reader that an algorithm with regret $O^*(\sqrt{n})$ has regret $O(\sqrt{T})$ under the first scaling method. In this section, we follow the literature and report the regret relative to the scaling $(n, nd(p))$ and report the regret in terms of n.

Early work on the constrained inventory model divided the horizon into a learning phase and an earning phase. Under these regime, it was possible to show that all policies have regret at least $\Omega(\sqrt{n})$, and that there exist a policy with regret $O(\sqrt{\log(n)}n^{2/3})$ for the parametric case, and regret $O(\sqrt{\log(n)}n^{3/4})$ for the non-parametric case. In this section, we review recent work that intertwines learning and earning and improves the regret to $O(\log(n)^{4.5}\sqrt{n})$ for both the parametric and non-parametric case, thus also achieving an $O^*(\sqrt{n})$ policy, which is equivalent to the results for the ample inventory model stated in terms of T.

The analysis is based on slightly different assumptions from the case with ample capacity. The most salient difference is that demand is assumed to be a Poisson process with rate $\lambda_t = d(p_t)$, where $d(p)$ is the expected demand rate that is strictly decreasing in p. This in contrast to the ample capacity case where the analysis is based on a binomial approximation to the Poisson. There is an inverse demand function given by $p = \gamma(\lambda)$, and a corresponding revenue rate $R(\lambda) = \lambda\gamma(\lambda)$ function written in terms of the sales rate instead of the price.

The analysis allows for both parametric and non-parametric models with constrained capacity and requires the following assumptions on the class $\mathcal{C}$ of admissible demand functions.

Assumption 1 Boundedness: $|d(p)| < M$ for all $p \in [0, 1] \cup \{p_\infty\}$ with $d(p_\infty) = 0$.

Assumption 2 Lipschitz continuity: $d(p)$ and $R(d(p)) = pd(p)$ are Lipschitz continuous functions with respect to p with constant K. The inverse demand function $\gamma(\lambda)$ is also Lipschitz continuous with constant K.

Assumption 3 Strict concavity and differentiability: The function $R(\lambda) = \lambda\gamma(\lambda)$ has a second derivative for all λ, and there are positive constants such that $-m_L \leq R''(\lambda) \leq -m_U < 0$ for all $p \in [0, 1]$.

The assumptions are all reasonable in light of our previous discussion of the case of ample capacity. The most significant difference here is the existence of a cutoff price p_∞ such that $d(p_\infty) = 0$. This is a modeling artifact that provides us a price to use when the system runs out of inventory.

For any d satisfying Assumptions 1–3, let $V_\pi(T, c; d)$ denote the expected revenue that can be obtained from c units of inventory over the selling horizon $[0, T]$ by applying an non-anticipating policy π. From our analysis of dynamic pricing, we know that for every demand function d,

$$V_\pi(T, c; d) \leq V(T, c; d) \leq \bar{V}(T, c : d),$$

where $V(T, c : d)$ is the maximum expected revenue under any non-anticipating policy and $\bar{V}(T, c; d)$ is the upper bound based on replacing demand by its expectations. Rather than measuring the regret of a policy π by $V(T, c; d) - V_\pi(T, c; d)$, in this section we measure the regret relative to the deterministic upper bound, resulting in

$$r_\pi(T, c; d) = \bar{V}(T, c; d) - V_\pi(T, c; d).$$

Normalizing $T = 1$, the regret is

$$r_\pi(c; d) = \bar{V}(c; d) - V_\pi(c; d),$$

where $T = 1$ is omitted for convenience. The goal is to minimize the worst-case regret, which is given by

$$\inf_\pi \sup_{d \in \mathcal{C}} r_\pi(nc; nd)$$

as n increases, where the infimum is taken over any non-anticipating policy and any demand function d satisfying Assumptions 1–3.

Learning and Dynamic Pricing (LDP)

Step 1. Initialization

(a) Consider a sequence of $\tau_i^u, \kappa_i^u, i = 1, 2, \ldots, N^u$ and $\tau_i^c, \kappa_i^c, i = 1, 2, \ldots, N^c$ Define $\underline{p}_1^u = \underline{p}_1^c = 0$ and $\overline{p}_1^u = \overline{p}_1^c = 1$. Define $t_i^u = \sum_{j=1}^i \tau_j^u$, for $i = 1$ to N^u and $t_i^c = \sum_{j=1}^i \tau_j^c$, for $i = 1$ to N^c;

Step 2. Learn p^u or Determine $p^c > p^u$

For $i = 1$ to N^u do

(a) Divide $[\underline{p}_i^u, \overline{p}_i^u]$ into κ_i^u equally spaced intervals and let $\{p_{i,j}^u, j = 1, 2, \ldots, \kappa_i^u\}$ be the left endpoints of these intervals;

(b) Divide the time interval $[t_{i-1}^u, t_i^u]$ into κ_i^u equal parts and define

$$\Delta_i^u = \frac{\tau_i^u}{\kappa_i^u}, \qquad t_{i,j}^u = t_{i-1}^u + j\Delta_i^u, \qquad j = 0, 1, \ldots, \kappa_i^u;$$

(c) For j from 1 to κ_i^u, apply $p_{i,j}^u$ from time $t_{i,j-1}^u$ to $t_{i,j}^u$. If inventory runs out, then apply p_∞ until time T and STOP;

(d) Compute

$$\hat{d}(p_{i,j}^u) = \frac{\text{total demand over } [t_{i,j-1}^u, t_{i,j}^u)}{\Delta_i^u}, \qquad j = 1, \ldots, \kappa_i^u;$$

(e) Compute

$$\hat{p}_i^u = \arg\max_{1 \le j \le \kappa_i^u} \{p_{i,j}^u \hat{d}(p_{i,j}^u)\} \quad \text{and} \quad \hat{p}_i^c = \arg\min_{1 \le j \le \kappa_i^u} \left|\hat{d}(p_{i,j}^u) - x/T\right|;$$

(f) If

$$\hat{p}_i^c > \hat{p}_i^u + 2\sqrt{\log n} \cdot \frac{\overline{p}_i^u - \underline{p}_i^u}{\kappa_i^u}$$

then break from Step 2, enter Step 3 and set $i_0 = i$;

Otherwise, set $\hat{p}_i = \max\{\hat{p}_i^c, \hat{p}_i^u\}$. The price range for the next iteration is given by

$$I_{i+1}^u = [\underline{p}_{i+1}^u, \overline{p}_{i+1}^u],$$

where

$$\underline{p}_{i+1}^u = \hat{p}_i - \frac{\log n}{3} \cdot \frac{\overline{p}_i^u - \underline{p}_i^u}{\kappa_i^u} \text{ and } \overline{p}_{i+1}^u = \hat{p}_i + \frac{2\log n}{3} \cdot \frac{\overline{p}_i^u - \underline{p}_i^u}{\kappa_i^u}.$$

We truncate the interval if it does not lie inside the feasible set [0, 1];

(g) If $i = N^u$, then enter Step 4(a);

Step 3. Learn p^c When $p^c > p^u$
For $i = i_0$ to N^c do

(a) Divide $[\underline{p}_i^c, \overline{p}_i^c]$ into κ_i^c equally spaced intervals and let $\{p_{i,j}^c, j = 1, 2, \ldots, \kappa_i^c\}$ be the left endpoints of these intervals;
(b) Define

$$\Delta_i^c = \frac{\tau_i^c}{\kappa_i^c}, \qquad t_{i,j}^c = t_{i-1}^c + j\Delta_i^c + t_{i_0}^u, \qquad j = 0, 1, \ldots, \kappa_i^c;$$

(c) For j from 1 to κ_i^c, apply $p_{i,j}^c$ from time $t_{i,j-1}^c$ to $t_{i,j}^c$. If inventory runs out, then apply p_∞ until time T and STOP;
(d) Compute

$$\hat{d}(p_{i,j}^c) = \frac{\text{total demand over } [t_{i,j-1}^c, t_{i,j}^c)}{\Delta_i^c}, \qquad j = 1, \ldots, \kappa_i^c;$$

(e) Compute

$$\hat{q}_i = \arg \min_{1 \le j \le \kappa_i^c} \left| \hat{d}(p_{i,j}^c) - x/T \right|.$$

The price range for the next iteration is given by

$$I_{i+1}^c = [\underline{p}_{i+1}^c, \overline{p}_{i+1}^c],$$

where

$$\underline{p}_{i+1}^c = \hat{q}_i - \frac{\log n}{2} \cdot \frac{\overline{p}_i^c - \underline{p}_i^c}{\kappa_i^c} \text{ and } \overline{p}_{i+1}^c = \hat{q}_i + \frac{\log n}{2} \cdot \frac{\overline{p}_i^c - \underline{p}_i^c}{\kappa_i^c},$$

and we truncate the interval if it doesn't lie inside the feasible set of [0, 1];

(f) If $i = N^c$, then enter Step 4(b);

Step 4. Apply the Learned Price

(a) Define $\tilde{p} = \hat{p}_{N^u} + 2\sqrt{\log n} \cdot \frac{\overline{p}^u_{N^u} - \underline{p}^u_{N^u}}{\kappa^u_{N^u}}$. Use $\tilde{p}$ for the rest of the selling season until the inventory runs out;

(b) Define $\tilde{q} = \hat{q}_{N^c}$. Use $\tilde{q}$ for the rest of the selling season until the inventory runs out.

A few comments about the LDP algorithm are in order. The selling season is divided into a set of time periods. In each time period, a set of a grid prices is tested within the current price interval. The intervals are then updated based on empirical observations at the end of each time interval, so the price intervals contain the optimal price with high probability. The process is repeated until the price interval is small enough so that the desired accuracy is achieved.

The optimal price is the largest between the unconstrained optimal price, say p^*, and the market clearing price, say p_{mc}. Finding these two prices require different shrinking strategies for the cases when $p^* > p_{mc}$ (Step 2) and $p_{mc} > p^*$ (Step 3). At the end of the algorithm, a fixed price is used for the remaining selling season (Step 4) until the inventory runs out.

The definitions of τ^u_i, κ^u_i, N^u, τ^c_i, κ^c_i, and N^c now follow, where $T = 1$ without loss of generality:

$$\left(\frac{\overline{p}^u_i - \underline{p}^u_i}{\kappa^u_i}\right)^2 \sim \sqrt{\frac{\kappa^u_i}{n\tau^u_i}}, \quad \forall i = 2, \ldots, N^u,$$

$$\overline{p}^u_{i+1} - \underline{p}^u_{i+1} \sim \log n \cdot \frac{\overline{p}^u_i - \underline{p}^u_i}{\kappa^u_i}, \quad \forall i = 1, \ldots, N^u - 1,$$

$$\tau^u_{i+1} \cdot \left(\frac{\overline{p}^u_i - \underline{p}^u_i}{\kappa^u_i}\right)^2 \cdot \sqrt{\log n} \sim \tau^u_1, \quad \forall i = 1, \ldots, N^u - 1,$$

$$N^u = \min_l \left\{ l : \left(\frac{\overline{p}^u_l - \underline{p}^u_l}{\kappa^u_l}\right)^2 \sqrt{\log n} < \tau^u_1 \right\}.$$

The main results for this section is the following.

Theorem 10.3 *For any function demand $d(p)$ satisfying Assumptions 1–3, the regret incurred by the LDP algorithm is bounded by*

$$\sup_{d \in \mathcal{C}} r_{\pi_{DPA}}(nc, nd) \leq K\sqrt{n} \log(n)^{4.5}$$

for a constant $K > 0$ that is independent of n for both the parametric and the non-parametric cases. For all non-anticipating polices, we have

$$\inf_{\pi} \sup_{f \in \mathcal{C}} r_{\pi}(nc, nd) \geq k\sqrt{n}$$

for a constant $k > 0$ that is independent of n.

Under mild assumptions, the results in this section can be extended to multiple market segments $d_m(p)$, $m = 1, \ldots, M$ using a primal-dual approach.

10.4 Bibliographical Remarks

There is a large and growing literature for parametric models that follows a dynamic programming formulation with Bayesian updating. Some examples in this stream of literature include the work by Aviv and Pazgal (2005), Bertsimas and Mersereau (2007), Araman and Caldentey (2009), Sen and Zhang (2009), Farias and Van Roy (2010), and Harrison et al. (2012). Bayesian methods require the specification of a prior distribution that belongs to a conjugate family, and the method is mostly use for the case of a single unknown parameter with a few notable exceptions. Alternatives to the Bayesian approach that are capable of dealing with a large number of parameters involve maximum likelihood methods and least squares as in Bertsimas and Perakis (2006) and Bertsimas and Misic (2019). Araman and Caldentey (2011) go over both Bayesian and non-parametric models. The survey paper by den Boer (2015) provides a comprehensive overview of this area.

There is an alternative stream of literature, closer to the results described in this chapter, that focus on the learning and earning problem to minimize the worst-case regret. This literature can be divided into the case of ample or constrained inventory. Some of the contributors to this literature come from the computer science literature, including Kleinberg and Leighton (2003), who provide analysis for an online posted-price auction for the case of ample capacity obtaining a regret of $O(\log T\sqrt{T})$ for a non-parametric model. The paper by Broder and Rusmevichientong (2012) provides an algorithm with regret $O(\log T\sqrt{T})$ for a parametric model based on maximum likelihood. They also show that the regret can be improved to $O(\log T)$ for situations where the demand functions can be separated. Cheung and Simchi-Levi (2017) also look into an infinite inventory model but restrict the class of demand functions to a finite set and allow a maximum of m price changes. They achieve a regret of $O((\log T)^m)$ under the assumption that exploration is done with informative prices. Besbes and Zeevi (2015) show that even if the demand is misspecified as linear, a regret of $O((\log T)^2\sqrt{T})$ can be surprisingly achieved under mild restrictions. None of the models mentioned allow for covariates.

The results presented in this chapter for the ample capacity case are due to Chen and Gallego (2018a). For the constrained capacity case, an important early reference is Besbes and Zeevi (2009), where learning and earning are separated into two phases. The section on constrained inventory is based on Wang et al. (2014), where more refined results are obtained by intertwining learning and earning. The extension of the constrained model to multiple market segments is due to Chen and Gallego (2018b). A related model that is applicable to multi-resource revenue management problems is given in Agrawal et al. (2014). Chen et al. (2019b, 2016a) discuss models for joint inventory and pricing decisions, when the price-demand relationship is unknown. There is also work on learning the customer preferences in assortment optimization problems, as exemplified by Saure and Zeevi (2013), Agrawal et al. (2018), Chen and Wang (2018), and Chen et al. (2018b,c).

Other works that focus on learning the price-demand relationship while making pricing decisions and earning revenues include Levina et al. (2009b), Besbes and Zeevi (2011), Kwon et al. (2012), Besbes and Saure (2014), Keskin and Zeevi (2014), and Nambiar et al. (2019). Ciocan and Farias (2012a) give bounds on the performance of a policy that is based on re-solving a mathematical program and updating the demand forecast. Ciocan and Farias (2014), Ban and Keskin (2017), Javanmard and Nazerzadeh (2018), and Cohen et al. (2018a) learn parameterized relationships for the demand for a product that are based on features. Keskin and Zeevi (2017) consider learning the price-demand relationship when this relationship is changing over time. A related model also appears in Besbes et al. (2015). The paper by den Boer and Keskin (2017a) focus on learning a discontinuous price-demand relationship, whereas den Boer and Keskin (2017b) study the case where there is an observable kink in the price-demand relationship. Keskin and Birge (2019) study a model where the firm learns the quality sensitivity of its customers and demonstrate that myopic policies can display near-optimal performance. A Bayesian approach based on Thompson sampling is given in Ferreira et al. (2018). Chen et al. (2019a) characterize the revenue loss of a policy that learns the multi-product demand function while making decisions. Afeche and Ata (2013) study a Bayesian learning model for a pricing problem in the queueing setting, where the proportion of patient customers needs to be learned. Lastly, Acemoglu et al. (2011), Crapis et al. (2017), and Ifrach et al. (2018) focus on learning problems within social networks.

Chapter 11
Competitive Assortment and Price Optimization

11.1 Introduction

In the models that we studied thus far, we considered the decisions made by a single firm. The implicit assumption in our development was that the other firms do not react to the decisions of each other. Naturally, this is almost never the case. When a firm decreases its prices, fearing loss of customers, its competitors may also decrease its prices. Both online and brick-and-mortar retail stores consider the assortments offered by the other stores when making planning their assortments. There is vast literature on modeling competition. Nevertheless, despite the fact that competition is the rule rather than an exception and there is vast literature on modeling competition, the development of operational models that can drive real-time decision making under competition is in its infancy. In most operational models, it is often the case that the competition is ignored or modeled rather simplistically. Perhaps, the most important reason for this is that explicitly modeling competition often times results in intractable models. Thus, for the sake of computational tractability, the reactions of the other firms are ignored. Furthermore, the data that drive the operational models are often collected in a competitive environment, and one usually naively hopes that building a noncompetitive model driven by data collected in a competitive environment will take care of the competition itself, but of course, this hope is not based on any scientific evidence.

Competition is a critical area for improvement for operational revenue management models, and we are starting to see more and more models in the literature that explicitly try to incorporate competition. In this chapter, we give a glimpse of two models. In Sect. 11.2, multiple firms compete in an environment where they choose the assortments they offer to their customers. The model here is static in the sense that there is no time dimension. In Sect. 11.3, multiple firms compete in their pricing decisions, there is limited inventory and the sales take place over time.

G. Gallego, H. Topaloglu, *Revenue Management and Pricing Analytics*, International Series in Operations Research & Management Science 279, https://doi.org/10.1007/978-1-4939-9606-3_11

11.2 Competitive Assortment Optimization

In this section, we consider a competitive assortment optimization problem between two firms, when the customers choose among the products offered by the firms according to the multinomial logit (MNL) model.

11.2.1 Problem Formulation

Consider two firms each of which has access to different sets of products. Among the set of products that a firm has access to, the firm chooses a subset, or an assortment, of products to offer to the customers. Considering all the products offered by *both* firms, a customer chooses among the products according to the MNL model. The goal of each firm is to choose an assortment of products to offer to maximize the expected revenue that it obtains from a customer. We index the firms by $\{1, -1\}$. For $i \in \{1, -1\}$, we use N_i to denote the set of products that firm i has access to. In other words, firm i offers an assortment within the set of products N_i. The set of all products is given by $N = N_1 \cup N_{-1}$. Let $v_j > 0$ be the attraction value of product $j \in N$, and v_0 be the attraction value of the no-purchase option. Let $V(S) := \sum_{j \in S} v_j$ denote the total attraction value of the products in set S. If the two firms offer subsets (S_1, S_{-1}) with $S_1 \subseteq N_1$ and $S_{-1} \subseteq N_{-1}$, then a customer chooses product $j \in S_1 \cup S_{-1}$ with probability

$$\pi_j(S_1, S_{-1}) := \frac{v_j}{v_0 + V(S_1) + V(S_{-1})}.$$

For $i \in \{1, -1\}$, we use $\mathcal{F}_i$ to denote the set of feasible assortments that can be offered by firm i. For example, each firm may be constrained by the number of products that they can display to their customers. Alternatively, each product may occupy a certain amount of space and the total space consumption of the products offered by a firm may have to be below a certain space limit. The revenue associated with product $j \in N$ is $p_j > 0$. Given that the two firms offer the assortments of products $(S_1, S_{-1}) \in \mathcal{F}_1 \times \mathcal{F}_{-1}$, the expected revenue that firm i obtains from a customer is

$$R_i(S_i, S_{-i}) := \sum_{j \in S_i} p_j \, \pi_j(S_i, S_{-i}) = \frac{\sum_{j \in S_i} p_j \, v_j}{v_0 + V(S_i) + V(S_{-i})}. \tag{11.1}$$

Therefore, if firm $-i$ offers the subset S_{-i} of products, then firm i maximizes its expected revenue by solving the problem

$$\max_{S_i \in \mathcal{F}_i} R_i(S_i, S_{-i}). \tag{11.2}$$

An optimal solution to the problem above is a best response of firm i to the assortment S_{-i} offered by firm $-i$. We say that the assortments $(S_1^*, S_{-1}^*) \in \mathcal{F}_1 \times \mathcal{F}_{-1}$ are a Nash equilibrium, if S_i^* is a best response to S_{-i}^* for all $i \in \{1, -1\}$. In the rest of our discussion, we show that a Nash equilibrium for competitive assortment optimization exists. We characterize a Pareto-dominating equilibrium in the sense that the expected revenue for each firm in the Pareto-dominating equilibrium is at least as large as its corresponding expected revenue in any other equilibria. Lastly, we compare the assortments in a Nash equilibrium with those chosen by a central planner to maximize the total expected revenue obtained by the two firms.

11.2.2 Existence of Equilibrium

Let $z_i^*(S_{-i})$ denote the optimal objective value of problem (11.2). In other words, $z_i^*(S_{-i})$ is the best expected revenue that firm i can achieve when firm $-i$ offers the assortment S_{-i}. Noting the expected revenue expression in (11.1), we have

$$z_i^*(S_{-i}) \geq \frac{\sum_{j \in S_i} p_j \, v_j}{v_0 + V(S_i) + V(S_{-i})} \qquad \forall \, S_i \in \mathcal{F}_i,$$

and the inequality above holds as equality at an optimal solution to problem (11.2). Since $V(S_i) = \sum_{j \in S_i} v_j$, this inequality is equivalent to

$$[v_0 + V(S_{-i})] \, z_i^*(S_{-i}) \geq \sum_{j \in S_i} (p_j - z_i^*(S_{-i})) \, v_j \qquad \forall \, S_i \in \mathcal{F}_i,$$

with equality holding at an optimal solution to (11.2), so

$$[v_0 + V(S_{-i})] \, z_i^*(S_{-i}) = \max_{S_i \in \mathcal{F}_i} \left\{ \sum_{j \in S_i} (p_j - z_i^*(S_{-i})) \, v_j \right\}.$$

Therefore, an optimal solution to problem (11.2) can be obtained by solving the problem:

$$\max_{S_i \in \mathcal{F}_i} \left\{ \sum_{j \in S_i} (p_j - z_i^*(S_{-i})) \, v_j \right\}. \tag{11.3}$$

Throughout, we assume that if problem (11.2) or (11.3) has multiple optimal solutions, then we choose a solution S_i that has the largest total attraction value $V(S_i)$. Note that problem (11.3) is not immediately useful to solve problem (11.2) because solving problem (11.3) requires knowing $z_i^*(S_{-i})$ and we do not know $z_i^*(S_{-i})$ before solving problem (11.2)! Nevertheless, we will use problem (11.3) to show the existence of Nash equilibria and to characterize the properties of such equilibria.

Consider two assortments $\hat{S}_{-i}$ and $\tilde{S}_{-i}$ that could be offered by firm $-i$. Let $\hat{S}_i$ be a best response of firm i to the assortment $\hat{S}_{-i}$ and $\tilde{S}_i$ be a best response of firm i to the assortment $\tilde{S}_{-i}$. In the next lemma, we present a key monotonicity result.

Lemma 11.1 *If* $V(\hat{S}_{-i}) \leq V(\tilde{S}_{-i})$, *then* $V(\hat{S}_i) \leq V(\tilde{S}_i)$.

The lemma above establishes a monotonicity property for the best response of each firm, where if firm $-i$ offers an assortment with a larger total attraction value, then firm i, in its best response, also offers an assortment with a larger total attraction value. By using this lemma, we will be able to show that a tatonnement process converges to a Nash equilibrium. In the process, we will also establish the existence of Nash equilibria. To describe the tatonnement process, we define the sequence of assortments $\{(\hat{S}_1^t, \hat{S}_{-1}^t) : t = 0, 1, \ldots\}$ offered by the two firms as follows. We start with $\hat{S}_1^0 = \varnothing$ and $\hat{S}_{-1}^0 = \varnothing$. Using $(\hat{S}_1^t, \hat{S}_{-1}^t)$, we compute $(\hat{S}_1^{t+1}, \hat{S}_{-1}^{t+1})$ as

$$\hat{S}_1^{t+1} \in \arg\max_{S_1 \in \mathcal{F}_1} R_1(S_1, \hat{S}_{-1}^t) \qquad \text{and} \qquad \hat{S}_{-1}^{t+1} \in \arg\max_{S_{-1} \in \mathcal{F}_{-1}} R_{-1}(S_{-1}, \hat{S}_1^{t+1}).$$

Thus, $\hat{S}_1^{t+1}$ is a best response of firm 1 to the assortment $\hat{S}_{-1}^t$ offered by firm -1, whereas $\hat{S}_{-1}^{t+1}$ is a best response of firm -1 to the assortment $\hat{S}_1^{t+1}$ offered by firm 1. In the next theorem, we use this tatonnement process to show that there exists a Nash equilibrium. In the proof, using Lemma 11.1, we argue that the sequence of the total attraction values in the assortments generated by the tatonnement process converges, in which case, we are able to construct a Nash equilibrium by using the limit of this sequence.

Theorem 11.2 *There exists a Nash equilibrium.*

In the proof of Theorem 11.2, we use Lemma 11.1 to argue that the sequence of assortments $\{(\hat{S}_1^t, \hat{S}_{-1}^t) : t = 0, 1, \ldots\}$ generated in the tatonnement process satisfies $V(\hat{S}_i^{t+1}) \geq V(\hat{S}_i^t)$ for all $i \in \{1, -1\}$. Thus, there exists an iteration counter $t_0 \geq 0$ in the tatonnement process such that $V(\hat{S}_1^{t_0}) = V(\hat{S}_1^{t_0+1}) = V(\hat{S}_1^{t_0+2}) = \ldots$ and $V(\hat{S}_{-1}^{t_0}) = V(\hat{S}_{-1}^{t_0+1}) = V(\hat{S}_{-1}^{t_0+2}) = \ldots$. In this case, we are able to show that $(\hat{S}_1^{t_0+1}, \hat{S}_{-1}^{t_0})$ is a Nash equilibrium. We refer to $(\hat{S}_1^{t_0+1}, \hat{S}_{-1}^{t_0})$ as a Nash equilibrium generated by the tatonnement process. In the tatonnement process, we started with the assortments $\hat{S}_1^0 = \hat{S}_{-1}^0 = \varnothing$, but the choice of $\hat{S}_1^0$ is irrelevant because we compute $\hat{S}_1^1$ as a best response to $\hat{S}_{-1}^0$ and we compute $\hat{S}_{-1}^1$ as a best response to $\hat{S}_1^1$. Thus, $\hat{S}_1^0$ does not play any role in the tatonnement process. Also, by using the same argument in the proof of Theorem 11.2, we can show that the tatonnement process would yield a Nash equilibrium even if we choose $\hat{S}_1^0$ and $\hat{S}_{-1}^0$ arbitrarily, but as we show in the next section, an equilibrium that we reach by choosing $\hat{S}_1^0 = \hat{S}_{-1}^0 = \varnothing$ Pareto dominates any equilibria. Therefore, when we say a Nash equilibrium generated by the tatonnement process, we will mean the one obtained by starting with $\hat{S}_1^0 = \hat{S}_{-1}^0 = \varnothing$.

11.2.3 Properties of Equilibrium

There can be multiple Nash equilibria in general, but it turns out that a Nash equilibrium generated by the tatonnement process will always Pareto dominate the others. In other words, the expected revenue of each firm in a Nash equilibrium generated by the tatonnement process is at least as large as its corresponding expected revenue in any other Nash equilibrium. We show this result in the next theorem.

Theorem 11.3 *A Nash equilibrium generated by the tatonnement process is Pareto dominant.*

The key to the result above is to show that the total attraction value of each assortment in a Nash equilibrium generated by the tatonnement process is no larger than its corresponding total attraction value in another Nash equilibria. Next, we compare the assortments offered in the absence of competition and the assortments offered by a central planner with the assortments offered in a Nash equilibrium.

In the absence of competition, firm i finds an assortment to offer by solving the problem $\max_{S_i \in \mathcal{F}_i} R_i(S_i, \varnothing)$. Let $(S_1^{\text{NC}}, S_{-1}^{\text{NC}})$ be the assortments offered by the two firms in the absence of competition, where the superscript NC stands for no competition. Also, if there were a central planner that chooses the assortments offered by the two firms to maximize the total expected revenue obtained by the two firms, then she would solve the problem

$$\max_{(S_1, S_{-1}) \in \mathcal{F}_1 \times \mathcal{F}_{-1}} \Big\{ R_1(S_1, S_{-1}) + R_{-1}(S_{-1}, S_1) \Big\}. \tag{11.4}$$

Let $(S_1^{\text{CP}}, S_{-1}^{\text{CP}})$ be the assortments offered by the central planner, where the superscript CP stands for central planner. In the next theorem, we show that the total attraction value of the products offered by each firm in any equilibrium is at least as large as the total attraction value of the products offered by the corresponding firm in the absence of competition. Furthermore, the total attraction value of the products offered by each firm in any equilibrium is also at least as large as the total attraction value of the products offered by the corresponding firm under the solution of the central planner. These results indicate that competition has the tendency to increase the total attraction values of the products offered by each firm. In other words, to deal with competition, the firms enlarge their assortments by offering assortments with larger total attraction values.

Theorem 11.4 *Let (S_1^*, S_{-1}^*) be any Nash equilibrium, $(S_1^{\text{NC}}, S_{-1}^{\text{NC}})$ be the assortments offered by the two firms in the absence of competition, and $(S_1^{\text{CP}}, S_{-1}^{\text{CP}})$ be the assortments offered by the central planner. Then, $V(S_i^{\text{NC}}) \le V(S_i^*)$ and $V(S_i^{\text{CP}}) \le V(S_i^*)$ for all $i \in \{1, -1\}$.*

Note that the result above holds for any Nash equilibrium.

11.3 Dynamic Pricing Under Competition

In this section, we consider dynamic pricing in an oligopolistic market with a mix of substitutable and complementary perishable products. Each firm has a fixed initial stock of items and competes in setting prices to sell them over a finite sales horizon. Customers sequentially arrive at the market, make a choice that includes the no-purchase alternative, and then leave the system. Assuming deterministic customer arrival rates, we show that any equilibrium strategy has a simple structure involving a finite set of time-invariant shadow prices measuring capacity externalities that firms exert on each other. This simple structure sheds light on dynamic revenue management problems under competition and demand uncertainty. Indeed, it turns out that the equilibrium solutions from the deterministic game provide precommitted and contingent heuristic policies that are asymptotic equilibria for the stochastic game when demand and supply are sufficiently large.

11.3.1 Problem Formulation

We consider a market of m competing firms selling differentiated perishable products over a finite horizon $[0, T]$. At time $t = 0$, each firm i has an initial inventory of c_i units of a unique product. We count the time forwards and use t for the elapsed time and $s = T - t$ for the remaining time. Let $p(t)$ be the vector of prices at time t, and let $d(t, p(t))$ be the vector of product demands at time t at prices $p(t)$, and let $r_i(t, p) = p_i\, d_i(t, p)$ be the revenue rate for firm i at time t when the price vector is $p = (p_i, p_{-i})$, where p_i is the priced offered by firm i and p_{-i} is the vector of prices from firms other than firm i. We make the following assumptions.

1. (a) The demand for firm i, $d_i(t, p)$ is continuously differentiable in p for all i and all t.
   (b) The aggregate demand $\int_0^T d_i(t, p(t))dt$ for firm i is pseudo-convex in its price path $p_i(t),\ t \in [0, T]$.
2. (a) The aggregate revenue $\int_0^T r_i(t, p(t))dt$ for firm i is pseudo-concave in its price path $p_i(t),\ t \in [0, T]$.
   (b) There exist a function $R_i(t)$ such that $r_i(t, p) \le R_i(t)$ and $\int_0^T R_i(t)dt < \infty$.
3. (a) There exist a choke price $p_i(t, p_{-i})$ such that

$$\lim_{p_i \to p_i(t,p_{-i})} d_i(t, p) = 0 \quad \text{and} \quad \lim_{p_i \to p_i(t,p_{-i})} r_i(t, p) = 0.$$

   Moreover, the choke price is always an available option for each firm.
   (b) Other than the choke price, firm i chooses prices from a compact and convex subset $\mathcal{P}_i(t, p_{-i})$ of $\{p_i \in \Re_+ : d_i(t, p_i, p_{-i}) \ge 0\}$.

(c) The salvage value of the products at the end of the horizon is zero, and all other costs are sunk.

4. All firms have perfect knowledge about the inventory levels of other firms at any time.

As examples of possible demand functions for the firms, consider the MNL demand function

$$d_i(t, p) = \lambda(t) \frac{\beta_i(t) \exp(-\alpha_i(t) p_i)}{a_0(t) + \sum_j \beta_j(t) \exp(-\alpha_j(t) p_j)},$$

where $\lambda(t), a_0(t), \alpha_i(t), \beta_i(t) > 0$ for all i and t. As a second example, consider now the linear demand function

$$d_i(t, p) = a_i(t) - b_i(t) p_i + \sum_{j \neq i} c_{ij}(t) \, p_j,$$

where $a_i(t), b_i(t) > 0$ for all i. These linear demand functions can arise from a representative consumer maximizing a quadratic utility function and can accommodate substitute and complementary products depending on whether $c_{ij}(t)$ is positive or negative. It can be shown that Assumptions 1 and 2 are satisfied both by the MNL model and the linear demand model.

Assumption 3(a) ensures that a firm immediately exits the market on a stockout. In this case, customers who originally prefer the stockout firm will spill over to the remaining firms that still have positive inventory. The spillover is endogenized from the demand model according to customers' preferences and product substitutability. Moreover, in view of Assumption 3(b), firms can use the choke price before it runs out of stock. The compactness assumption of $\mathcal{P}_i(t, p_{-i})$ fails to hold for some models, like the MNL. Fortunately, for the MNL model there are ways around that avoid compactness.

Assumption 3(c) is without loss of generality. Assumption 4 is standard in game theory and it is realistic in an airline setting as major airlines offer a feature of previewing seat availability from their websites.

Let $x(t) \in [0, c]$ be the joint inventory at time t. A joint open-loop strategy $p(t)$ depends only on time t and the initial inventory $x(0) = c$. In contrast, a feedback strategy $p(t, x(t))$ depends on t and the current inventory $x(t)$. The set of all open-loop strategies is denoted by $\mathcal{P}_0$, and the set of all feedback strategies is denoted by $\mathcal{P}_F$.

Let $D[0, T]$ denote the set of all right-continuous real-valued functions with left limits defined on $[0, T]$, where the left discontinuities allow for price jumps after a sale. Given a price control path $p \in D[0, T]^m$, we denote the total profit for firm i by

$$J_i[p] = \int_0^T r_i(t, p_i(t)) dt.$$

Moreover, under $p \in D[0, T]^m$, the inventory of product i evolves according to

$$\dot{x}_i(t) = -d_i(t, p(t)) \quad 0 \le t \le T$$

starting from $x_i(0) = c_i$.

The objective of each firm is to maximize its own total revenue over the sales horizon subject to *all* capacity constraints over the entire sales horizon. Thus, firm's i problem is

$$\max_{p_i(t), t \in [0,T]} \int_0^T r_i(t, p(t))dt$$

subject to

$$x_j(t) = c_j - \int_0^t d_j(v, p(v))dv \ge 0 \quad \forall t \in [0, T], \; j = 1, \dots, m.$$

Firms simultaneously solve their own revenue maximization problems subject to a joint set of constraints, giving rise to a game with coupled strategy constraints for all firms. These are known as generalized Nash games with coupled constraints. If some pricing policy results in negative inventory at some time, then it will be eliminated from the joint feasible strategy space. In other words, all firms face a joint set of constraints, $x(t) \ge 0$ for all $t \in [0, T]$ in selecting feasible strategies.

A generalized open-loop Nash equilibrium (OLNE) is an open-loop control path $p \in \mathcal{P}_0$ such that $p \in D[0, T]^m$, and $p_i(t), 0 \le t \in T$ solves firm i's problem for all i. Likewise, a feedback loop Nash Equilibrium (FNE) is a feedback control path $p \in \mathcal{P}_F$ such that $p \in D[0, T]^m$, and $p_i(t, x(t))$ solves firm i's problem for all i.

In a nonzero-sum differential game, open-loop and feedback strategies are generally different. However, re-solving the OLNE with the current time and inventory level continuously over time results in an FNE, which generates the same price path and inventory trajectory as those of the OLNE with the same initial time and inventory level. Because of this relationship, we call an OLNE, an equilibrium strategy.

11.3.2 Equilibrium Results

The following theorem gives important existence results.

Theorem 11.5 *If the choke price $p_i^\infty(t, p_{-i})$ is in the convex and compact set $\mathcal{P}_i(t, p_{-i})$ for each i, then an equilibrium exists. For the MNL model, an equilibrium exists where firms never use the choke price.*

The first part of Theorem 11.5 applies to the linear demand model, but it does not apply to the MNL model, because for the MNL model, the sets $\mathcal{P}_i(t, p_{-i})$ are not compact. Nevertheless, the second part guarantees the existence of an equilibrium that does not involve the choke price for any of the firms.

Necessary Conditions

From Pontryagin's maximum principle for constrained set space, the following are necessary conditions for an OLNE.

Theorem 11.6 *If an open-loop pricing policy p^* is an OLNE, then there exists a non-negative, $m \times m$ matrix of non-negative shadow prices μ_{ij}, such that for any t such that $x_i^*(t) > 0$, the open-loop policy maximizes*

$$r_i(t, p_i, p_{-i}^*(t)) - \sum_j \mu_{ij} d_j(t, p_i, p_{-i}^*(t)).$$

Moreover, $\mu_{ij} x_j^*(T) = 0$ for all j. Let $E_i = \{t \in [0, T] : x_i^*(t) = 0\}$, and if E_i is non-empty, define $t_i = \inf E_i$. For all $t \in [t_i, T]$, firm i uses the choke price $p^\infty(t, p_{-i}^*(t))$. There exists a decreasing shadow price process $\mu_{ij}(t) \in [0, \mu_{ij}]$ for all j and $t \in [t_i, T]$ such that the choke price $p_i^\infty(t)$ maximizes

$$r_i(t, p_i, p_{-i}^*(t)) - \sum_j \mu_{ij}(t) d_j(t, p_i, p_{-i}^*(t)).$$

As a result, the OLNE has a simple structure. First, there exist an $m \times m$ matrix of finite, non-negative, time invariant, shadow prices μ_{ij}. At time t, let $S(t) = \{i : x_i^*(t) > 0\}$ be the set of firms with positive inventories. Then every firm in $S(t)$ simultaneously solves the problem

$$\max_{p_i} \left\{ r_i(t, p_i, p_{-i}) - \sum_{j \in S(t)} \mu_{ij} d_j(t, p_i, p_{-i}) \right\},$$

where $p_i \in \mathcal{P}_i(t, p_{-i}) \cup p_i^\infty(t, p_{-i})$ and all firms $i \notin S(t)$ use their choke price. Notice that the set of allowable prices for firms in $S(t)$ include the choke price, thus a firm may use its choke price even if it has positive inventory.

We now illustrate how capacity externalities influence the equilibrium pricing. Fix an arbitrary time t. If firms i and j offer substitutable products, then firm j's scarce capacity exerts an externality on firm i by pushing up firm i's price: Since firm j has limited capacity, it has a tendency to increase its own price due to the self-inflicted capacity externality. Because of the substitutability between products from firms i and j, the price competition between the two firms will be alleviated so that firm i can also post a higher price. On the other hand, if firms i and j offer complementary products, then firm j's scarce capacity exerts an externality on firm i by pushing down firm i's price: While firm j has a tendency to increase its own price, due to the complementarity between products from firms i and j, firm i has to undercut its price to compensate for the price increase of firm j. By a similar reasoning, on stockout, a product's market exit by posting choke prices will be a boon for its substitutable products and a bane for its complementary products.

An important special case is that of time invariant demands $d(t, p) = d(p)$. In this case, the price trajectories and the available products in the market remain constant before the first stockout event, between any two consecutive stockout events, and after the final stockout event until the end of the sales horizon.

Sufficient Conditions

Consider now a bounded rational OLNE where the matrix of shadow prices is diagonal, so $\mu_{ii} \geq 0$ and $\mu_{ij} = 0$ for all $i \neq j$. Such a bounded rational equilibrium may arise if firms only care about their own capacity constraint. This bounded rational equilibrium may also arise if firms do not have inventory information for their competitors, and equilibrium outcomes emerge from repeated best responses. Moreover, it can also arise when firms proceed under the assumption that the competitors have sufficiently large capacities as if they would never stock out.

Theorem 11.7 *If $r_i(t, p)$ is concave in p_i and $d_i(t, p)$ is convex in p_j for all i, j, t, then the necessary conditions are also sufficient. Moreover, if $\int_0^T [r_i(t, p(t)) - \mu(t)d_i(t, p(t))]dt$ is pseudo-concave in $p_i(t), 0 \leq t \leq T$ for all $\mu(t) \geq 0$, $0 \leq t \leq T$, then the necessary conditions together with $\mu_{ij}(t) = 0$ for all $i \neq j$ and t are also sufficient for a bounded rational OLNE.*

The first part of Theorem 11.7 applies to the linear demand model, but fails for the MNL model. On the other hand, second set of sufficient conditions apply to the MNL model.

11.3.3 Comparative Statics

If all products are substitutable such that the price competition is (log-)supermodular, then a decrease in the initial capacity level of any firm leads to higher equilibrium prices at any time for all firms in a bounded rational OLNE. Consider now a duopoly selling complementary products. If the price competition is (log-)submodular, then a decrease in the initial capacity level of one firm leads to higher equilibrium prices at any time for the firm itself and lower equilibrium prices at any time for the other firm in a bounded rational OLNE.

Uniqueness

A normalized OLNE has a matrix of constant shadow prices where all the rows are the same, so $\mu_{ij} = \mu_j$ independent of i for all i, j. In essence, all firms use the same set of shadow prices for a firm's capacity constraint in their best-response problems. Suppose that $d_i(t, p)$ is twice continuously differentiable in p for all i and t. If $d_i(t, p)$ is convex in p_j for all i, j, t, and

$$\frac{\partial^2 r_i(t, p)}{\partial p_i^2} + \sum_{j \neq i} \left| \frac{\partial^2 r_i(t, p)}{\partial p_i \partial p_j} \right| < 0$$

for all i, t, then there exist a unique normalized OLNE.

Unfortunately, this result does not apply to the MNL model as the demand function is not convex in p. For this reason, we present alternative conditions for uniqueness that apply to the MNL model. Assume that $d_i(t, p)$ is twice continuously differentiable in p for all i and t. If $\partial d_i(t, p)/\partial p_i < 0$ for all i, t, and the Jacobian and Hessian matrix of the demand function $d(t, p)$ with respect to p are negative semidefinitive for all $p \in \mathcal{P}$, then there exist at most one bounded rational OLNE for any vector of diagonal shadow prices. Moreover, there exists a unique bounded rational OLNE for some vector of diagonal shadow prices. It is possible to verify that the MNL model satisfies these latter set of conditions.

Coupled with the existence results, we know that there exists at least one bounded rational OLNE, and that there exists at least one vector of diagonal shadow prices such that its corresponding bounded rational OLNE is unique.

Applications
In dynamic Bertrand–Edgeworth models, firms may avoid head-to-head competition and take turns acting as monopolists. This can also happen in our model depending on the inter-temporal demand structure; however, it is possible to show that this can never happen under the MNL model. On the other hand, examples exist where firms run out of stock before the end of the sales horizon in equilibrium even if demands are stationary. However, among all such equilibria, the one using the whole sales horizon Pareto dominates all others and it is the unique bounded rational OLNE.

11.3.4 Asymptotic Optimality for the Stochastic Case

We extend the differential game to account for demand uncertainty by considering its stochastic-game counterpart in continuous time. We show that the solutions suggested by the differential game capture the essence and provide a good approximation to the stochastic game. Given a feasible pricing policy u, we denote the revenue for firm i as $G_i(u)$. A policy u^* is a Markovian equilibrium if $G(u_i, u^*_{-i}) \leq G(u^*_i, u^*_{-i})$ for all i. An equilibrium can be found, in theory, by simultaneously solving the corresponding Hamilton–Jacobi–Bellman equations for all the firms. It can be shown that using an affine functional approximation for the value functions of all the firms coincides with the differential game we have studied earlier.

Using k as an index, we consider a sequence of problems with demand rate $d^k(t, p) = kd(t, p)$ and capacity $c^k = kc$. Let $\bar{G}^k_i(u) = G^k(u)/k$ be the revenue form firm i. In a stochastic game, a feasible policy u^* is called an asymptotic Nash equilibrium in the limiting regime of the sequence of scaled stochastic games, if for any $\epsilon > 0$ and all i, there exists an l such that for all $k > l$, $\bar{G}^l_i(u_i, u^*_{-i}) \leq \bar{G}^l(u^*) + \epsilon$ for all feasible policies (u_i, u^*_{-i}).

Theorem 11.8 *Any OLNE heuristic corresponding to an OLNE of the differential game is an asymptotic Nash equilibrium in the limiting regime of the sequence of scaled stochastic games.*

Under the stochastic regime, firms may prefer to use a re-solving feedback strategy that updates prices continuously based on the state of the system as a potentially better heuristic. The following theorem tells us that this feedback policy is also asymptotically optimal.

Theorem 11.9 *The re-solving feedback heuristic is an asymptotic Nash equilibrium in the limiting regime of the sequence of scaled stochastic games.*

The last two results are of practical value to capacity providers that can be assured that heuristics based on the differential game have good asymptotic properties.

11.4 End of Chapter Problems

1. Consider a version of the Markov chain (MC) choice model that we studied in the chapters on choice modeling and assortment optimization, but we make pricing decisions, instead of assortment offer decisions. With probability λ_i, a consumer arriving into the system is primarily interested in product i and checks its price p_i. The consumer makes a purchase with $e^{-\alpha_i p_i}$ and leaves the system generating a revenue equal to p_i. Otherwise, the consumer rejects product i, and transitions to product j with probability ρ_{ij} and checks its price p_j. Assume that $\sum_{j \in N} \rho_{ij} < 1$ for all $i \in N$ so that a customer visiting product i and not purchasing this product transitions to the no-purchase option and leaves the system with probability $1 - \sum_{j \in N} \rho_{ij}$. The customer transitions among the products until she makes a purchase decision or she decides to leave without a purchase.
   (a) Given that we charge the prices $p = \{p_i : i \in N\}$, write a system of equations that we can solve to compute the purchase probability of each product.
   (b) Let g_i be the optimal expected revenue from a customer currently visiting product i. Write a dynamic program that can be used to compute $\{g_i : i \in N\}$.
2. We continue with the MC choice model setup discussed in the previous problem. Assume that the set of products N are partitioned into the sets N^1 and N^{-1}. There are two firms, firm 1 and firm -1. Firm i owns and sets the prices of products in the set N^i. The customers make a choice over the whole set of products N according to the MC choice model. If a customer purchases a product owned by firm i, then firm i generates a revenue. Each firm is interested in maximizing its own expected revenue.
   (a) Assume that firm -1 charges the prices $\{p_i^{-1} : i \in N^{-1}\}$ for the products that it owns. Let g_i^1 be the optimal expected revenue of firm 1 from a customer that is currently visiting product i. Write a dynamic program that can be used to compute $\{g_i^1 : i \in N^1 \cup N^{-1}\}$. Note that g_i^1 is nonzero for $i \in N^{-1}$.

In particular, firm 1 can generate nonzero revenue from a customer visiting a product owned by firm -1 because this customer may decide not to purchase the product and subsequently transition to a product owned by firm 1.

(b) What is the optimal price that firm 1 should charge for product $i \in N^1$ as a function of $\{g_j^1 : j \in N^1 \cup N^{-1}\}$?

3. In the second part of the previous problem, we derived how to compute the best response of firm 1 to the prices $\{p_i^{-1} : i \in N^{-1}\}$ charged by firm -1. Let $\hat{p}^1 = \{\hat{p}_i^1 : i \in N^1\}$ be the best response of firm 1 to the prices $\hat{p}^{-1} = \{\hat{p}_i^{-1} : i \in N^{-1}\}$ charged by firm -1. Let $\tilde{p}^1 = \{\tilde{p}_i^1 : i \in N^1\}$ be the best response of firm 1 to the prices $\tilde{p}^{-1} = \{\tilde{p}_i^{-1} : i \in N^{-1}\}$ charged by firm -1.

   (a) Show that if $\hat{p}_i^{-1} \geq \tilde{p}_i^{-1}$ for all $i \in N^{-1}$, then $\hat{p}_i^1 \geq \tilde{p}_i^1$ for all $i \in N^1$.
   (b) Using the previous part, show that there exists a Nash equilibrium for the two firms.

4. Consider a symmetric duopoly market with firms 1 and 2 each selling one product by setting prices p_1 and p_2, respectively. The two products are substitutable. The demand system that governs the market has a linear form:

$$d_1(p_1, p_2) = a - p_1 + \gamma p_2,$$

$$d_2(p_1, p_2) = a - p_2 + \gamma p_1,$$

   where $a > 0$ and $\gamma \in [0, 1)$. Both firms have the identical marginal cost of z to procure, produce, and distribute their products. Note that though the two firms are symmetric, nothing prevents them from adopting asymmetric decisions.

   (a) As γ increases, how does the total sales $d_1(p_1, p_2) + d_2(p_1, p_2)$ change for a given price vector (p_1, p_2)? Is there any issue with this monotonicity property?
   (b) If both firms simultaneously set *prices* to maximize their profit, what is the price equilibrium?
   (c) If both firms simultaneously make decisions on the *sales quantity* to maximize their profit, what is the market equilibrium outcome in terms of prices?
   (d) Compare the equilibrium outcomes in parts (b) and (c). Provide an intuitive explanation why the comparison you observe hold.
   (e) Consider a two-stage sequential game in which firms maximize their profit. In the first stage, both firms simultaneously decide on the *capacity* of their production and distribution. As a result, how many each can sell will be capped by the capacity level determined by themselves. In the second stage, given the capacity level they build in the first stage, both firms simultaneously set *prices* to maximize their profit. For this two-stage game, what is the price equilibrium in the second stage given the equilibrium capacity level set in the first stage?

(f) Compare the equilibrium outcomes in parts (c) and (e). Provide an intuitive explanation why the comparison you observe hold.
(g) On top of the two-stage sequential game of part (e), suppose in the second stage, firms can produce, distribute, and sell more than the capacity level each sets in the first stage. The downside is the additional quantity produced, distributed, and sold beyond the capacity level incurs an *additional* cost z' per unit beyond z. What is the equilibrium market outcome you expect to see from this modified two-stage sequential game?

11.5 Bibliographical Remarks

The competitive assortment optimization model that we discuss in this chapter is based on Besbes and Saure (2016). The authors extend the results that we discuss in this chapter to the case where there are common products that can be offered by both firms. If a customer chooses such a common product, then she makes the purchase decision from either of the firms with equal probabilities. Also, the authors analyze the setting where the firms choose the assortment of products to offer, as well as the prices of the products in the offered assortment. The competitive pricing model presented in this chapter is due to Gallego and Hu (2014). We refer the reader to that paper for the details of the analysis. Related models also appear in Federgruen and Hu (2015) and Federgruen and Hu (2018).

Hopp and Xu (2008) study a dynamic price and assortment optimization problem under competition. The authors adopt a fluid approximation framework, where the demand takes on its expected value. They establish the existence of a Nash equilibrium and provide conditions for uniqueness. Gallego et al. (2006a) study competitive pricing problems under the MNL model. Chen and Chen (2017) incorporate the network effects into the problem in a duopoly setting. Anderson and de Palma (1992) and Li and Huh (2011) study competitive pricing problems under the nested logit (NL) model. In the first paper, the authors assume that all products have the same price sensitivity, whereas in the second paper, the authors assume that the products in the same nest have the same price sensitivity. Gallego and Wang (2014) extend these results to the case where the products can have arbitrary price sensitivities. Cachon and Kok (2007b) use the NL model to analyze the decisions made by category managers, who focus on the expected revenue obtained from a customer purchasing a product in their own category. The authors characterize the potential revenue loss and provide remedies to attain expected revenues close to those that can be obtained by a central planner. Kok and Xu (2011) study the structural properties of the best-response dynamics when the customers choose according to the NL model. Cooper et al. (2015) develop a model to understand the consequences of ignoring the competition while estimating the customer demand. Feng and Hu (2017) study a competitive product investment model to understand the customer herding behavior.

There is a large and still growing body of literature on competitive models in network revenue management. There is work focusing on dynamic pricing models with competition; see Perakis and Sood (2006), Gallego et al. (2006b), Xu and Hopp (2006), Kachani et al. (2007), Levin et al. (2009), Adida and Perakis (2010a), Martinez-de-Albeniz and Talluri (2011), Caro and Martinez-de-Albeniz (2012) and Kirshner et al. (2018). There is also work on competitive assortment models; see Heese and Martinez-de-Albeniz (2018). Lastly, there is work on studying price or quantity competition in static problems that include either a single or two time periods; see Farahat and Perakis (2009), Nalca et al. (2010, 2013), Farahat and Perakis (2010), Martinez-de-Albeniz and Roels (2011), Afeche et al. (2014), Wang and Hu (2014), Cho and Tang (2014), Bazhanov et al. (2015), Nazerzadeh and Perakis (2016), Aviv et al. (2017, 2018) and Cachon and Feldman (2017).

Appendix

Proof of Lemma 11.1 If $V(\hat{S}_{-i}) = V(\tilde{S}_{-i})$, then since the objective function of problem (11.2) depends on S_{-i} only through $V(S_{-i})$, a best response of firm i to the assortment $\hat{S}_{-i}$ is also a best response of firm i to the assortment $\tilde{S}_{-i}$. In this case, the result follows immediately. Assume without loss of generality that $V(\hat{S}_{-i}) < V(\tilde{S}_{-i})$, and assume for a contradiction that $V(\hat{S}_i) > V(\tilde{S}_i)$. From (11.1), for any $S_i \in \mathcal{F}_i$ and $S_i \neq \varnothing$, $R_i(S_i, \hat{S}_{-i}) > R_i(S_i, \tilde{S}_{-i})$. Consequently,

$$z_i^*(\hat{S}_{-i}) = \max_{S_i \in \mathcal{F}_i} R_i(S_i, \hat{S}_{-i}) > \max_{S_i \in \mathcal{F}_i} R_i(S_i, \tilde{S}_{-i}) = z_i^*(\tilde{S}_{-i}),$$

so that $z_i^*(\hat{S}_{-i}) > z_i^*(\tilde{S}_{-i})$, where we implicitly assume that there exists a nonempty feasible solution to the two maximization problems above; otherwise, $\hat{S}_i = \tilde{S}_i = \varnothing$ and the result trivially holds. On the other hand, by the discussion before the lemma, a best response of firm i to the assortment S_{-i} is given by an optimal solution to problem (11.3). Therefore, $\hat{S}_i$ is an optimal solution to problem (11.3) after replacing $z_i^*(S_{-i})$ with $z_i^*(\hat{S}_{-i})$. In other words, $\hat{S}_i$ is an optimal solution to the problem

$$\max_{S_i \in \mathcal{F}_i} \left\{ \sum_{j \in S_i} (p_j - z_i^*(\hat{S}_{-i}))\, v_j \right\}.$$

Since $\tilde{S}_i$ is a feasible but not necessarily an optimal solution to the problem above, it follows that

$$\sum_{j \in \hat{S}_i} (p_j - z_i^*(\hat{S}_{-i}))\, v_j \geq \sum_{j \in \tilde{S}_i} (p_j - z_i^*(\hat{S}_{-i}))\, v_j.$$

Interchanging the roles of $\hat{S}_{-i}$ and $\tilde{S}_{-i}$ and following the same argument, we see that

$$\sum_{j\in\tilde{S}_i}(p_j - z_i^*(\tilde{S}_{-i}))\, v_j \geq \sum_{j\in\hat{S}_i}(p_j - z_i^*(\tilde{S}_{-i}))\, v_j.$$

Adding the two inequalities yields, it follows that $z_i^*(\hat{S}_{-i})\,[V(\tilde{S}_i) - V(\hat{S}_i)] \geq z_i^*(\tilde{S}_{-i})\,[V(\tilde{S}_i) - V(\hat{S}_i)]$. The last inequality contradicts the fact that $z_i^*(\hat{S}_{-i}) > z_i^*(\tilde{S}_{-i})$ and $V(\hat{S}_i) > V(\tilde{S}_i)$. □

Proof of Theorem 11.2 We use induction over the iterations to show that $V(\hat{S}_1^t) \geq V(\hat{S}_1^{t-1})$ and $V(\hat{S}_{-1}^t) \geq V(\hat{S}_{-1}^{t-1})$ for all $t = 1, 2, \ldots$. The result trivially holds for $t = 1$, since we have $\hat{S}_1^0 = \hat{S}_{-1}^0 = \varnothing$ so that $V(\hat{S}_1^0) = V(\hat{S}_{-1}^0) = 0$. Assuming that $V(\hat{S}_1^t) \geq V(\hat{S}_1^{t-1})$ and $V(\hat{S}_{-1}^t) \geq V(\hat{S}_{-1}^{t-1})$, we proceed to showing that $V(\hat{S}_1^{t+1}) \geq V(\hat{S}_1^t)$ and $V(\hat{S}_{-1}^{t+1}) \geq V(\hat{S}_{-1}^t)$. By definition $\hat{S}_1^t$ is a best response of firm 1 to the assortment $\hat{S}_{-1}^{t-1}$, whereas $\hat{S}_1^{t+1}$ is a best response of firm 1 to the assortment $\hat{S}_{-1}^t$. Since $V(\hat{S}_{-1}^{t-1}) \leq V(\hat{S}_{-1}^t)$ by the induction hypothesis, by Lemma 11.1, it follows that $V(\hat{S}_1^t) \leq V(\hat{S}_1^{t+1})$. Similarly, by definition, $\hat{S}_{-1}^t$ is a best response of firm -1 to the assortment $\hat{S}_1^t$, whereas $\hat{S}_{-1}^{t+1}$ is a best response of firm -1 to the assortment $\hat{S}_1^{t+1}$. We just established that $V(\hat{S}_1^t) \leq V(\hat{S}_1^{t+1})$, so by Lemma 11.1, we obtain $V(\hat{S}_{-1}^t) \leq V(\hat{S}_{-1}^{t+1})$. This discussion completes the induction argument, so that we have $V(\hat{S}_1^t) \geq V(\hat{S}_1^{t-1})$ and $V(\hat{S}_{-1}^t) \geq V(\hat{S}_{-1}^{t-1})$ for all $t = 1, 2, \ldots$. Since the number of possible assortments is finite and the sequences $\{V(\hat{S}_1^t) : t = 0, 1, \ldots\}$ and $\{V(\hat{S}_{-1}^t) : t = 0, 1, \ldots\}$ are increasing, these sequences converge. Therefore, there exists $t_0 \geq 0$ such that $V(\hat{S}_1^{t_0}) = V(\hat{S}_1^{t_0+1}) = V(\hat{S}_1^{t_0+2}) = \ldots$ and $V(\hat{S}_{-1}^{t_0}) = V(\hat{S}_{-1}^{t_0+1}) = V(\hat{S}_{-1}^{t_0+2}) = \ldots$.

We claim that $(\hat{S}_1^{t_0+1}, \hat{S}_{-1}^{t_0})$ is a Nash equilibrium. By definition of the tatonnement process, $\hat{S}_1^{t_0+1}$ is a best response of firm 1 to the assortment $\hat{S}_{-1}^{t_0}$ offered by firm -1. It only remains to argue that $\hat{S}_{-1}^{t_0}$ is a best response of firm -1 to the assortment $\hat{S}_1^{t_0+1}$ offered by firm 1. By the definition of the tatonnement process, note that $\hat{S}_{-1}^{t_0}$ is a best response of firm -1 to the assortment $\hat{S}_1^{t_0}$. The best response of firm -1 to the assortment $\hat{S}_1^{t_0}$ is computed by solving the problem $\max_{S_{-1}\in\mathcal{F}_{-1}} R_{-1}(S_{-1}, \hat{S}_1^{t_0})$. By (11.1), $R_{-1}(S_{-1}, \hat{S}_1^{t_0})$ depends on $\hat{S}_1^{t_0}$ only through $V(\hat{S}_1^{t_0})$. Since $V(\hat{S}_1^{t_0}) = V(\hat{S}_1^{t_0+1})$, it follows that $\hat{S}_{-1}^{t_0}$ is also a best response of firm -1 to the assortment $\hat{S}_1^{t_0+1}$, establishing the claim. □

Proof of Theorem 11.3 Let (S_1^*, S_{-1}^*) be any Nash equilibrium. Assume that the sequence of assortments $\{(\hat{S}_1^t, \hat{S}_{-1}^t) : t = 0, 1, \ldots\}$ is generated by the tatonnement process. We use induction over the iterations of the tatonnement process to show that $V(\hat{S}_1^t) \leq V(S_1^*)$ and $V(\hat{S}_{-1}^t) \leq V(S_{-1}^*)$ for all $t = 0, 1, \ldots$. The result trivially holds for $t = 0$, since we have $\hat{S}_1^0 = \hat{S}_{-1}^0 = \varnothing$. Assuming that $V(\hat{S}_1^t) \leq V(S_1^*)$ and $V(\hat{S}_{-1}^t) \leq V(S_{-1}^*)$, we proceed to showing that $V(\hat{S}_1^{t+1}) \leq V(S_1^*)$ and $V(\hat{S}_{-1}^{t+1}) \leq V(S_{-1}^*)$. By definition of the tatonnement process, $\hat{S}_1^{t+1}$ is a best response of firm 1 to the assortment $\hat{S}_{-1}^t$. Also, by the definition of a Nash equilibrium, S_1^* is a best response of firm 1 to the assortment S_{-1}^*. In this case, since we have $V(\hat{S}_{-1}^t) \leq V(S_{-1}^*)$, by Lemma 11.1, we obtain $V(\hat{S}_1^{t+1}) \leq V(S_1^*)$. Similarly, by the definition of the tatonnement process, $\hat{S}_{-1}^{t+1}$ is a best response of firm -1 to the assortment $\hat{S}_1^{t+1}$. By the definition of a Nash equilibrium, S_{-1}^* is a best response of firm -1 to the assortment S_1^*. Since we have just shown that $V(\hat{S}_1^{t+1}) \leq V(S_1^*)$, by Lemma 11.1, we obtain $V(\hat{S}_{-1}^{t+1}) \leq V(S_{-1}^*)$, completing the induction argument.

By the preceding discussion, the sequence $\{(\hat{S}_1^t, \hat{S}_{-1}^t) : t = 0, 1, \ldots\}$ of assortments generated by the tatonnement process satisfies $V(\hat{S}_1^t) \leq V(S_1^*)$ and $V(\hat{S}_{-1}^t) \leq V(S_{-1}^*)$ for all $t = 0, 1, \ldots$. In this case, letting $(\hat{S}_1^{t_0+1}, \hat{S}_{-1}^{t_0})$ be a Nash equilibrium generated by the tatonnement process, we have $V(\hat{S}_1^{t_0+1}) \leq V(S_1^*)$ and $V(\hat{S}_{-1}^{t_0}) \leq V(S_{-1}^*)$. By (11.1), $R_i(S_i, S_{-i})$ is decreasing in $V(S_{-i})$. Therefore, since $V(\hat{S}_{-1}^{t_0}) \leq V(S_{-1}^*)$, we get $R_1(S_1, \hat{S}_{-1}^{t_0}) \geq R_1(S_1, S_{-1}^*)$ for all $S_1 \in \mathcal{F}_1$. Also, since $(\hat{S}_1^{t_0+1}, \hat{S}_{-1}^{t_0})$ and (S_1^*, S_{-1}^*) are Nash equilibria, we have $\hat{S}_1^{t_0+1} \in \arg\max_{S_1 \in \mathcal{F}_1} R_1(S_1, \hat{S}_{-1}^{t_0})$ and $S_1^* \in \arg\max_{S_1 \in \mathcal{F}_1} R_1(S_1, S_{-1}^*)$, because the assortments offered by firm 1 must be a best response to the assortments offered by firm -1 in any Nash equilibrium. In this case, we obtain

$$R_1(\hat{S}_1^{t_0+1}, \hat{S}_{-1}^{t_0}) = \max_{S_1 \in \mathcal{F}_1} R_1(S_1, \hat{S}_{-1}^{t_0}) \geq \max_{S_1 \in \mathcal{F}_1} R_1(S_1, S_{-1}^*) = R_1(S_1^*, S_{-1}^*),$$

where the inequality uses the fact that $R_1(S_1, \hat{S}_{-1}^{t_0}) \geq R_1(S_1, S_{-1}^*)$ for all $S_1 \in \mathcal{F}_1$. The chain of inequalities above shows that the expected revenue of firm 1 in the equilibrium $(\hat{S}_1^{t_0+1}, \hat{S}_{-1}^{t_0})$ is at least as large as its expected revenue in the equilibrium (S_1^*, S_{-1}^*). We can use a similar argument to show that the same statement holds for firm -1 as well. □

Proof of Theorem 11.4 By its definition, S_i^{NC} is a best response of firm i to the empty assortment. Also, by the definition of a Nash equilibrium, S_i^* is a best response of firm i to the assortment S_{-i}^*. Since $V(\varnothing) = 0 \leq V(S_{-i}^*)$, by Lemma 11.1, we obtain $V(S_i^{\text{NC}}) \leq V_i(S_i^*)$, establishing the first inequality in the theorem. To show the second inequality in the theorem, let z^* be the optimal objective value of problem (11.4) and $z_i^*(S_{-i}^*)$ be the optimal objective value of problem (11.2) after replacing S_{-i} with S_{-i}^*. Because (S_1^*, S_{-1}^*) is a feasible but not necessarily an optimal solution to problem (11.4), we have

$$z^* = R_1(S_1^{\text{CP}}, S_{-1}^{\text{CP}}) + R_{-1}(S_{-1}^{\text{CP}}, S_1^{\text{CP}})$$
$$\geq R_1(S_1^*, S_{-1}^*) + R_{-1}(S_{-1}^*, S_1^*) = z_1^*(S_{-1}^*) + z_{-1}^*(S_1^*),$$

where the last equality follows from the fact that S_i^* is a best response of firm i to the assortment S_{-i}^*. In problem (11.2), each firm can trivially obtain a strictly positive expected revenue by offering any product. Therefore, $z_1^*(S_{-1}^*) > 0$ and $z_{-1}^*(S_1^*) > 0$, in which case, the chain of inequalities above implies that $z^* > z_1^*(S_{-1}^*)$ and $z^* > z_{-1}^*(S_1^*)$.

Note that in problem (11.4), if we fix the assortment S_{-1} at its optimal value S_{-1}^{CP} and optimize only over the assortment S_1, then setting $S_1 = S_1^{\text{CP}}$ would still yield an optimal solution. Therefore, S_1^{CP} is an optimal solution to the problem

$$\max_{S_1 \in \mathcal{F}_1} \left\{ R_1(S_1, S_{-1}^{\text{CP}}) + R_{-1}(S_{-1}^{\text{CP}}, S_1) \right\} = \max_{S_1 \in \mathcal{F}_1} \frac{\sum_{j \in S_1} p_j\, v_j + \sum_{j \in S_{-1}^{\text{CP}}} p_j\, v_j}{v_0 + V(S_1) + V(S_{-1}^{\text{CP}})}$$

yielding the optimal objective value z^*. In this case, we have

$$z^* \geq \frac{\sum_{j \in S_1} p_j\, v_j + \sum_{j \in S_{-1}^{\text{CP}}} p_j\, v_j}{v_0 + V(S_1) + V(S_{-1}^{\text{CP}})} \qquad \forall\, S_1 \in \mathcal{F}_1,$$

and the inequality above holds as equality at the optimal solution S_1^{CP}. Following the same sequence of steps that we used to obtain problem (11.3), it follows that

$$[v_0 + V(S_{-1}^{\text{CP}})]\, z^* \geq \sum_{j \in S_1} (p_j - z^*)\, v_j + \sum_{j \in S_{-1}^{\text{CP}}} p_j\, v_j \qquad \forall\, S_1 \in \mathcal{F}_1,$$

and the inequality holds as equality at the optimal solution S_1^{CP}. Therefore, S_1^{CP} is an optimal solution to the problem

$$\max_{S_1 \in \mathcal{F}_1} \left\{ \sum_{j \in S_1} (p_j - z^*)\, v_j \right\},$$

but since S_1^* is a feasible but not necessarily an optimal solution to the problem above, we obtain

$$\sum_{j \in S_1^{\text{CP}}} (p_j - z^*)\, v_j \geq \sum_{j \in S_1^*} (p_j - z^*)\, v_j.$$

Also, since S_1^* is a best response of firm 1 to the assortment S_{-1}^*, S_1^* is an optimal solution to problem (11.3) with $i = 1$ and $S_{-i} = S_{-i}^*$. However, since S_1^{CP} is a feasible but not necessarily an optimal solution to this problem, we get

$$\sum_{j \in S_1^*} (p_j - z_1^*(S_{-1}^*))\, v_j \geq \sum_{j \in S_1^{\text{CP}}} (p_j - z_1^*(S_{-1}^*))\, v_j.$$

Adding the last two inequalities, we obtain

$$z^* \, [V(S_1^*) - V(S_1^{\text{CP}})] \geq z_1^*(S_{-1}^*) \, [V(S_1^*) - V(S_1^{\text{CP}})].$$

If $V(S_1^*) < V(S_1^{\text{CP}})$, then the last inequality implies that $z^* \leq z_1^*(S_{-1}^*)$, which contradicts the fact that $z^* > z_1^*(S_{-1}^*)$. Therefore, we must have $V(S_1^*) \geq V(S_1^{\text{CP}})$. A symmetric argument also shows that $V(S_{-1}^*) \geq V(S_{-1}^{\text{CP}})$, establishing the second inequality in the theorem. □

References

D. Acemoglu, M. Dahleh, I. Lobel, A. Ozdaglar, Bayesian learning in social networks. Econ. Stud. **78**(4), 1201–1236 (2011)

D. Adelman, Dynamic bid-prices in revenue management. Oper. Res. **55**(4), 647–661 (2007)

E. Adida, O. Ozer, Why markdown as a pricing modality? Manag. Sci. **65**(5), 2161–2178 (2019)

E. Adida, G. Perakis, A robust optimization approach to dynamic pricing and inventory control with no backorders. Math. Program. **107**(1–2), 97–129 (2006)

E. Adida, G. Perakis, A nonlinear continuous time optimal control model of dynamic pricing and inventory control with no backorders. Nav. Res. Logist. **54**(7), 767–795 (2007)

E. Adida, G. Perakis, Dynamic pricing and inventory control with no backorders: uncertainty and competition. Oper. Res. **58**(2), 289–302 (2010a)

E. Adida, G. Perakis, Dynamic pricing and inventory control: robust vs. stochastic uncertainty models – a computational study. Ann. Oper. Res. **181**(1), 125–157 (2010b)

P. Afeche, B. Ata, Bayesian dynamic pricing in queueing systems with unknown delay cost characteristics. Manuf. Serv. Oper. Manag. **15**(2), 292–304 (2013)

P. Afeche, M. Hu, Y. Li, The downside of reorder flexibility under price competition. Technical report, University of Toronto, Toronto, ON (2014)

S. Agrawal, Z. Wang, Y. Ye, A dynamic near-optimal algorithm for online linear programming. Oper. Res. **62**(4), 876–890 (2014)

S. Agrawal, V. Avadhanula, V. Goyal, A. Zeevi, MNL-Bandit: a dynamic learning approach to assortment selection. Technical report, Columbia University, New York, NY (2018)

S.D. Ahipasaoglu, X. Li, K. Natarajan, A convex optimization approach for computing correlated choice probabilities with many alternatives. IEEE Trans. Autom. Control **64**(1), 190–205 (2019)

M. Akan, B. Ata, Bid-price controls for network revenue management: martingale characterization of optimal bid prices. Math. Oper. Res. **64**(4), 912–936 (2009)

M. Akan, B. Ata, R.C. Savaskan-Ebert, Dynamic pricing of remanufacturable products under demand substitution: a product life cycle model. Ann. Oper. Res. **211**(1), 1–25 (2013)

M. Akan, B. Ata, J.D. Dana, Revenue management by sequential screening. J. Econ. Theory **159** (B), 728–774 (2015)

A. Alptekinoglu, A. Grasas, When to carry eccentric products? Optimal retail assortment under consumer returns. Prod. Oper. Manag. **23**(5), 877–892 (2014)

A. Alptekinoglu, J.H. Semple, The exponomial choice model: a new alternative for assortment and price optimization. Oper. Res. **64**(1), 79–93 (2016)

G. Gallego, H. Topaloglu, *Revenue Management and Pricing Analytics*,
International Series in Operations Research & Management Science 279,
https://doi.org/10.1007/978-1-4939-9606-3

K. Amaruchkul, W.L. Cooper, D. Gupta, Single-leg air-cargo revenue management. Transp. Sci. **41**(4), 457–469 (2007)

K. Amaruchkul, W.L. Cooper, D. Gupta, A note on air-cargo capacity contracts. Prod. Oper. Manag. **20**(1), 152–162 (2011)

T. Amornpetchkul, H.-S. Ahn, O. Sahin, Conditional promotions and consumer overspending. Prod. Oper. Manag. **27**(8), 1455–1475 (2018)

S.P. Anderson, A. de Palma, Multiproduct firms: a nested logit approach. J. Ind. Econ. **40**(3), 261–276 (1992)

S.P. Anderson, A. de Palma, J.G. Thisse, *Discrete Choice Theory of Product Differentiation* (MIT Press, Cambridge, 1992)

A. Aouad, D. Segev, Display optimization for vertically differentiated locations under multinomial logit preferences. Technical report, London School of Business, London, UK (2018)

A. Aouad, V.F. Farias, R. Levi, Assortment optimization under consider-then-choose choice models. Technical report, MIT, Cambridge, MA (2016)

A. Aouad, V.F. Farias, R. Levi, D. Segev, The approximability of assortment planning under ranking preferences. Oper. Res. **66**(6), 1661–1669 (2018a)

A. Aouad, J.B. Feldman, D. Segev, The exponomial choice model: assortment optimization and application to public transit choice prediction in San Francisco. Technical report, Washington University, St. Louis, MO (2018b)

A. Aouad, R. Levi, D. Segev, Greedy-like algorithms for dynamic assortment planning under multinomial logit preferences. Oper. Res. **66**(5), 1321–1345 (2018c)

A. Aouad, R. Levi, D. Segev, Approximation algorithms for dynamic assortment optimization models. Math. Oper. Res. **44**(2), 487–511 (2019)

V.F. Araman, R. Caldentey, Dynamic pricing for nonperishable products with demand learning. Oper. Res. **57**(5), 1169–1188 (2009)

V.F. Araman, R. Caldentey, Revenue management with incomplete demand information, in *Wiley Encyclopedia of Operations Research and Management Science*, ed. by J.J. Cochran, L.A. Cox Jr., P. Keskinocak, J.P. Kharoufeh, J.C. Smith (Wiley, New York, 2011)

B. Ata, M. Akan, On bid-price controls for network revenue management. Stoch. Syst. **5**(2), 268–323 (2015)

B. Ata, J.D. Dana, Price discrimination on booking time. Int. J. Ind. Organ. **43**(November), 175–181 (2015)

Y. Aviv, A. Pazgal, A partially observed Markov decision process for dynamic pricing. Manag. Sci. **51**(9), 1400–1416 (2005)

Y. Aviv, Y. Levin, M. Nediak, Counteracting strategic consumer behavior in dynamic pricing systems, in *Consumer-Driven Demand and Operations Management Models*, ed. by S. Netessine, C.S. Tang (Springer, New York, 2009)

Y. Aviv, A. Bazhanov, Y. Levin, M. Nediak, Quantity competition when most-favored customers are strategic. Prod. Oper. Manag. **26**(11), 2107–2121 (2017)

Y. Aviv, A. Bazhanov, Y. Levin, M. Nediak, Price-matching competition in the presence of strategic consumers and retail price maintenance. Technical report, Queen's University, Kingston, Canada (2018)

G. Aydin, S. Ziya, Pricing promotional products under upselling. Manuf. Serv. Oper. Manag. **10** (3), 360–376 (2008)

G. Aydin, S. Ziya, Technical note – personalized dynamic pricing of limited inventories. Oper. Res. **57**(6), 1523–1531 (2009)

N. Aydin, S.I. Birbil, J.B.G. Frenk, N. Noyan, Single-leg airline revenue management with overbooking. Transp. Sci. **47**(4), 560–583 (2013)

L. Baardman, M.C. Cohen, K. Panchamgam, G. Perakis, D. Segev, Scheduling promotion vehicles to boost profits. Manag. Sci. **65**(1), 55–70 (2019)

M.O. Ball, M. Queyranne, Toward robust revenue management: competitive analysis of online booking. Oper. Res. **57**(4), 950–963 (2009)

S. Balseiro, C. Gocmen, R. Phillips, G. Gallego, Revenue management of consumer options for tournaments. Technical report, Columbia University, New York, NY (2010)

G.-Y. Ban, N.B. Keskin, Personalized dynamic pricing with machine learning. Technical report, Duke University, Durham, NC (2017)

M. Bansal, C. Maglaras, Dynamic pricing when customers strategically time their purchase: asymptotic optimality of a two-price policy. J. Revenue Pricing Manag. **8**(1), 42–66 (2009)

C. Barz, D. Adelman, C. Uckun, Power approximations for network revenue management, in *26th European Conference on Operational Research* (2013)

A. Bassamboo, S. Kumar, R.S. Randhawa, Dynamics of new product introduction in closed rental systems. Oper. Res. **57**(6), 1347–1359 (2009)

A.V. Bathia, S.C. Prakesh, Optimal allocation of seats by fare, in *AGIFORS Reservations and Yield Management Study Group Annual Meeting Proceedings* (1973)

A. Bazhanov, Y. Levin, M. Nediak, Quantity competition in the presence of strategic consumers and retail price maintenance. Technical report, Queen's University, Kingston, Canada (2015)

A. Belkaid, V. Martinez-de-Albeniz, Here comes the sun: fashion goods retailing under weather shocks. Technical report, University of Navarra, Barcelona, Spain (2017)

P.P. Belobaba, Airline yield management: an overview of seat inventory control. Transp. Sci. **21** (2), 63–73 (1987)

P.P. Belobaba, Application of a probabilistic decision model to airline seat inventory control. Oper. Res. **37**(2), 183–197 (1989)

P.P. Belobaba, A. Farkas, Yield management impacts on airline spill estimation. Transp. Sci. **33** (2), 217–232 (1999)

P.P. Belobaba, L.R. Weatherford, Comparing decision rules that incorporate customer diversion in perishable asset revenue management situations. Decis. Sci. **27**(2), 343–363 (1996)

M. Ben-Akiva, S.R. Lerman, *Discrete Choice Analysis Theory and Application to Travel Demand: Theory and Application to Travel Demand* (MIT Press, Cambridge, 1985)

G. Berbeglia, Discrete choice models based on random walks. Oper. Res. Lett. **44**(2), 234–237 (2016)

G. Berbeglia, The generalized stochastic preference choice model. Technical report, Melbourne Business School, Melbourne, Australia (2018)

G. Berbeglia, G. Joret, Assortment optimisation under a general discrete choice model: a tight analysis of revenue-ordered assortments, in *Proceedings of the 2017 ACM Conference on Economics and Computation*, Cambridge, MA (2017), pp. 345–346

G. Berbeglia, A. Garrassino, G. Vulcano, A comparative empirical study of discrete choice models in retail operations. Technical report, Universidad Torcuato di Tella, Buenos Aires, Argentina (2018)

F. Bernstein, V. Martinez-de-Albeniz, Dynamic product rotation in the presence of strategic customers. Manag. Sci. **63**(7), 2092–2107 (2017)

F. Bernstein, A.G. Kok, L. Xie, Dynamic assortment customization with limited inventories. Manuf. Serv. Oper. Manag. **17**(4), 538–553 (2015)

D. Bertsimas, S. de Boer, Dynamic pricing and inventory control for multiple products. J. Revenue Pricing Manag. **3**(4), 303–319 (2004)

D. Bertsimas, S. de Boer, Simulation-based booking limits for airline revenue management. Oper. Res. **53**(1), 90–106 (2005)

D. Bertsimas, A. Mersereau, A learning approach for interactive marketing to a customer segment. Oper. Res. **55**(6), 1120–1135 (2007)

D. Bertsimas, V. Misic, Robust product line design. Oper. Res. **65**(1), 19–37 (2017)

D. Bertsimas, V. Misic, Exact first-choice product line optimization. Oper. Res. **67**(3), 651–670 (2019)

D. Bertsimas, G. Perakis, Dynamic pricing: a learning approach, in *Mathematical and Computational Models for Congestion Charging*, ed. by S. Lawphongpanich, D.W. Hearn, M.J. Smith (Springer US, Boston, 2006)

D. Bertsimas, I. Popescu, Revenue management in a dynamic network environment. Transp. Sci. **37**(3), 257–277 (2003)

D. Bertsimas, R. Shioda, Restaurant revenue management. Oper. Res. **51**(3), 472–486 (2003)

D. Bertsimas, J. Hawkins, G. Perakis, Optimal bidding in online auctions. J. Revenue Pricing Manag. **8**(1), 21–41 (2009)

O. Besbes, I. Lobel, Intertemporal price discrimination: structure and computation of optimal policies. Manag. Sci. **61**(1), 92–110 (2015)

O. Besbes, C. Maglaras, Revenue optimization for a make-to-order queue in an uncertain market environment. Oper. Res. **57**(6), 1438–1450 (2009)

O. Besbes, C. Maglaras, Dynamic pricing with financial milestones: feedback-form policies. Manag. Sci. **58**(9), 1715–1731 (2012)

O. Besbes, D. Saure, Dynamic pricing in the presence of demand shifts. Manuf. Serv. Oper. Manag. **16**(4), 513–528 (2014)

O. Besbes, D. Saure, Product assortment and price competition under multinomial logit demand. Prod. Oper. Manag. **25**(1), 114–127 (2016)

O. Besbes, A. Zeevi, Dynamic pricing without knowing the demand function: risk bounds and near optimal algorithms. Oper. Res. **57**(6), 1407–1420 (2009)

O. Besbes, A. Zeevi, On the minimax complexity of pricing in a changing environment. Oper. Res. **59**(1), 66–79 (2011)

O. Besbes, A. Zeevi, Blind network revenue management. Oper. Res. **60**(6), 1537–1550 (2012)

O. Besbes, A. Zeevi, On the (surprising) sufficiency of linear models for dynamic pricing with demand learning. Manag. Sci. **61**(4), 723–739 (2015)

O. Besbes, R. Phillips, A. Zeevi, Testing the validity of a demand model: an operations perspective. Manuf. Serv. Oper. Manag. **12**(1), 162–183 (2010)

O. Besbes, Y. Gur, A. Zeevi, Non-stationary stochastic optimization. Oper. Res. **63**(5), 1227–1244 (2015)

O. Besbes, D. Iancu, N. Trichakis, Dynamic pricing under debt: spiraling distortions and efficiency losses. Manag. Sci. **64**(10), 4572–4589 (2018)

G. Bitran, R. Caldentey, An overview of pricing models and revenue management. Manuf. Serv. Oper. Manag. **5**(3), 203–229 (2003)

G. Bitran, R. Caldentey, S. Mondschein, Coordinating clearance markdown sales of seasonal products in retail chains. Oper. Res. **46**(5), 609–624 (1998)

J. Blanchet, G. Gallego, V. Goyal, A Markov chain approximation to choice modeling. Oper. Res. **64**(4), 886–905 (2016)

H.D. Block, Random orderings and stochastic theories of responses, in *Economic Information, Decision, and Prediction* (Springer Netherlands, Dordrecht, 1974), pp. 172–217

C. Borgs, O. Candogan, J. Chayes, I. Lobel, H. Nazerzadeh, Optimal multiperiod pricing with service guarantees. Manag. Sci. **60**(7), 1792–1811 (2014)

T. Boyaci, Y. Akcay, Pricing when customers have limited attention. Manag. Sci. **64**(7), 2995–3014 (2018)

T. Boyaci, G. Gallego, Coordinating pricing and inventory replenishment policies for one wholesaler and one or more geographically dispersed retailers. Int. J. Prod. Econ. **77**(2), 95–111 (2002)

T. Boyaci, O. Ozer, Information acquisition for capacity planning via pricing and advance selling: when to stop and act? Oper. Res. **58**(5), 1328–1349 (2010)

P. Bremaud, *Point Processes and Queues: Martingale Dynamics* (Springer Verlag, New York, 1980)

J. Broder, P. Rusmevichientong, Dynamic pricing under a general parametric choice model. Oper. Res. **60**(4), 965–980 (2012)

J.J.M. Bront, I. Mendez Diaz, G. Vulcano, A column generation algorithm for choice-based network revenue management. Oper. Res. **57**(3), 769–784 (2009)

S.L. Brumelle, J.I. McGill, Airline seat allocation with multiple nested fare classes. Oper. Res. **41** (1), 127–137 (1993)

S. Brumelle, D. Walczak, Dynamic airline revenue management with multiple semi-Markov demand. Oper. Res. **51**(1), 137–148 (2003)

S.L. Brumelle, J.I. McGill, T.H. Oum, M.W. Tretheway, K. Sawaki, Allocation of airline seats between stochastically dependent demands. Transp. Sci. **24**(3), 183–192 (1990)

G. Cachon, P. Feldman, Pricing services subject to congestion: charge per-use fees or sell subscriptions. Manuf. Serv. Oper. Manag. **13**(2), 244–260 (2011)

G. Cachon, P. Feldman, Price commitments with strategic consumers: why it can be optimal to discount more frequently... than optimal. Manuf. Serv. Oper. Manag. **17**(3), 399–410 (2015)

G.P. Cachon, P. Feldman, Is advance selling desirable with competition? Mark. Sci. **36**(2), 214–231 (2017)

G. Cachon, A.G. Kok, Implementation of the newsvendor model with clearance pricing: how to (and how not to) estimate a salvage value. Manuf. Serv. Oper. Manag. **9**(3), 276–290 (2007a)

G.P. Cachon, A.G. Kok, Category management and coordination in retail assortment planning in the presence of basket shopping consumers. Manag. Sci. **53**(6), 934–951 (2007b)

G.P. Cachon, R. Swinney, Purchasing, pricing, and quick response in the presence of strategic consumers. Manag. Sci. **55**(3), 497–511 (2009)

G.P. Cachon, R. Swinney, The value of fast fashion: quick response, enhanced design, and strategic consumer behavior. Manag. Sci. **57**(4), 778–795 (2011)

G. Cachon, C. Terwiesch, Y. Xu, Retail assortment planning in the presence of consumer search. Manuf. Serv. Oper. Manag. **7**(4), 330–346 (2005)

G. Cachon, K.M. Daniels, R. Lobel, The role of surge pricing on a service platform with self-scheduling capacity. Manuf. Serv. Oper. Manag. **19**(3), 368–384 (2017)

R. Caldentey, G. Vulcano, Online auction and list price revenue management. Manag. Sci. **53**(5), 795–813 (2007)

R. Caldentey, L.M. Wein, Revenue management of a make-to-stock queue. Oper. Res. **54**(5), 859–875 (2006)

R. Caldentey, Y. Liu, I. Lobel, Intertemporal pricing under minimax regret. Oper. Res. **65**(1), 104–129 (2017)

A. Calmon, D.F. Ciocan, G. Romero, Revenue management with repeated customer interactions. Technical report, University of Toronto, Toronto, ON (2019)

A. Caplin, B. Nalebuff, Aggregation and imperfect competition: on the existence of equilibrium. Econometrica **59**(1), 25–59 (1991)

A. Caplin, M. Dean, Revealed preference, rational inattention, and costly information acquisition. Working paper, National Bureau of Economic Research, Cambridge, MA (2014)

F. Caro, J. Gallien, Dynamic assortment with demand learning for seasonal consumer goods. Manag. Sci. **53**(2), 276–292 (2007)

F. Caro, J. Gallien, Inventory management of a fast-fashion retail network. Oper. Res. **58**(2), 257–273 (2010)

F. Caro, J. Gallien, Clearance pricing optimization for a fast-fashion retailer. Oper. Res. **60**(6), 1404–1422 (2012)

F. Caro, V. Martinez-de-Albeniz, The effect of assortment rotation on consumer choice, and its impact on competition, in *Operations Management Models with Consumer-Driven Demand*, ed. by S. Netessine, C. Tang (Springer, New York, 2009)

F. Caro, V. Martinez-de-Albeniz, Product and price competition with satiation effects. Manag. Sci. **58**(7), 1357–1373 (2012)

F. Caro, V. Martinez-de-Albeniz, How fast fashion works: can it work for you, too? IESE Insight Rev. **21**, 58–65 (2014)

F. Caro, V. Martinez-de-Albeniz, Fast fashion: business model overview and research opportunities, in *Retail Supply Chain Management: Quantitative Models and Empirical Studies*, ed. by N. Agrawal, S.A. Smith (Springer, New York, 2015)

F. Caro, J. Gallien, M. Diaz, J. Garcia, J.M. Corredoira, M. Montes, J.A. Ramos, J. Correa, Zara uses operations research to reengineer its global distribution process. Interfaces **40**(1), 71–84 (2010)

F. Caro, V. Martinez-de-Albeniz, P. Rusmevichientong, The assortment packing problem: multiperiod assortment planning for short-lived products. Manag. Sci. **60**(11), 2701–2721 (2014)

J. Carrasco, J. de Ortuzar, Review and assessment of the nested logit model. Transp. Rev. **22**(2), 197–218 (2002)

L.E. Celis, G. Lewis, M. Mobius, H. Nazerzadeh, Buy-it-now or take-a-chance: price discrimination through randomized auctions. Manag. Sci. **60**(12), 2927–2948 (2014)

O. Ceryan, O. Sahin, I. Duenyas, Dynamic pricing of substitutable products in the presence of capacity flexibility. Manuf. Serv. Oper. Manag. **15**(1), 86–101 (2013)

O. Ceryan, I. Duenyas, O. Sahin, Dynamic pricing and replenishment with consumer upgrades. Prod. Oper. Manag. **27**(4), 663–679 (2018)

L.M.A. Chan, D. Simchi-Levi, J. Swann, Pricing, production, and inventory policies for manufacturing with stochastic demand and discretionary sales. Manuf. Serv. Oper. Manag. **8**(2), 149–168 (2006)

S. Chandler, S. Ja, Revenue opportunity modeling at American Airlines, in *AGIFORS Reservations and Yield Management Study Group Annual Meeting Proceedings*, Jeju (2007)

J. Chaneton, G. Vulcano, Computing bid-prices for revenue management under customer choice behavior. Manuf. Serv. Oper. Manag. **13**(4), 452–470 (2011)

R.E. Chatwin, Multiperiod airline overbooking with a single fare class. Oper. Res. **46**(6), 805–819 (1998)

R.E. Chatwin, Continuous-time airline overbooking with time-dependent fares and refunds. Transp. Sci. **33**(2), 182–191 (1999)

N. Chen, Y.-J. Chen, Duopoly competition with network effects in discrete choice models. Technical report, Hong Kong University of Science and Technology, Hong Kong (2017)

Y. Chen, V.F. Farias, Simple policies for dynamic pricing with imperfect forecasts. Oper. Res. **61** (3), 612–624 (2013)

Y. Chen, V.F. Farias, On the efficacy of static prices for revenue management in the face of strategic customers. Technical report, MIT, Cambridge, MA (2017)

Y. Chen, V.F. Farias, Robust dynamic pricing with strategic customers. Math. Oper. Res. **43**(4), 1119–1142 (2018)

N. Chen, G. Gallego, Nonparametric learning and optimization with covariates. Technical report, Hong Kong University of Science and Technology, Hong Kong (2018a)

N. Chen, G. Gallego, A primal-dual learning algorithm for personalized dynamic pricing with an inventory constraint. Technical report, Hong Kong University of Science and Technology, Hong Kong (2018b)

N. Chen, G. Gallego, Welfare analysis of dynamic pricing. Manag. Sci. **65**(1), 139–151 (2019)

X. Chen, P. Hu, Joint pricing and inventory management with deterministic demand and costly price adjustment. Oper. Res. Lett. **40**(5), 385–389 (2012)

Y. Chen, S. Jasin, Power of dynamic pricing in revenue management with strategic (forward-looking) customers. Technical report, University of Michigan, Ann Arbour, MI (2018)

N. Chen, J. Nasiry, Does loss aversion preclude price variation? Manuf. Serv. Oper.Manag. (2019, forthcoming)

Y. Chen, C. Shi, Joint pricing and inventory management with strategic customers. Oper. Res. (2019, forthcoming)

X. Chen, D. Simchi-Levi, Coordinating inventory control and pricing strategies with random demand and fixed ordering cost: the infinite horizon case. Math. Oper. Res. **29**(3), 698–723 (2004a)

X. Chen, D. Simchi-Levi, Coordinating inventory control and pricing strategies with random demand and fixed ordering cost: the finite horizon case. Oper. Res. **52**(6), 887–896 (2004b)

X. Chen, D. Simchi-Levi, Coordinating inventory control and pricing strategies with random demand and fixed ordering cost: the continuous review model. Oper. Res. Lett. **34**(3), 323–332 (2006)

X. Chen, D. Simchi-Levi, Pricing and inventory management, in *The Oxford Handbook of Pricing Management*, ed. by R. Phillips, O. Ozer (Oxford University Press, Oxford, 2012), pp. 784–822

X. Chen, Y. Wang, A note on tight lower bound for MNL-bandit assortment selection models. Oper. Res. Lett. **46**(5), 534–537 (2018)

X. Chen, S. Zhou, F. Chen, Integration of inventory and pricing decisions with costly price adjustments. Oper. Res. **59**(5), 1144–1158 (2011)

X. Chen, Z. Pang, L. Pan, Coordinating inventory control and pricing strategies for perishable products. Oper. Res. **62**(2), 284–300 (2014)

B. Chen, X. Chao, C. Shi, Nonparametric algorithms for joint pricing and inventory control with lost-sales and censored demand. Technical report, University of Michigan, Ann Arbour, MI (2016a)

Q. Chen, S. Jasin, I. Duenyas, Real-time dynamic pricing with minimal and flexible price adjustments. Manag. Sci. **62**(8), 2437–2455 (2016b)

X. Chen, P. Hu, S. Shum, Y. Zhang, Dynamic stochastic inventory management with reference price effects. Oper. Res. **64**(6), 1529–1536 (2016c)

X. Chen, W. Ma, D. Simchi-Levi, L. Xin, Assortment planning for recommendations at checkout under inventory constraints. Technical report, MIT, Cambridge, MA (2016d)

H. Chen, M. Hu, G. Perakis, Distribution-free pricing. Technical report, University of Toronto, Toronto, ON (2017a)

X. Chen, P. Hu, Z. Hu, Efficient algorithms for dynamic pricing problem with reference price effect. Manag. Sci. **63**(12), 4389–4408 (2017b)

Y. Chen, R. Levi, C. Shi, Revenue management of reusable resources with advanced reservations. Prod. Oper. Manag. **26**(5), 836–859 (2017c)

H. Chen, M. Hu, J. Wu, Intertemporal price discrimination via randomized pricing. Technical report, University of Toronto, Toronto, ON (2018a)

X. Chen, Y. Wang, Y. Zhou, Dynamic assortment optimization with changing contextual information. Technical report, New York University, New York, NY (2018b)

X. Chen, Y. Wang, Y. Zhou, Dynamic assortment selection under the nested logit models. Technical report, New York University, New York, NY (2018c)

Q.G. Chen, S. Jasin, I. Duenyas, Nonparametric self-adjusting control for joint learning and optimization of multiproduct pricing with finite resource capacity. Math. Oper. Res. **44**(2), 601–631 (2019a)

B. Chen, X. Chao, H.-S. Ahn, Coordinating pricing and inventory replenishment with nonparametric demand learning. Oper. Res. (2019b, forthcoming)

W.C. Cheung, D. Simchi-Levi, Technical note – dynamic pricing and demand learning with limited price experimentation. Oper. Res. **65**(6), 1722–1731 (2017)

S.-H. Cho, C.S. Tang, Technical note – capacity allocation under retail competition: uniform and competitive allocations. Oper. Res. **62**(1), 72–80 (2014)

J.-K. Chong, T.-H. Ho, C.S. Tang, A modeling framework for category assortment planning. Manuf. Serv. Oper. Manag. **3**(3), 191–210 (2001)

H. Chung, H.-S. Ahn, S. Jasin, (Rescaled) multi-attempt approximation of choice model and its application to assortment optimization. Prod. Oper. Manag. **8**(2), 341–353 (2019)

E. Cinar, V. Martinez-de-Albeniz, A closed-loop approach to dynamic assortment planning. Technical report, University of Navarra, Barcelona (2014)

D.F. Ciocan, V. Farias, Model predictive control for dynamic resource allocation. Math. Oper. Res. **37**(3), 501–525 (2012a)

D.F. Ciocan, V.F. Farias, Dynamic allocation problems with volatile demand. Math. Oper. Res. **37** (3), 501–525 (2012b)

D.F. Ciocan, V.F. Farias, Fast demand learning for display advertising revenue management. Technical report, MIT, Boston, MA (2014)

M. Cohen, S. Gupta, J. Kalas, G. Perakis, An efficient algorithm for dynamic pricing using a graphical representation. Technical report, MIT, Cambridge, MA (2016)

M. Cohen, J. Kalas, G. Perakis, Optimizing promotions for multiple items in supermarkets. Technical report, MIT, Cambridge, MA (2017a)

M. Cohen, R.S. Pindyck, G. Perakis, Pricing with limited knowledge of demand. Technical report, MIT, Cambridge, MA (2017b)

M.C. Cohen, N.-H.Z. Leung, K. Panchamgam, G. Perakis, A. Smith, The impact of linear optimization on promotion planning. Oper. Res. **65**(2), 446–468 (2017c)

M. Cohen, I. Lobel, R.P. Leme, Feature-based dynamic pricing. Technical report, New York University, New York, NY (2018a)

M. Cohen, R. Lobel, G. Perakis, Dynamic pricing through data sampling. Prod. Oper. Manag. **27** (6), 1074–1088 (2018b)

P.B. Collado, V. Martinez-de-Albeniz, Estimating and optimizing the impact of inventory on consumer choices in a fashion retail setting. Technical report, University of Navarra, Barcelona (2017)

W. Cooper, Asymptotic behavior of an allocation policy for revenue management. Oper. Res. **50** (4), 720–727 (2002)

W.L. Cooper, D. Gupta, Stochastic comparisons in airline revenue management. Manuf. Serv. Oper. Manag. **8**(3), 221–234 (2006)

W.L. Cooper, T. Homem de Mello, Some decomposition methods for revenue management. Transp. Sci. **41**(3), 332–353 (2007)

W.L. Cooper, L. Li, On the use of buy up as a model of customer choice in revenue management. Prod. Oper. Manag. **21**(5), 833–850 (2012)

W. Cooper, T. Homem de Mello, A. Kleywegt, Models of the spiral-down effect in revenue management. Oper. Res. **54**(5), 968–987 (2006)

W.L. Cooper, T. Homem de Mello, A.J. Kleywegt, Learning and pricing with models that do not explicitly incorporate competition. Oper. Res. **63**(1), 86–103 (2015)

P. Courty, J. Nasiry, Loss aversion and the uniform pricing puzzle for media and entertainment products. Econ. Theory **66**(1), 105–140 (2018)

D. Crapis, B. Ifrach, C. Maglaras, M. Scarsini, Monopoly pricing in the presence of social learning. Manag. Sci. **63**(11), 3531–3997 (2017)

Y. Cui, I. Duenyas, O. Sahin, Unbundling of ancillary service: how does price discrimination of main service matter? Manuf. Serv. Oper. Manag. **20**(3), 455–466 (2018)

R.E. Curry, Optimal airline seat allocation with fare nested by origins and destinations. Transp. Sci. **24**(3), 193–204 (1990)

C. Daganzo, *Multinomial Probit: The Theory and Its Application to Demand Forecasting* (Academic, London, 1979)

J. Dai, W. Ding, A. Kleywegt, X. Wang, Y. Zhang, Choice based revenue management for parallel flights. Technical report, Georgia Institute of Technology, Atlanta, GA (2015)

J. Dai, A. Kleywegt, Y. Xiao, Network revenue management with cancellations and no-shows. Prod. Oper. Manag. **8**(2), 292–318 (2019)

J. Davis, G. Gallego, H. Topaloglu, Assortment planning under the multinomial logit model with totally unimodular constraint structures. Technical report, Cornell University, School of Operations Research and Information Engineering (2013)

J.M. Davis, G. Gallego, H. Topaloglu, Assortment optimization under variants of the nested logit model. Oper. Res. **62**(2), 250–273 (2014)

S.V. de Boer, R. Freling, N. Piersma, Mathematical programming for network revenue management revisited. Eur. J. Oper. Res. **137**(1), 72–92 (2002)

A.V. den Boer, Dynamic pricing and learning: historical origins, current research, and new directions. Surv. Oper. Res. Manag. Sci. **20**(1), 1–18 (2015)

A. den Boer, N.B. Keskin, Discontinuous demand functions: estimation and pricing. Technical report, Duke University, Durham, NC (2017a)

A. den Boer and N.B. Keskin, Dynamic pricing with demand learning and reference effects. Technical report, Duke University, Durham, NC (2017b)

A. Desir, V. Goyal, An FPTAS for capacity constrained assortment optimization. Technical report, Columbia University, New York, NY (2013)

A. Desir, V. Goyal, D. Segev, C. Ye, Capacity constrained assortment optimization under the Markov chain based choice model. Technical report, Columbia University, New York, NY (2015)

A. Desir, V. Goyal, S. Jagabathula, D. Segev, A Mallows-smoothed distribution over rankings approach for modeling choice. Technical report, Columbia University, New York, NY (2018)

S. Diwan. Performance of dynamic programming methods in airline revenue management. Master's thesis, Massachusetts Institute of Technology, Cambridge, MA (2010)

T.A. Domencich, D. McFadden, *Urban Travel Demand : A Behavioral Analysis* (North Holland Publishing, Amsterdam, 1975)

C. Du, W.L. Cooper, Z. Wang, Optimal pricing for a multinomial logit choice model with network effects. Oper. Res. **64**(2), 2016 (2016)

C. Du, W.L. Cooper, Z. Wang, Optimal worst-case pricing for a logit demand model with network effects. Oper. Res. Lett. **46**(3), 2018 (2018)

D. Dzyabura, S. Jagabathula, Offline assortment optimization in the presence of an online channel. Manag. Sci. **64**(6), 2767–2786 (2018)

A.N. Elmachtoub, M.L. Hamilton, The power of opaque products in pricing. Technical report, Columbia University, New York, NY (2017)

A.N. Elmachtoub, Y. Wei, Retailing with opaque products. Technical report, Columbia University, New York, NY (2015)

A.N. Elmachtoub, V. Gupta, M.L. Hamilton, The value of personlized pricing. Technical report, Columbia University, New York, NY (2018)

W. Elmahraby, P. Keskinocak, Dynamic pricing in the presence of inventory considerations: research overview, current practices, and future directions. Manag. Sci. **49**(10), 1287–1309 (2003)

A. Erdelyi, H. Topaloglu, Separable approximations for joint capacity control and overbooking decisions in network revenue management. J. Revenue Pricing Manag. **8**(1), 3–20 (2009)

A. Erdelyi, H. Topaloglu, A dynamic programming decomposition method for making overbooking decisions over an airline network. INFORMS J. Comput. **22**(3), 443–456 (2010)

S.S. Eren, C. Maglaras, Monopoly pricing with limited demand information. J. Revenue Pricing Manag. **9**(1–2), 23–48 (2010)

A. Farahat, G. Perakis, Profit loss in differentiated oligopolies. Oper. Res. Lett. **37**(1), 43–46 (2009)

A. Farahat, G. Perakis, A nonnegative extension of the affine demand function and equilibrium analysis for multiproduct price competition. Oper. Res. Lett. **38**(4), 280–286 (2010)

V.F. Farias, B. Van Roy, An approximate dynamic programming approach to network revenue management. Technical report, MIT, Cambridge, MA (2007)

V. Farias, B. Van Roy, Dynamic pricing with a prior on market response. Oper. Res. **58**(1), 16–29 (2010)

V.F. Farias, S. Jagabathula, D. Shah, A non-parametric approach to modeling choice with limited data. Manag. Sci. **59**(2), 305–322 (2013)

V.F. Farias, S. Jagabathula, D. Shah, Building optimized and hyperlocal product assortments: a nonparametric choice approach. Technical report, MIT, Cambridge, MA (2016)

A. Federgruen, A. Heching, Combined pricing and inventory control under uncertainty. Oper. Res. **47**(3), 454–475 (1999)

A. Federgruen, M. Hu, Multi-product price and assortment competition. Oper. Res. **63**(3), 572–584 (2015)

A. Federgruen, M. Hu, Global and robust stability in a general price and assortment competition model. Technical report, University of Toronto, Toronto, ON (2018)

J. Feldman, Space constrained assortment optimization under the paired combinatorial model. Technical report, Washington University, St. Louis, MO (2018)

J. Feldman, A. Paul, Relating the approximability of the fixed cost and space constrained assortment problems. Prod. Oper. Manag. **8**(5), 1238–1255 (2019)

J. Feldman, H. Topaloglu, Bounding optimal expected revenues for assortment optimization under mixtures of multinomial logits. Prod. Oper. Manag. **24**(10), 1598–1620 (2015a)

J.B. Feldman, H. Topaloglu, Technical note – capacity constraints across nests in assortment optimization under the nested logit model. Oper. Res. **63**(4), 812–822 (2015b)

J.B. Feldman, H. Topaloglu, Revenue management under the Markov chain choice model. Oper. Res. **65**(5), 1322–1342 (2017)

J. Feldman, H. Topaloglu, Technical note – capacitated assortment optimization under the multinomial logit model with nested consideration sets. Oper. Res. **66**(2), 380–391 (2018)

J. Feldman, D. Zhang, X. Liu, N. Zhang, Taking assortment optimization from theory to practice: evidence from large field experiments on alibaba. Technical report, Washington University, St. Louis, MO (2018)

J. Feldman, A. Paul, H. Topaloglu, Technical note – assortment optimization with small consideration sets. Oper. Res. (2019, forthcoming)

Y. Feng, G. Gallego, Optimal starting times of end-of-season sales and optimal stopping times for promotional fares. Manag. Sci. **41**(8), 1371–1391 (1995)

Y. Feng, G. Gallego, Perishable asset revenue management with Markovian time-dependent demand intensities. Manag. Sci. **46**(7), 941–956 (2000)

Y. Feng, M. Hu, Blockbuster or niche? Competitive strategy under network effects. Technical report, University of Toronto, Toronto, ON (2017)

Y. Feng, B. Xiao, Optimal policies of yield management with multiple predetermined prices. Oper. Res. **48**(2), 332–343 (2000)

B. Feng, Y. Li, Z.-J.M. Shen, Air cargo operations: literature review and comparison with practices. Transp. Res. C Emerg. Technol. **56**(July), 263–280 (2015)

G. Feng, X. Li, Z. Wang, Technical note – on the relationship between several discrete choice models. Oper. Res. **65**(6), 1516–1525 (2017)

G. Feng, X. Li, Z. Wang, On the substitutabiliy and complementarity in discrete choice models. Oper. Res. Lett. **46**(1), 141–146 (2018)

K. Ferreira, J. Goh, Assortment rotation and the value of concealment. Technical report, Harvard Business School, Cambridge, MA (2018)

K.J. Ferreira, B.H.A. Lee, D. Simchi-Levi, Analytics for an online retailer: demand forecasting and price optimization. Manuf. Serv. Oper. Manag. **18**(1), 69–88 (2016)

K.J. Ferreira, D. Simchi-Levi, H. Wang, Online network revenue management using Thompson sampling. Oper. Res. **66**(6), 1586–1602 (2018)

T. Fiig, K. Isler, C. Hopperstad, P. Belobaba, Optimization of mixed fare structures: theory and applications. J. Revenue Pricing Manag. **9**(1–2), 152–170 (2010)

G. Gallego, A min-max distribution-free procedure for the (Q, r) inventory model. Oper. Res. Lett. **11**(1), 55–60 (1992)

G. Gallego, M. Hu, Dynamic pricing of perishable assets under competition. Manag. Sci. **60**(5), 1241–1259 (2014)

G. Gallego, A. Li, Attention, consideration then selection choice model. Technical report, Hong Kong University of Science and Technology, Hong Kong (2016)

G. Gallego, I. Moon, The distribution free newsboy problem: review and extensions. J. Oper. Res. Soc. **44**(8), 824–834 (1993)

G. Gallego, O. Sahin, Revenue management with partially refundable fares. Oper. Res. **58**(4), 817–833 (2010)

G. Gallego, C. Stefanescu, Upgrades, upsells and pricing in revenue management. Technical report, Columbia University, New York, NY (2009)

G. Gallego, C. Stefanescu, Services engineering: design and pricing of service features, in *The Oxford Handbook of Pricing Management*, ed. by R. Phillips, Ö. Özer (Oxford University Press, Oxford, 2012), pp. 711–733

G. Gallego, H. Topaloglu, Constrained assortment optimization for the nested logit model. Manag. Sci. **60**(10), 2583–2601 (2014)

G. Gallego, G. van Ryzin, Optimal dynamic pricing of inventories with stochastic demand over finite horizons. Manag. Sci. **40**(8), 999–1020 (1994)

G. Gallego, G.J. van Ryzin, A multiple product dynamic pricing problem and its applications to network yield management. Oper. Res. **45**(1), 24–41 (1997)

G. Gallego, R. Wang, Multiproduct price optimization and competition under the nested logit model with product-differentiated price sensitivities. Oper. Res. **62**(2), 450–461 (2014)

G. Gallego, R. Wang, Reservation value choice models. Technical report, Hong Kong University of Science and Technology (2019)

G. Gallego, G. Iyengar, R. Phillips, A. Dubey, Managing flexible products on a network. Computational Optimization Research Center Technical Report TR-2004-01, Columbia University (2004)
G. Gallego, W.T. Huh, W. Kang, R. Phillips, Price competition with the attraction demand model: existence of unique equilibrium and its stability. Manuf. Serv. Oper. Manag. **8**(4), 359–375 (2006a)
G. Gallego, S. Krishnamoorthy, R. Phillips, Dynamic revenue management games with forward and spot markets. J. Revenue Pricing Manag. **5**(1), 10–31 (2006b)
G. Gallego, S. Kou, R. Phillips, Revenue management of callable products. Manag. Sci. **54**(3), 550–564 (2008a)
G. Gallego, R. Phillips, O. Sahin, Strategic management of distressed inventories. Prod. Oper. Manag. **17**(4), 402–415 (2008b)
G. Gallego, L. Lin, R. Ratliff, Choice based EMSR methods for single-resource revenue management with demand dependencies. J. Revenue Pricing Manag. **8**(2–3), 207–240 (2009a)
G. Gallego, L. Lin, R. Ratliff, Demand arrival order and revenue management controls, in *AGIFORS Cargo and RM Study Group* (2009b)
G. Gallego, R. Ratliff, S. Shebalov, A general attraction model and sales-based linear program for network revenue management under consumer choice. Oper. Res. **63**(1), 212–232 (2015)
G. Gallego, A. Li, V.A. Truong, X. Wang, Approximation algorithms for product framing and pricing. Technical report, Columbia University, New York, NY (2016a)
G. Gallego, A. Li, V.A. Truong, X. Wang, Online bipartite matching with customer choice. Technical report, Columbia University, New York, NY (2016b)
G. Gallego, A. Li, V.A. Truong, X. Wang, Online personalized resource allocation with customer choice. Technical report, Columbia University, New York, NY (2016c)
G. Gallego, M.Z.F. Li, Y. Liu, Dynamic nonlinear pricing of inventories over finite sales horizons. Technical report, Hong Kong University of Science and Technology, Hong Kong (2018)
J. Gallien, Dynamic mechanism design for online commerce. Manag. Sci. **54**(2), 291–310 (2006)
V. Gaur, D. Honhon, Assortment planning and inventory decisions under a locational choice model. Manag. Sci. **52**(10), 1528–1543 (2006)
Y. Ge, C. Pan, Study on overbooking management with a choice model of consumer behavior, in *2010 International Conference on Management and Service Science* (2010), pp. 1–4
I.I. Gihman, A.V. Skorohod, *Controlled Stochastic Processes* (Springer Verlag, New York, 1979)
N. Golrezaei, H. Nazerzadeh, Auctions with dynamic costly information acquisition. Oper. Res. **65**(1), 130–144 (2017)
N. Golrezaei, H. Nazerzadeh, P. Rusmevichientong, Real-time optimization of personalized assortments. Manag. Sci. **60**(6), 1532–1551 (2014)
N. Golrezaei, H. Nazerzadeh, R. Randhawa, Dynamic pricing for heterogeneous time-sensitive customers. Technical report, University of Southern California, Los Angeles, CA (2017)
V. Goyal, R. Levi, D. Segev, Near-optimal algorithms for the assortment planning problem under dynamic substitution and stochastic demand. Oper. Res. **64**(1), 219–235 (2016)
D. Gupta, W.L. Cooper, Stochastic comparisons in product yield management. Oper. Res. **53**(2), 377–384 (2005)
J.M. Harrison, B.N. Keskin, A. Zeevi, Bayesian dynamic pricing policies: learning and earning under a binary prior distribution. Manag. Sci. **58**(3), 570–586 (2012)
A. Heching, G. Gallego, G.J. van Ryzin, Markdown pricing: an empirical analysis of policies and revenue potential at one apparel retailer. J. Pricing Revenue Manag. **1**(2), 139–160 (2002)
H.S. Heese, V. Martinez-de-Albeniz, Effects of assortment breadth announcements on manufacturer competition. Manuf. Serv. Oper. Manag. **20**(2), 302–316 (2018)
T.-H. Ho, C.S. Tang, D.R. Bell, Rational shopping behavior and the option value of variable pricing. Manag. Sci. **44**(12–2), S145–S160 (1998)
D. Honhon, S. Seshadri, Fixed vs. random proportions demand models for the assortment planning problem under stockout-based substitution. Manuf. Serv. Oper. Manag. **15**(3), 378–386 (2013)
D. Honhon, V. Gaur, S. Seshadri, Assortment planning and inventory decisions under stockout-based substitution. Oper. Res. **58**(5), 1364–1379 (2010)

D. Honhon, S. Jonnalagedda, X.A. Pan, Optimal algorithms for assortment selection under ranking-based consumer choice models. Manuf. Serv. Oper. Manag. **14**(2), 279–289 (2012)
W.J. Hopp, X. Xu, A static approximation for dynamic demand substitution with applications in a competitive market. Oper. Res. **56**(3), 630–645 (2008)
X. Hu, W.J. Hopp, Price trends in a dynamic pricing model with heterogeneous customers: a martingale perspective. Oper. Res. **57**(5), 1298–1302 (2009)
Z. Hu, J. Nasiry, Are markets with loss-averse consumers more sensitive to losses? Manag. Sci. **64**(3), 1384–1395 (2018)
M. Hu, M. Pavlin, M. Shi, When gray markets have silver linings: all-unit discounts, gray markets, and channel management. Manuf. Serv. Oper. Manag. **15**(2), 250–262 (2013a)
X. Hu, R. Caldentey, G. Vulcano, Revenue sharing in airline alliances. Manag. Sci. **59**(5), 1177–1195 (2013b)
M. Hu, J. Milner, J. Wu, Liking and following and the newsvendor: operations and marketing policies under social influence. Manag. Sci. **62**(3), 867–879 (2016a)
Z. Hu, X. Chen, P. Hu, Technical note – dynamic pricing with gain-seeking reference price effects. Oper. Res. **64**(1), 150–157 (2016b)
F. Huettner, T. Boyaci, Y. Akcay, Consumer choice under limited attention when options have different information costs. Technical report, ESMT, Berlin (2018)
W.T. Huh, H. Li, Technical note – pricing under the nested attraction model with a multi-stage choice structure. Oper. Res. **63**(4), 840–850 (2015)
B. Ifrach, C. Maglaras, M. Scarsini, Bayesian social learning from consumer reviews. Technical report, Columbia University, New York, NY (2018)
S. Jagabathula, Assortment optimization under general choice. Technical report, New York University, New York, NY (2016)
S. Jagabathula, P. Rusmevichientong, A nonparametric joint assortment and price choice model. Manag. Sci. **63**(9), 3128–3145 (2017)
S. Jagabathula, P. Rusmevichientong, The limit of rationality in choice modeling: formulation, computation, and implications. Manag. Sci. **65**(5), 2196–2215 (2019)
S. Jagabathula, G. Vulcano, A partial-order-based model to estimation individual preferences using panel data. Manag. Sci. **64**(4), 1609–1628 (2018)
S. Jagabathula, L. Subramanian, A. Venkataraman, A conditional gradient approach for nonparametric estimation of mixtures of choice models. Technical report, New York University, New York, NY (2018)
S. Jasin, Re-optimization and self-adjusting price control for network revenue management. Oper. Res. **62**(5), 1168–1178 (2014)
S. Jasin, Performance of an LP-based control for revenue management with unknown demand parameters. Oper. Res. **63**(4), 909–915 (2015)
S. Jasin, S. Kumar, A re-solving heristic with bounded revenue loss for network revenue management with customer choice. Math. Oper. Res. **37**(2), 313–345 (2012)
S. Jasin, S. Kumar, Analysis of deterministic LP-based heuristics for network revenue management. Oper. Res. **61**(6), 1312–1320 (2013)
A. Javanmard, H. Nazerzadeh, Dynamic pricing in high dimensions. Technical report, University of Southern California, Los Angeles, CA (2018)
C.R. Johnson, Positive definite matrices. Am. Math. Mon. **77**(3), 259–264 (1970)
S. Kachani, G. Perakis, C. Simon, Modeling the transient nature of dynamic pricing with demand learning in a competitive environment, in *Network Science, Nonlinear Science and Infrastructure Science*, ed. by T. Friesz (Springer, Boston, 2007)
S.M. Kakade, I. Lobel, H. Nazerzadeh, Optimal dynamic mechanism design and the virtual-pivot mechanism. Oper. Res. **61**(4), 837–854 (2013)
Y. Kanoria, H. Nazerzadeh, Dynamic reserve prices for repeated auctions: learning from bids, in *Web and Internet Economics*, ed. by T.-Y. Lie, Q. Qi (Springer International Publishing, Beijing, 2014), pp. 232–232
I. Karaesmen, G. van Ryzin, Overbooking with substitutable inventory classes. Oper. Res. **52**(1), 83–104 (2004a)

I. Karaesmen, G. van Ryzin, Coordinating overbooking and capacity control decisions on a network. Technical report, Columbia Business School (2004b)

P.W. Keller, R. Levi, G. Perakis, Efficient formulations for pricing under attraction demand models. Math. Program. **14**(1–2), 223–261 (2014)

N.B. Keskin, J. Birge, Dynamic selling mechanisms for product differentiation and learning. Oper. Res. (2019, forthcoming)

N.B. Keskin, A. Zeevi, Dynamic pricing with an unknown demand model: asymptotically optimal semi-myopic policies. Oper. Res. **62**(5), 1142–1167 (2014)

N.B. Keskin, A. Zeevi, Chasing demand: learning and earning in a changing environment. Math. Oper. Res. **42**(2), 277–307 (2017)

W.K. Kincaid, D.A. Darling, An inventory pricing problem. J. Math. Anal. Appl. **7**(2), 183–208 (1963)

S.N. Kirshner, M. Nediak, Scalable dynamic bid prices for network revenue management in continuous time. Prod. Oper. Manag. **24**(10), 1621–1635 (2015)

S. Kirshner, Y. Levin, M. Nediak, Quantity competition in C2C exchange markets with strategic consumers and dynamic preferences. Technical report, Queen's University, Kingston (2018)

R. Kleinberg, T. Leighton, The value of knowing a demand curve: bounds on regret for online posted-price auctions, in *44th Annual IEEE Symposium on Foundations of Computer Science*, Cambridge, MA (2003), pp. 594–605

A. Kleywegt, An optimal control problem of dynamic pricing. Technical report, Georgia Institute of Technology, Atlanta, GA (2001)

A.J. Kleywegt, J.D. Papastavrou, The dynamic and stochastic knapsack problem. Oper. Res. **46** (1), 17–35 (1998)

A.G. Kok, M. Fisher, Demand estimation and assortment optimization under substitution: methodology and application. Oper. Res. **55**(6), 1001–1021 (2007)

A.G. Kok, Y. Xu, Optimal and competitive assortments with endogenous pricing under hierarchical consumer choice models. Manag. Sci. **57**(9), 1546–1563 (2011)

A.G. Kok, M. Fisher, R. Vaidyanathan, Assortment planning: review of literature and industry practice, in *Retail Supply Chain Management*, ed. by N. Agrawal, S.A. Smith (Springer, New York, 2008)

F.S. Koppelman, C.-H. Wen, The paired combinatorial logit model: properties, estimation and application. Transp. Res. B **34**(2), 75–89 (2000)

V. Kostami, D. Kostamis, S. Ziya, Pricing and capacity allocation for shared services. Manuf. Serv. Oper. Manag. **19**(2), 230–245 (2017)

S. Kunnumkal, Randomization approaches for network revenue management with customer choice behavior. Prod. Oper. Manag. **23**(9), 1617–1633 (2014)

S. Kunnumkal, On upper bounds for assortment optimization under the mixture of multinomial logit models. Oper. Res. Lett. **43**(2), 189–194 (2015)

S. Kunnumkal, V. Martinez-de-Albeniz, Tractable approximations for assortment planning with product costs. Oper. Res. **67**(2), 436–452 (2019)

S. Kunnumkal, K. Talluri, A strong Lagrangian relaxation for general discrete-choice network revenue management. Technical report, Indian School of Business, Hyderabad (2015a)

S. Kunnumkal, K. Talluri, Choice network revenue management based on new compact formulations. Technical report, Indian School of Business, Hyderabad (2015b)

S. Kunnumkal, K. Talluri, Technical note – a note on relaxations of the choice network revenue management dynamic program. Oper. Res. **64**(1), 158–166 (2016a)

S. Kunnumkal, K. Talluri, On a piecewise-linear approximation for network revenue management. Math. Oper. Res. **41**(1), 72–91 (2016b)

S. Kunnumkal, H. Topaloglu, A refined deterministic linear program for the network revenue management problem with customer choice behavior. Nav. Res. Logist. **55**(6), 563–580 (2008)

S. Kunnumkal, H. Topaloglu, A stochastic approximation method for the revenue management problem on a single flight leg with discrete demand distributions. Math. Meth. Oper. Res. **70** (3), 477–504 (2009)

S. Kunnumkal, H. Topaloglu, A new dynamic programming decomposition method for the network revenue management problem with customer choice behavior. Prod. Oper. Manag. **19**(5), 575–590 (2010a)

S. Kunnumkal, H. Topaloglu, A stochastic approximation algorithm for making pricing decisions in network revenue management problems. J. Revenue Pricing Manag. **9**(5), 419–442 (2010b)

S. Kunnumkal, H. Topaloglu, Computing time-dependent bid prices in network revenue management problems. Transp. Sci. **44**(1), 38–62 (2010c)

S. Kunnumkal, H. Topaloglu, A randomized linear program for the network revenue management problem with customer choice behavior. J. Revenue Pricing Manag. **10**(5), 455–470 (2011a)

S. Kunnumkal, H. Topaloglu, A stochastic approximation algorithm to compute bid prices for joint capacity allocation and overbooking over an airline network. Nav. Res. Logist. **58**(4), 323–343 (2011b)

S. Kunnumkal, K. Talluri, H. Topaloglu, A randomized linear programming method for network revenue management with product-specific no-shows. Transp. Sci. **46**(1), 90–108 (2012)

C.W. Kuo, H.S. Ahn, G. Aydin, Dynamic pricing of limited inventories when customers negotiate. Oper. Res. **59**(4), 882–897 (2011)

H.D. Kwon, S.A. Lippman, C.S. Tang, Optimal markdown pricing strategy with demand learning. Probab. Eng. Inf. Sci. **26**(1), 77–104 (2012)

T.L. Lai, H. Robbins, Maximally dependent random variables. Proc. Natl. Acad. Sci. **73**(2), 286–288 (1976)

M.A. Larriviere, E.L. Porteus, Selling to the newsvendor: an analysis of price-only contracts. Manuf. Serv. Oper. Manag. **3**(4), 293–305 (2001)

C.J. Lautenbacher, S. Stidham, The underlying Markov decision process in the single-leg airline yield management problem. Transp. Sci. **33**(2), 136–146 (1999)

C.T. Lee, M. Hersh, A model for dynamic airline seat inventory control with multiple seat bookings. Transp. Sci. **27**(3), 252–265 (1993)

Y. Lei, S. Jasin, Real-time dynamic pricing for revenue management with reusable resources, advance reservation, and deterministic service time requirements. Technical report, University of Michigan, Ann Arbour, MI (2018)

Y. Lei, S. Jasin, A. Sinha, Joint dynamic pricing and order fulfillment for e-commerce retailers. Manuf. Serv. Oper. Manag. **20**(2), 269–284 (2018)

R. Levi, A. Radovanovic, Provably near-optimal LP-based policies for revenue management of reusable resources. Oper. Res. **58**(2), 503–507 (2010)

R. Levi, G. Perakis, J. Uichanco, The data-driven newsvendor problem: new bounds and insights. Oper. Res. **63**(6), 1294–1306 (2015)

Y. Levin, J. McGill, M. Nediak, Price guarantees in dynamic pricing and revenue management. Oper. Res. **55**(1), 75–97 (2007)

Y. Levin, J. McGill, M. Nediak, Risk in revenue management and dynamic pricing. Oper. Res. **56** (2), 326–343 (2008)

Y. Levin, J. McGill, M. Nediak, Dynamic pricing in the presence of strategic consumers and oligopolistic competition. Manag. Sci. **55**(1), 32–46 (2009)

Y. Levin, J. McGill, M. Nediak, Optimal dynamic pricing of perishable items by a monopolist facing strategic consumers. Prod. Oper. Manag. **19**(1), 40–60 (2010)

Y. Levin, M. Nediak, H. Topaloglu, Cargo capacity management with allotments and spot market demand. Oper. Res. **60**(2), 351–365 (2012)

Y. Levin, M. Nediak, A. Bazhanov, Quantity premiums and discounts in dynamic pricing. Oper. Res. **62**(4), 846–863 (2014)

T. Levina, Y. Levin, J. McGill, M. Nediak, Dynamic pricing with online learning and strategic consumers: an application of the aggregating algorithm. Oper. Res. **57**(2), 327–341 (2009a)

T. Levina, Y. Levin, J. McGill, M. Nediak, Dynamic pricing with online learning and strategic consumers: an application of the aggregating algorithm. Oper. Res. **57**(2), 327–341 (2009b)

T. Levina, Y. Levin, J. McGill, M. Nediak, Network cargo capacity management. Oper. Res. **59** (4), 1008–1023 (2011)

T. Levina, Y. Levin, J. McGill, M. Nediak, Strategic consumer cooperation in a name-your-own-price channel. Prod. Oper. Manag. **24**(12), 1883–1900 (2015)

H. Li, W.T. Huh, Pricing multiple products with the multinomial logit and nested logit models: concavity and implications. Manuf. Serv. Oper. Manag. **13**(4), 546–563 (2011)

H. Li, W.T. Huh, Optimal pricing for short life-cycle products when customer price sensitivity varies over time. Nav. Res. Logist. **59**(7), 552–576 (2012)

G. Li, P. Rusmevichientong, H. Topaloglu, The d-level nested logit model: assortment and price optimization problems. Oper. Res. **62**(2), 325–342 (2015)

W.S. Lim, C.S. Tang, An auction model arising from an internet search service provider. Eur. J. Oper. Res. **172**(3), 956–970 (2006)

W.S. Lim, C.S. Tang, Optimal dynamic pricing strategies in the presence of speculators and forward looking consumers: are speculators friends or foes? Prod. Oper. Manag. **22**(3), 571–587 (2013)

K. Littlewood, Forecasting and control of passenger bookings, in *AGIFORS Reservations and Yield Management Study Group Annual Meeting Proceedings*, Nathanya (1972)

Y. Liu, W.L. Cooper, Optimal dynamic pricing with patient customers. Oper. Res. **63**(6), 1307–1319 (2015)

Q. Liu, G. van Ryzin, On the choice-based linear programming model for network revenue management. Manuf. Serv. Oper. Manag. **10**(2), 288–310 (2008a)

Q. Liu, G.J. van Ryzin, Strategic capacity rationing to induce early purchases. Manag. Sci. **54**(6), 1115–1131 (2008b)

Q. Liu, G.J. van Ryzin, Strategic capacity rationing when customers learn. Manuf. Serv. Oper. Manag. **13**(1), 89–107 (2011)

I. Lobel, Dynamic pricing with heterogeneous patience levels. Technical report, New York University, New York, NY (2017)

Y. Lu, D. Simchi-Levi, On the unimodality of the profit function of the pricing newsvendor. Prod. Oper. Manag. **22**(3), 615–625 (2013)

R.D. Luce, *Individual Choice Behavior: A Theoretical Analysis* (Wiley, New York, 1959)

R.D. Luce, The choice axiom after twenty years. J. Math. Psychol. **15**(3), 215–233 (1977)

W. Ma, D. Simchi-Levi, Online resource allocation under arbitrary arrivals: optimal algorithms and tight competitive ratios. Technical report, MIT, Cambridge, MA (2017)

W. Ma, D. Simchi-Levi, Learning valuation distributions from bundle sales. Technical report, MIT, Cambridge, MA (2018)

W. Ma, D. Simchi-Levi, C.-P. Teo, On policies for single-leg revenue management with limited demand information. Technical report, MIT, Cambridge, MA (2018)

C. Maglaras, J. Meissner, Dynamic pricing strategies for multiproduct revenue management problems. Manuf. Serv. Oper. Manag. **8**(2), 135–148 (2006)

C. Maglaras, A. Zeevi, Pricing and design of differentiated services: approximate analysis and structural insights. Oper. Res. **53**(2), 242–262 (2005)

S. Mahajan, G.J. van Ryzin, Stocking retail assortments under dynamic consumer substitution. Oper. Res. **49**(3), 334–351 (2001)

P. Manzini, M. Mariotti, Stochastic choice and consideration sets. Econometrica **82**(3), 1153–1176 (2014)

V. Martinez-de-Albeniz, G. Roels, Competing for shelf space. Prod. Oper. Manag. **20**(1), 32–46 (2011)

V. Martinez-de-Albeniz, A. Saez-de-Tejada, Dynamic attraction models for cultural choice. Technical report, University of Navarra, Barcelona (2014)

V. Martinez-de-Albeniz, K. Talluri, Dynamic price competition with fixed capacities. Manag. Sci. **27**(6), 1078–1093 (2011)

F. Matejka, A. McKay, Rational inattention to discrete choices: a new foundation for the multinomial logit model. Am. Econ. Rev. **105**(1), 272–298 (2015)

P.R. McAfee, Coarse matching. Econometrica **75**(5), 2025–2034 (2004)

P.R. McAfee, V. te Velde, Dynamic pricing with constant demand elasticity. Prod. Oper. Manag. **17**(4), 432–438 (2008)

D. McFadden, Conditional logit analysis of qualitative choice behavior, in *Frontiers in Econometrics*, ed. by P. Zarembka (Academic, New York, 1974), pp. 105–142

D. McFadden, Modeling the choice of residential location, in *Spatial Interaction Theory and Planning Models*, ed. by A. Karlqvist, L. Lundqvist, F. Snickars, J. Weibull (Amsterdam, North-Holland, 1978), pp. 75–96

D. McFadden, Econometric models for probabilistic choice among products. J. Bus. **53**(3), 13–29 (1980)

D. McFadden, K. Train, Mixed MNL models for discrete response. J. Appl. Econ. **15**, 447–470 (2000)

J.I. McGill, G.J. van Ryzin, Revenue management: research overview and prospects. Transp. Sci. **33**(2), 233–256 (1999)

J. Meissner, A. Strauss, Network revenue management with inventory-sensitive bid prices and customer choice. Eur. J. Oper. Res. **216**(2), 459–468 (2012)

J. Meissner, A. Strauss, K. Talluri, An enhanced concave program relaxation for choice network revenue management. Prod. Oper. Manag. **22**(1), 71–87 (2013)

I. Mendez-Diaz, J.M. Bront, G. Vulcano, P. Zabala, A branch-and-cut algorithm for the latent-class logit assortment problem. Discrete Appl. Math. **164**(1), 246–263 (2014)

A. Mersereau, D. Zhang, Markdown pricing with unknown fraction of strategic customers. Manuf. Serv. Oper. Manag. **14**(3), 355–370 (2012)

V.K. Mishra, K. Natarajan, H. Tao, C.-P. Teo, Choice prediction with semidefinite optimization when utilities are correlated. IEEE Trans. Autom. Control **57**(10), 2450–2463 (2012)

V.K. Mishra, K. Natarajan, D. Padmanabhan, C.-P. Teo, X. Li, On theoretical and empirical aspects of marginal distribution choice models. Manag. Sci. **60**(4), 1511–1531 (2014)

A. Nalca, T. Boyaci, S. Ray, Competitive price-matching guarantees under imperfect store availability. Quant. Mark. Econ. **8**(3), 275–300 (2010)

A. Nalca, T. Boyaci, S. Ray, Competitive price-matching guarantees: equilibrium analysis of the availability verification clause under demand uncertainty. Manag. Sci. **59**(4), 971–986 (2013)

M. Nambiar, D. Simchi-Levi, H.Wang, Dynamic learning and pricing with model misspecification. Manag. Sci. (2019, forthcoming)

J. Nasiry, I. Popescu, Advance selling when consumers regret. Manag. Sci. **58**(6), 1160–1177 (2012)

K. Natarajan, M. Song, C.-P. Teo, Persistency model and its applications in choice modeling. Manag. Sci. **55**(3), 453–469 (2009)

H. Nazerzadeh, G. Perakis, Technical note – nonlinear pricing competition with private capacity information. Oper. Res. **64**(2), 329–340 (2016)

S. Netessine, R.A. Shumsky, Revenue management games: horizontal and vertical competition. Manag. Sci. **51**(5), 813–831 (2005)

N. Osadchiy, G. Vulcano, Selling with binding reservations in the presence of strategic consumers. Manag. Sci. **56**(12), 2173–2190 (2010)

O. Ozer, R. Phillips (eds.), *The Oxford Handbook of Pricing Management* (Oxford University Press, Oxford, 2012)

O. Ozer, Y. Zheng, Markdown or everyday low price: the role of behavioral motives. Manag. Sci. **62**(2), 326–346 (2016)

X.A. Pan, D. Honhon, Assortment planning for vertically differentiated products. Prod. Oper. Manag. **21**(2), 253–275 (2012)

Z. Pang, O. Berman, M. Hu, Up then down: bid-price trends in revenue management. Prod. Oper. Manag. **24**(7), 1135–1147 (2014)

A. Paul, J. Feldman, J.M. Davis, Assortment optimization and pricing under a nonparametric tree choice model. Manuf. Serv. Oper. Manag. **20**(3), 550–565 (2018)

G. Perakis, G. Roels, Regret in the newsvendor problem with partial information. Oper. Res. **56** (1), 188–203 (2008)

G. Perakis, G. Roels, Robust controls for network revenue management. Manuf. Serv. Oper. Manag. **12**(1), 56–76 (2010)

G. Perakis, A. Sood, Competitive multi-period pricing for perishable products: a robust optimization approach. Math. Program. **107**(1–2), 295–335 (2006)

N.C. Petruzzi, M. Dada, Pricing and the newsvendor problem: a review with extensions. Oper. Res. **47**(2), 183–194 (1999)

R.L. Phillips, *Pricing and Revenue Optimization* (Stanford University Press, Stanford, 2005)

R. Phillips, Optimizing prices for consumer credit. J. Revenue Pricing Manag. **12**(4), 360–377 (2013)

R. Phillips, A.S. Simsek, G.J. van Ryzin, The effectiveness of field price discretion: empirical evidence from auto lending. Manag. Sci. **61**(8), 1741–4759 (2015)

R. Phillips, A.S. Simsek, G. van Ryzin, Estimating buyer willingness-to-pay and seller reserve prices from negotiation data and the implications for pricing. Technical report, University of Texas at Dallas, Dallas, TX (2016)

I.P.L. Png, Reservations: customer insurance in the marketing of capacity. Mark. Sci. **8**(3), 248–264 (1989)

R. Ratliff, Revenue management demand distributions, in *AGIFORS Reservations and Yield Management Study Group Annual Meeting Proceedings*, Cape Town, South Africa (2005)

W.Z. Rayfield, P. Rusmevichientong, H. Topaloglu, Approximation methods for pricing problems under the nested logit model with price bounds. INFORMS J. Comput. **27**(2), 335–357 (2015)

M.I. Reiman, Q. Wang, An asymptotically optimal policy for a quantity-based network revenue management problem. Math. Oper. Res. **33**(2), 257–282 (2008)

H. Richter, The differential revenue method to determine optimal seat allotments by fare type, in *Proceedings 22nd AGIFORS Symposium*, Lagonissi (1982), pp. 339–362

L.W. Robinson, Optimal and approximate control policies for airline booking with sequential nonmonotonic fare classes. Oper. Res. **43**(3), 252–263 (1995)

G. Roels, K. Fridgeirsdottir, Dynamic revenue management for online display advertising. J. Revenue Pricing Manag. **8**(5), 452–466 (2009)

M. Rothstein, An airline overbooking model. Transp. Sci. **5**(2), 180–192 (1971)

P. Rusmevichientong, H. Topaloglu, Robust assortment optimization under the multinomial logit choice model. Oper. Res. **60**(4), 865–882 (2012)

P. Rusmevichientong, B. Van Roy, P.W. Glynn, A nonparametric approach to multiproduct pricing. Oper. Res. **54**(1), 82–98 (2006)

P. Rusmevichientong, Z.-J.M. Shen, D.B. Shmoys, A PTAS for capacitated sum-of-ratios optimization. Oper. Res. Lett. **37**(4), 230–238 (2009)

P. Rusmevichientong, Z.-J.M. Shen, D.B. Shmoys, Dynamic assortment optimization with a multinomial logit choice model and capacity constraint. Oper. Res. **58**(6), 1666–1680 (2010)

P. Rusmevichientong, D.B. Shmoys, C. Tong, H. Topaloglu, Assortment optimization under the multinomial logit model with random choice parameters. Prod. Oper. Manag. **23**(11), 2023–2039 (2014)

D. Saure, A. Zeevi, Optimal dynamic assortment planning with demand learning. Manuf. Serv. Oper. Manag. **15**(3), 387–404 (2013)

D. Segev, Assortment planning with nested preferences: dynamic programming with distributions as states? Algorithmica **81**(1), 393–417 (2019)

A. Sen, A.X. Zhang, Style goods pricing with demand learning. Eur. J. Oper. Res. **196**(3), 1058–1075 (2009)

S. Shugan, J. Xie, Advance pricing of services and other implications of separating purchase and consumption. J. Serv. Res. **2**(3), 227–239 (2000)

R. Shumsky, The Southwest effect, airline alliances and revenue management. J. Revenue Pricing Manag. **5**(1), 83–89 (2006)

R.A. Shumsky, F. Zhang, Dynamic capacity management with substitution. Oper. Res. **57**(3), 671–684 (2009)

D.D. Sierag, G.M. Koole, R.D. van der Mei, J.I. van der Rest, B. Zwart, Revenue management under customer choice behaviour with cancellations and overbooking. Eur. J. Oper. Res. **246** (1), 170–185 (2015)

J.L. Simon, An almost practical solution to airline overbooking. J. Trans. Econ. Policy **2**(2), 201–202 (1968)

R. Simpson, Theoretical concepts for capacity/yield management, in *Proceedings 25th AGIFORS Symposium* (1985), pp. 281–293

A.S. Simsek, H. Topaloglu, Technical note – an expectation-maximization algorithm to estimate the parameters of the Markov chain choice model. Oper. Res. **66**(3), 748–760 (2018)

A.G. Sogomonian, C.S. Tang, A modeling framework for coordinating promotion and production decisions within a firm. Manag. Sci. **39**(2), 191–203 (1993)

Y. Song, S. Ray, T. Boyaci, Technical note: optimal dynamic joint inventory-pricing control for multiplicative demand with fixed order costs and lost sales. Oper. Res. **57**(1), 245–250 (2009)

C. Stein, V.A. Truong, X. Wang, Advance reservations with heterogeneous customers. Manag. Sci. (2019, forthcoming)

A.K. Strauss, K. Talluri, Tractable consideration set structures for assortment optimization and network revenue management. Prod. Oper. Manag. **26**(7), 1359–1368 (2017)

N. Surasvadi, C. Tang, G. Vulcano, Using contingent markdown with reservation to profit from strategic consumer behavior. Prod. Oper. Manag. **26**(12), 2226–2246 (2017)

K. Talluri, New formulations for choice network revenue management. INFORMS J. Comput. **26** (2), 401–413 (2014)

K. Talluri, G. van Ryzin, An analysis of bid-price controls for network revenue management. Manag. Sci. **44**(11), 1577–1593 (1998)

K. Talluri, G. van Ryzin, A randomized linear programming method for computing network bid prices. Transp. Sci. **33**(2), 207–216 (1999)

K. Talluri, G. van Ryzin, Revenue management under a general discrete choice model of consumer behavior. Manag. Sci. **50**(1), 15–33 (2004a)

K. Talluri, G. van Ryzin, *The Theory and Practice of Revenue Management* (Springer Science+Business Media, New York, 2004b)

K. Talluri, F. Castejon, B. Codina, J. Magaz, Proving the performance of a new revenue management system. J. Revenue Pricing Manag. **9**(4), 300–312 (2010)

C.S. Tang, R. Yin, Joint ordering and pricing strategies for managing substitutable products. Prod. Oper. Manag. **16**(1), 138–153 (2007)

C.S. Tang, K. Rajaram, A. Alptekinoglu, J. Ou, The benefits of advance booking discount programs: model and analysis. Manag. Sci. **50**(4), 465–478 (2004)

C. Temath, S. Polt, L. Suhi, On the robustness of the network-based revenue opportunity model. J. Revenue Pricing Manag. **9**(4), 341–355 (2010)

C. Tong, H. Topaloglu, On approximate linear programming approaches for network revenue management problems. INFORMS J. Comput. **26**(1), 121–134 (2014)

H. Topaloglu, A stochastic approximation method to compute bid prices in network revenue management problems. INFORMS J. Comput. **20**(4), 596–610 (2008)

H. Topaloglu, On the asymptotic optimality of the randomized linear program for network revenue management. Eur. J. Oper. Res. **197**(3), 884–896 (2009a)

H. Topaloglu, Using Lagrangian relaxation to compute capacity-dependent bid prices in network revenue management. Oper. Res. **57**(3), 637–649 (2009b)

H. Topaloglu, Joint stocking and product offer decisions under the multinomial logit model. Prod. Oper. Manag. **22**(5), 1182–1199 (2013)

K. Train, *Discrete Choice Methods with Simulation* (Cambridge University Press, New York, 2002)

V.-A. Truong, Optimal selection of medical formularies. J. Revenue Pricing Manag. **13**(2), 113–132 (2014)

C. Ulu, D. Honhon, A. Alptekinoglu, Learning consumer tastes through dynamic assortments. Oper. Res. **60**(4), 833–849 (2012)

G.J. van den Berg, On the uniqueness of optimal prices set by monopolistic sellers. J. Econ. **141** (2), 482–491 (2007)

G.J. van Ryzin, Models of demand. J. Revenue Pricing Manag. **4**(2), 204–210 (2005)

G.J. van Ryzin, S. Mahajan, On the relationship between inventory costs and variety benefits in retailassortments. Manag. Sci. **45**(11), 1496–1509 (1999)

G.J. van Ryzin, J. McGill, Yield management without forecasting or optimization: an adaptive algorithm for protection level optimization. Manag. Sci. **46**(6), 568–573 (2000)

G.J. van Ryzin, K.T. Talluri, Revenue management, in *Handbook of Transportation*, ed. by R. Hall (Springer, New York, 2003)

G.J. van Ryzin, K.T. Talluri, An introduction to revenue management, in *Tutorials in Operations Research* (INFORMS, New Orleans, 2005)

G.J. van Ryzin, G. Vulcano, Optimal auctioning and ordering in an infinite horizon inventory-pricing system. Oper. Res. **52**(3), 346–367 (2004)

G.J. van Ryzin, G. Vulcano, Simulation-based optimization of virtual nesting controls for network revenue management. Oper. Res. **56**(4), 865–880 (2008a)

G.J. van Ryzin, G. Vulcano, Computing virtual nesting controls for network revenue management under customer choice behavior. Manuf. Serv. Oper. Manag. **10**(3), 448–467 (2008b)

G. van Ryzin, G. Vulcano, A market discovery algorithm to estimate a general class of non-parametric choice models. Manag. Sci. **61**(2), 281–300 (2015)

X. Vives, *Oligopoly Pricing: Old Ideas and New Tools* (MIT Press, Cambridge, 2001)

T. Vossen, D. Zhang, A dynamic disaggregation approach to approximate linear programs for network revenue management. Prod. Oper. Manag. **24**(3), 469–487 (2015)

G. Vulcano, G.J. van Ryzin, Estimating primary demand for substitutable products from sales transaction data. Oper. Res. **60**(2), 313–334 (2012)

G. Vulcano, G.J. van Ryzin, C. Maglaras, Optimal dynamic auctions for revenue management. Manag. Sci. **48**(11), 1388–1407 (2002)

G. Vulcano, G.J. van Ryzin, W. Chaar, Choice-based revenue management: an empirical study of estimation and optimization. Manuf. Serv. Oper. Manag. **12**(3), 371–392 (2010)

D. Walczak, S. Mardan, R. Kallesen, Customer choice, fare adjustments and the marginal expected revenue data transformation: a note on using old yield management techniques in the brave new world of pricing. J. Revenue Pricing Manag. **9**, 94–109 (2010)

R. Wang, Capacitated assortment and price optimization under the multinomial logit choice model. Oper. Res. Lett. **40**(6), 492–497 (2012)

R. Wang, Assortment management under the generalized attraction model with a capacity constraint. J. Revenue Pricing Manag. **12**(3), 254–270 (2013)

Z. Wang, Technical note – intertemporal price discrimination via reference price effects. Oper. Res. **64**(2), 290–296 (2016)

R. Wang, Consumer choice and market expansion: modeling, optimization and implementation. Technical report, Johns Hopkins University, Baltimore, MD (2018a)

R. Wang, When prospect theory meets consumer choice models: assortment and pricing management with reference prices. Manuf. Serv. Oper. Manag. **20**(3), 583–600 (2018b)

Z. Wang, M. Hu, Committed versus contingent pricing under competition. Prod. Oper. Manag. **23** (11), 1919–1936 (2014)

R. Wang, O. Sahin, The impact of consumer search cost on assortment planning and pricing. Manag. Sci. **64**(8), 3649–3666 (2018)

Y. Wang, Z.-J.M. Shen, Joint optimization of capacitated assortment and pricing problem under the tree logit model. Technical report, University of California, Berkeley, CA (2017)

R. Wang, Z. Wang, Consumer choice models with endogenous network effects. Manag. Sci. **63** (11), 3944–3960 (2017)

Z. Wang, Y. Ye, Hidden-city ticketing: the cause and impact. Transp. Sci. **50**(1), 288–305 (2016)

Z. Wang, S. Deng, Y. Ye, Close the gaps: a learning-while-doing algorithm for single-product revenue management problems. Oper. Res. **62**(2), 318–331 (2014)

J. Wang, Y. Levin, M. Nediak, Selling passes to strategic customers. Technical report, Queen's University, Kingston (2015)

X. Wang, V.A. Truong, D. Bank, Online advance admission scheduling for services with customer preferences. Technical report, Columbia University, New York, NY (2018)

R. Wang, M. Dada, O. Sahin, Pricing ancillary service subscriptions. Manag. Sci. (2019, forthcoming)

L.R. Weatherford, S.E. Bolidy, A taxonomy and research overview of perishable-asset revenue management: yield management, overbooking, and pricing. Oper. Res. **40**(5), 831–844 (1992)

L.R. Weatherford, S.E. Bolidy, P.E. Pfeifer, Modeling the customer arrival process and comparing decision rules in perishable asset revenue management situations. Transp. Sci. **27**(3), 239–251 (1993)

C.-H. Wen, F.S. Koppelman, The generalized nested logit model. Transp. Res. B **35**, 627–641 (2001)

E.L. Williamson, Airline network seat inventory control: Methodologies and revenue impacts. PhD thesis, Flight Transportation Laboratory, Massachusetts Institute of Technology, Cambridge, MA, 1992

R.D. Wollmer, An airline seat management model for a single leg route when lower fare classes book first. Oper. Res. **40**(1), 26–37 (1992)

C.P. Wright, H. Groenevelt, R.A. Shumsky, Dynamic revenue management in airline alliances. Transp. Sci. **44**(1), 15–37 (2010)

J. Xie, S. Shugan, Electronic tickets, smart cards, and online prepayments: when and how to advance sell. Mark. Sci. **20**, 219–243 (2001)

X. Xu, W.J. Hopp, A monopolistic and oligopolistic stochastic flow revenue management model. Oper. Res. **54**(6), 1098–1109 (2006)

Z. Yan, C. Cheng, K. Natarajan, C.-P. Teo, Marginal estimation price optimization: data driven multi-product pricing problem. Technical report, Singapore University of Technology and Design, Singapore (2017)

R. Yin, Y. Aviv, A. Pazgal, and C.S. Tang. Optimal markdown pricing: implications of inventory display formats in the presence of strategic customers. Manag. Sci. **55**(8), 1391–1422 (2009)

Y. Yu, X. Chen, F. Zhang, Dynamic capacity management with general upgrading. Oper. Res. **63** (6), 1372–1389 (2015)

D. Zhang, An improved dynamic programming decomposition approach for network revenue management. Manuf. Serv. Oper. Manag. **13**(1), 35–52 (2011)

D. Zhang, D. Adelman, An approximate dynamic programming approach to network revenue management with customer choice. Transp. Sci. **43**(3), 381–394 (2009)

D. Zhang, W. Cooper, Revenue management for parallel flights with customer choice behavior. Oper. Res. **53**(3), 415–431 (2005)

D. Zhang, W.L. Cooper, Managing clearance sales in the presence of strategic customers. Prod. Oper. Manag. **17**(4), 416–431 (2008)

D. Zhang, W. Cooper, Pricing substitutable flights in airline revenue management. Eur. J. Oper. Res. **197**(3), 848–861 (2009)

D. Zhang, R. Kallesen, Incorporating competitive price information in revenue management. J. Revenue Pricing Manag. **7**, 17–26 (2008)

D. Zhang, Z. Lu, Assessing the value of dynamic pricing in network revenue management. INFORMS J. Comput. **25**(1), 102–115 (2013)

D. Zhang, T. Vossen, Reductions of approximate linear programs for network revenue management. Oper. Res. **63**(6), 1352–1371 (2015)

D. Zhang, L. Weatherford, Dynamic pricing for network revenue management: a new approach and application in the hotel industry. INFORMS J. Comput. **29**(1), 18–35 (2017)

H. Zhang, C. Shi, C. Qin, C. Hua, Stochastic regret minimization for revenue management problems with nonstationary demands. Nav. Res. Logist. **63**(6), 433–448 (2016)

H. Zhang, P. Rusmevichientong, H. Topaloglu, Assortment optimization under the paired combinatorial logit model. Technical report, Cornell University, Ithaca, NY (2017)

H. Zhang, P. Rusmevichientong, H. Topaloglu, Technical note – multiproduct pricing under the generalized extreme value models with homogeneous price sensitivity parameters. Oper. Res. **66**(6), 1559–1570 (2018)

W. Zhao, Y.-S. Zheng, A dynamic model for airline seat allocation with passenger diversion and no-shows. Transp. Sci. **35**(1), 80–98 (2001)

W. Zhuang, M. Gumus, D. Zhang, A single-resource revenue management problem with random resource consumptions. J. Oper. Res. Soc. **63**, 1213–1227 (2012)

P. Zipkin, *Foundations of Inventory Management* (McGraw-Hill/Irwin, Boston, 2000)

S. Ziya, H. Ayhan, R.D. Folley, Relationships among three assumptions in revenue management. Oper. Res. **52**(5), 804–809 (2004)

Index

G. Gallego, H. Topaloglu, *Revenue Management and Pricing Analytics*,
International Series in Operations Research & Management Science 279,
https://doi.org/10.1007/978-1-4939-9606-3

Printed by Printforce, the Netherlands